CALIFORN
D0778854

Ecology of Heathlands

Ecology of Heathlands

C. H. GIMINGHAM
Professor, Department of Botany, University of Aberdeen

London
CHAPMAN AND HALL

First published 1972
by Chapman and Hall Ltd
11 New Fetter Lane, London, EC4P 4EE

Printed in Great Britain by
Cox & Wyman Ltd., Fakenham, Norfolk

SBN *412 10460 1*

Published in the U.S.A.
by Halsted Press, a Division
of John Wiley & Sons, Inc.
New York

'The warm breeze during its leisurely passage over moor and heath has become imbued with their delicious aroma. Gently and steadily it sweeps across the immense expanses, which offer it no resistance. Peace and quiet reign supreme. . . .'

C. Raunkiaer, 1910; translated
by H. Gilbert-Carter, 1934.

'Callunetum is at once the glory and the tragedy of the Scottish Highlands.'

A. S. Watt, 1961.

'If we are not careful to preserve a considerable area of heath as a memorial, then I do not doubt that our descendants will censure us for our short-sightedness and lack of feeling.'

C. Raunkiaer, 1913; translated
by H. Gilbert-Carter, 1934.

Preface

Heathlands have always attracted the attention of ecologists. Particularly in recent years, however, research has been focused on the heathland ecosystem from so many different angles (Fig. 1) that a remarkably comprehensive picture is beginning to emerge. An attempt is made here to present the outlines of this picture, the details of which have hitherto been scattered in a great variety of European ecological journals. The wealth of information available suggests that the time is ripe for integration, even though additions to the literature continue to appear in almost every new issue of the journals. Several reasons may be suggested for this concentration of interest upon the dwarf-shrub heaths. On the one hand, they offer considerable scope for research in ecological history, phytosociology, community structure and dynamics, while on the other they raise important questions concerning the ecological consequences of various forms of land use and management. Furthermore, the fact that in parts of west Europe they are fast disappearing establishes an urgent need for conservation, in connection with which it is essential to establish appropriate codes of management.

Probably more than in any other branch of biology, ecological theory and methodology often appear to become detached from reality unless firmly linked to practical examples. In this context, few examples could be more illuminating than heathland. Almost every aspect of ecological theory and technique finds a place in the interpretation of dwarf-shrub heaths. The whole, therefore, amounts to more than just an analysis of heathland vegetation: it offers an example of the application of ecological principles to a particular practical problem. Starting with the analysis of the heath ecosystem, the problem extends to consideration of the use, management and conservation of the vegetation and the findings are therefore of interest to agriculturalists, foresters, land-owners, land consultants and wild life conservationists.

In selecting material for inclusion, these considerations have been uppermost and for this reason a purely descriptive account of heaths has been avoided. Attention has been concentrated on the results of research into processes, particularly those which are applicable throughout the west European heath region. On the grounds that a writer deals more competently with subject matter familiar to him, there is emphasis on work carried out in Britain, particularly in the north. The space devoted to discussion of the ecology of *Calluna vulgaris* may at first sight seem somewhat disproportionate, but this species dominates a very large proportion of the total area of heathland. More than any other it controls and influences the processes taking place in heath vegetation, and embodies many of the characteristics of heath communities. Furthermore, *Calluna vulgaris* is a species of great ecological interest in its own right, as will emerge from almost every chapter. In much the same way as numerous avenues of approach have impinged on heathland ecology, so anatomists, morphologists, physiologists and ecologists have unravelled many aspects of the organization and functioning of this species, demonstrating its remarkable versatility. In so doing they have contributed much to an understanding of its role in heathland ecosystems.

The animal components of heathland communities have been considered only in so far as they affect processes taking place in the vegetation, and all too little attention has been given to the invertebrate fauna. Much current research is being devoted to the animal ecology of heaths. Although it might have been desirable to incorporate more of this in the present account, limitations both of space and of the author's qualifications have determined otherwise.

To a considerable degree, the convergence of numerous lines of investigation on to one object of study multiplies the interest and significance of the results, and attracts yet more workers to the field. This book has been concerned with integrating the findings of numerous workers, in a variety of different centres and organizations. To many of them I am indebted for most valuable exchanges of ideas. In particular, I owe a great deal to the stimulus of discussions and collaboration in the field with Dr J. B. Kenworthy, Dr G. R. Miller and Mr I. A. Nicholson, who will doubtless recognize that they sowed the seeds of many ideas which have grown in these pages. To them, and to others who have also read and commented on sections of the manuscript – Professor J. D. Matthews, Dr E. A. Fitzpatrick, Dr D. N. McVean, Mr J. Stone – my thanks are offered. Colleagues who have answered my inquiries and taken me to localities which they know intimately or where their work is in progress are too many to name individually, but their help is greatly appreciated. In this connection I am especially grateful to Dr S. B. Chapman, Dr D. T. Streeter, Dr J. J. Barkman and Dr E. Preising. Dr Chapman and Dr Miller have also both kindly allowed me to use data and illustrations prior to publication elsewhere, for which I am most grateful. The same privilege has been extended to me by Dr P. B. Bridgewater, Mr H. M. Hinshiri, Dr T. H. Keatinge, Dr D. K. L. MacKerron, Mr D. A. Smart, Dr C. F. Summers and Dr S. D.

Ward, while Mrs S. M. Anderson, Mrs E. M. Birse and Dr S. Y. Landsberg have allowed me to base several of the Figures on their data or drawings. Plate 2 has been prepared from a photograph by Dr P. S. Ashton, and part of Table 3 is quoted by kind permission of Professor Dr H. Ellenberg.

In the preparation of diagrams invaluable help has been given by Miss A. M. Slater and Mrs L. Forbes. A number of the photographs were taken by Mr I. Moir and Mr E. F. Middleton, and Mr Middleton prepared all the photographic illustrations for reproduction. To all of these, and to Mrs H. Murray, Miss L. Smith and Miss P. A. Grinsted who typed the manuscript, and Mr D. R. Paterson who helped with proof reading, I am extremely grateful for their essential and expert assistance.

For permission to reproduce a number of published Figures and Tables I am indebted to the authors named in the Legends, and to the following:

The Editor of the Journal of Ecology and Blackwell Scientific Publications Ltd (Figs. 3, 5, 7, 14, 15, 18, 20, 26, 28, 30, 34, 35, 39, 40, 45).

Section CT of the International Biological Programme and Blackwell Scientific Publications Ltd (Part of Table 3).

The Editor of the Transactions of the Botanical Society of Edinburgh (Fig. 4).

The Editor of Botaniska Notiser (Fig. 6).

Dr W. Junk N. V., Publishers of Vegetatio (Figs. 21, 22 and Table 7).

The Editors of The New Phytologist (Fig. 29).

The Editors of Oikos (Fig. 38).

Edinburgh University Press (Table 32).

Finally, it is a special pleasure to thank my wife who not only translated many passages in Scandinavian languages for me, but also read and made valuable comments on the whole manuscript.

C.H.G.

Aberdeen, 1971.

Nomenclature of plants belonging to the British flora follows Clapham, A. R., Tutin, T. G., and Warburg, E. F., (2nd Edition, 1962), *Flora of the British Isles*, Cambridge; Richards, P. W., and Wallace, E. C., (1950), 'An annotated list of British mosses,' *Trans. Brit. Bryol. Soc.*, **1**, i–xxxi; and James, P. W., (1965), 'A new check-list of British lichens', *Lichenologist*, **3**, 95–153.

Following common practice, *Calluna vulgaris* (L.) Hull (heather) is referred to simply as *Calluna*.

Contents

Plates

1 Introduction

The heaths of western Europe

Heathlands have been known to the peoples of west European countries for many generations. In certain parts they have been so widespread as to constitute an environment in which a characteristic way of life has evolved, with its distinctive patterns of agriculture and domestic architecture. To some, heathlands have meant home and livelihood, becoming a loved symbol of nation or region to which the mind of the exile constantly returns. To many others they have been a source of delight and interest, appealing especially to those who appreciate wide expanses of open country and the plants and animals which inhabit them. At times, it is true, heathland gives an impression of barren waste and windswept monotony, but when in late summer the rich purple of heath plants in flower spreads across the country, native and visitor alike respond to the beauty of the landscape.

Later in this chapter mention will be made of similar vegetation in other parts of the world, but the Atlantic zone of Europe is universally regarded as the 'type-locality' for heaths. Their origins and development, the many variations in composition of plant and animal communities, and their relationships with climate and soil offer numerous important problems to the ecologist, plant geographer and the general student of natural history and wild life. Particularly instructive opportunities are presented for investigating the influence of man and his domestic animals upon vegetation and habitat.

To the agricultural scientist heathlands are of significance not only in regard to the rather marginal type of farming traditionally practised upon them, but also because of their potentialities for improvement, or for reclamation and conversion into more productive vegetation. Similarly, the forester, whose task it has often been to establish productive forest on heathland sites, is deeply concerned with their plant communities and habitats. In several west European countries, notably Sweden, Denmark, the Netherlands, Belgium and Germany, the extent of heathland has been or is at present being greatly reduced as large areas are converted to more profitable farm-land or forest. This process inevitably raises controversies and tensions, which underline the importance of a proper ecological appraisal of the best uses for any particular area of land. At the same time, the urgency of conservation becomes apparent, since the wealth and diversity of scenery, wild life and game represent an inheritance cherished by scientists and naturalists, walkers, tourists and sportsmen. While it may be agreed that representative examples of heath should be retained, the conservationist is faced with the problem of deciding precisely what to conserve and how to manage heathland reserves.

It is therefore hoped that a book in which some of the fruits of ecological research on heath vegetation are gathered will prove of value to readers coming to the subject along any one of a variety of approaches (Fig. 1). For the ecologist and geographer, a unified account of the many facets of a single vegetational formation may be useful, not least because it represents a synthesis of numerous lines of investigation. Studies in historical ecology, phytosociology, physiological ecology, community structure and dynamics, microclimate, soil, primary production, nutrient cycling, the effects of man's activities, and many other aspects, all contribute towards an understanding of the one object of inquiry. For the agriculturalist and forester, it may be of service to consider the bearing of ecological findings upon the choice of procedures for improvement and reclamation. The future of our heathlands is a matter of concern to all who feel that land should be used wisely, with the aim of combining productivity with the maintenance of fertility and public enjoyment. Whether heaths are to be replaced by crops or retained under more or less traditional forms of management, it is essential that decisions should be based upon knowledge of the ecological consequences of past, present and future activities in these areas. Only with this knowledge can principles be established upon which to strike a balance between intensified production from the land, and the need for conservation of a thing of value, all too easily lost. Economic and sociological factors, as well as scientific considerations, affect this balance, but the time is perhaps ripe to review some of the botanical and ecological investigations which in recent years have contributed so much to a better understanding of the problem. A great deal remains to be accomplished, and perhaps future efforts may be stimulated by some assessment of the present position.

In the early days of the scientific study of vegetation, some of the classics of ecological literature were concerned with heaths. Among the best known is 'Die Heide Norddeutschlands' (1st edition 1901; 2nd edition, 1925) in which P. Graebner gathered the results of his work during the late nineteenth century. This treatment was among the first of its kind to demonstrate effectively the inter-relationships between climate, soil, vegetation and historical processes in determining the characteristics of a particular vegetation type, and it remains an important source of information

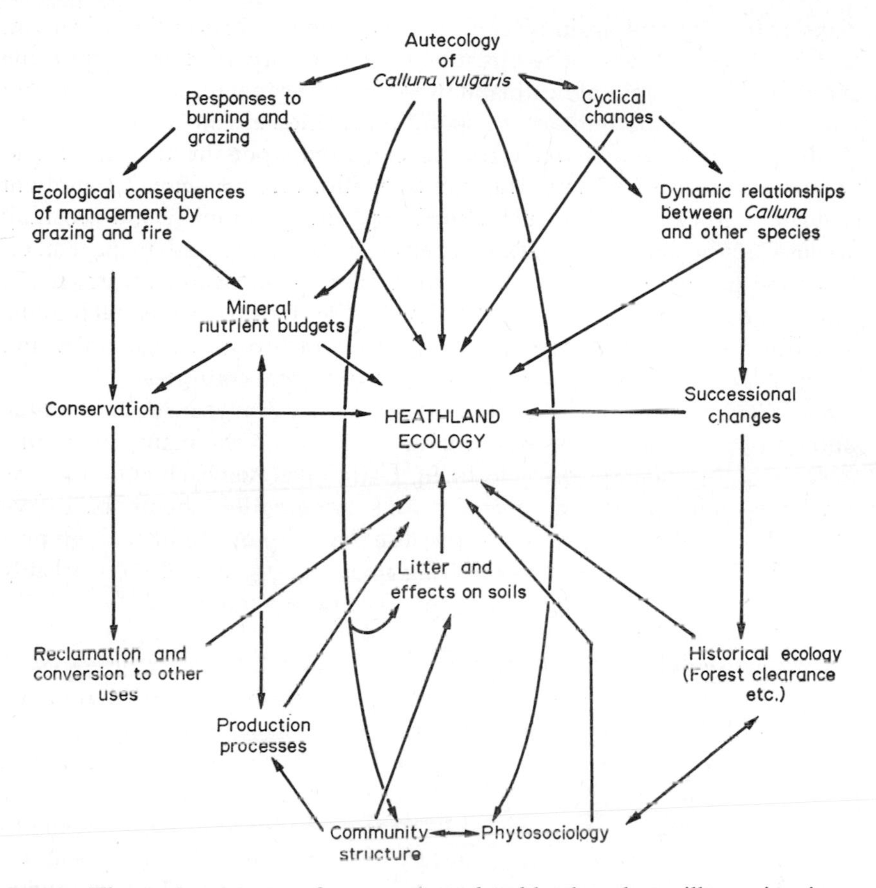

Fig. 1. The main avenues of approach to heathland ecology, illustrating interactions between the various aspects of the subject.

on the plants and soils of west European heaths, their affinities with related communities and the factors influencing their use or cultivation.

About the same time, P. E. Müller (1884, 1924) was publishing his views on the development of Danish heath vegetation and soils, and especially on the formation of mor humus, which laid the foundation for more recent research and discussion of these topics. In Britain, interest in the survey and mapping of vegetation sprang up in the years around the turn of the century, deriving its inspiration partly from workers on the continent of Europe. Among the first to contribute substantially to this movement were the Scottish brothers R. and W. G. Smith, and not unnaturally their attention was soon focused upon the origin and development of Scottish heaths and moors, their excellent descriptions of which (1900, 1902, 1905, 1911) long remained the accepted account. Other valuable data on British heaths contained in regional ecological surveys were summarized by

Tansley in 1911 and again more fully, with some later additions, in 1939. In retrospect, however, it is a matter for some surprise that, despite the extent and importance of heathland in western Europe, for some years after 1911 rather little was built on the sound foundation already laid.

More recently renewed interest has centred upon heathlands, firstly because of the excellent opportunities they afford for the scientific study of ecological processes and of plant sociology and geography, and secondly because of the increasing rate of change in their use and management. Attention has been drawn to the low productivity of heathland in general and to signs of decrease in an already low level of fertility. Research into the practical problems of improvement, in terms of forestry, agriculture and other purposes, has met with much success in several countries.

As a result, during the course of the past twenty-five years a considerable amount of new information has become available on the plant and animal ecology of heathlands, the effects of traditional management, and the principles upon which improvement and conservation should be based. While there is much still to be learnt, the diversity of the interests represented in these investigations is leading to the emergence of a remarkably comprehensive ecological picture of this type of vegetation.

Dwarf-shrub heaths in general

In this book the word 'heathland' is used to describe territories in which trees or tall shrubs are sparse or absent, and in which the dominant life-form is that of the evergreen dwarf-shrub, particularly as represented in Ericaceae (Plate 1). The word 'heath' and its equivalents in Germanic languages (German: *Heide*; Swedish: *hed*) probably referred originally to this kind of vegetation, being transferred also in the form 'heather' (German: *Heidekraut*) to the most abundant species of the plant communities. In common English usage this term 'heath' has become applicable more widely, and less precisely, to almost any kind of 'barren waste ground' particularly on acid soils, and even in botanical contexts has been employed in the compounds 'grass heath', 'moss heath' and 'lichen heath', designating respectively certain types of grassland on acid soils, and arctic, alpine or coastal bryophyte- or lichen-dominated communities (Tansley, 1939; Watt, 1940; McVean and Ratcliffe, 1962; McVean, 1964). Difficulties have been increased by indiscriminate use of the words 'heath' and 'moor', which are sometimes applied to one and the same area. Some difference of meaning seems to attach to these words, for 'moor' (or, in Scots, 'muir') generally conveys an impression of northern or upland regions and peaty soils. The equivalent word in Germanic languages signifies wet peatlands (German: *Moor*; Swedish: *myr*) whereas 'heath' suggests sandy podsolic ground in more southern and lowland districts.

Despite these apparent distinctions it is clearly better for ecological purposes to characterize a vegetation-type or 'formation' in terms of structural features, and a formation dominated by ericoid dwarf-shrubs emerges as a fairly clear entity. To this it seems appropriate to apply the original, and in Europe most widely used, name 'heath', restricting its sphere of reference by using the phrase 'dwarf-shrub heath' as the descriptive title.

For a brief designation of this formation it is difficult to improve on the definitions given by Graebner (1901) and Warming (1909). The former referred to 'true heath' as 'open country without tree growth', where the 'woody plants consist essentially of undershrubs and low shrubs' and continuous grass turf is absent. Warming wrote 'By the term heath, so far as it concerns northern Europe, is meant a treeless tract that is mainly occupied by evergreen, slow-growing, small-leaved, dwarf-shrubs and creeping shrubs which are largely Ericaceae'. Reference was also made to the height of the vegetation which 'often rises to thirty centimetres or more, but often only to ten or twenty'. The caespitose growth-form characteristic of many of the dwarf-shrubs concerned is emphasized by other writers.

In its ecological context, 'heath' is thus more precisely defined and limited than in common usage, and while closely related to acid grassland, peat bog and other types of vegetation, is differentiated from them on grounds of community structure. The incorporation of most of what are commonly called 'moors' into the category of dwarf-shrub heaths (with the obvious exception of 'grass moors', 'sedge moors', etc.) follows from even a cursory botanical analysis. This was accepted by Tansley (1939), although he maintained for convenience the broad distinction used by R. Smith (1900), W. G. Smith, (1911) and Elgee, (1914), between heaths (lowland and upland) and heather moors, the latter being differentiated by their situation on the deeper peats rather than on floristic grounds.

To define the formation in terms of the dominance of ericoid dwarf-shrubs is not, of course, to confine it to the area of distribution of the family Ericaceae, and much less to treat it as unique to Europe. The word 'ericoid' is used to denote low, rather densely-branched woody shrubs, normally with the branches radiating from a common base, microphyllous or with small rather sclerophyllous leaves and often, though not invariably, evergreen. In addition to the heaths of Europe structurally similar communities have been described from other parts of the world, wherever conditions favouring the dominance of an ericoid dwarf-shrub type of life-form obtain.

Except in a few special cases, these conditions involve the combination of a relatively cool temperature regime, high atmospheric humidity throughout much of the year, and rather freely drained substrata not conducive to deep peat formation. In addition, some factor, whether climatic, edaphic, biotic or anthropogenic, must operate to exclude the development of taller shrubs and trees. This type of environment may be expected in three main categories of situation:

(*i*) where for any reason forest is excluded in the more strongly oceanic regions of the cool-temperate belt;
(*ii*) in certain parts of sub-arctic and sub-antarctic territory;
(*iii*) where adequate humidity prevails at sub-alpine or low-alpine altitudes on mountains.

Brief reference to each follows:

(*i*) The first type of situation is most widely represented in western Europe, where numerous species of ericoid dwarf-shrub are native, and where, as already mentioned, the heath formation has played an important part in lowland vegetation. There are some habitats in this region permanently occupied by heath rather than woodland, on account of edaphic or, in certain northern and coastal localities, climatic factors. However, although climate and soil are widely favourable, heaths could not have played so large a part in the vegetation of western Europe if the native forests had not suffered at the hand of man (Chapter 2).

Similar environments have arisen in other regions, as for example in Nova Scotia, eastern Canada, where the vegetation of some of the so-called 'pine barrens' strongly resembles certain European heaths. Ericaceae are prominent, although *Erica* and *Calluna* are lacking since the natural distribution of both genera is confined to the Old World. *Vaccinium*, however, is prominently represented, particularly by the polymorphic *V. angustifolium* which gives an appearance to the community reminiscent of that of European *V. myrtillus* heaths. Other important Ericaceous dwarf-shrubs include species of *Gaylussacia*, *Gaultheria* and *Arctostaphylos*.

These pine barrens occur on boulder-strewn ground surrounded usually by woodland. They are dotted with somewhat stunted trees (for example, *Pinus strobus*, *Acer rubrum*) and there is evidence to suggest that, like their European counterparts, they have been derived from former forest. Whether this change was, perhaps, a consequence of the climatic shift at the onset of the sub-Atlantic period, or whether man played a significant part remains an open question. The soils are podsolic, closely resembling in appearance and properties those of the more northerly heaths in Europe.

(*ii*) Dwarf-shrub heaths replace forest and scrub vegetation as the climate becomes increasingly severe in sub-arctic and arctic regions. The communities concerned are virtually circumpolar in distribution, many of the most characteristic species such as *Vaccinium uliginosum*, *V. vitis-idaea*, *Empetrum hermaphroditum*, *Rhododendron lapponicum*, *Cassiope tetragona*, etc., appearing, for example, in Canada, Greenland and Scandinavia. Similar communities extend to more southerly latitudes on exposed oceanic islands such as the Faroe Islands and even Shetland (*Empetrum hermaphroditum* heath, Spence 1960).

Arctic heaths give place in still more exacting environments to tundra, or to the more open vegetation of fell-field. These formations are distinguished generally by the prominence of Cyperaceae, Gramineae, prostrate woody plants such as species of *Salix*, *Dryas*, etc., and mosses or lichens, although ericoid dwarf-shrubs extend into many of the communities.

In the southern hemisphere, land surfaces in the climatic zone suitable for this type of vegetation are much more restricted in area. However, a type of heath based on the genus *Empetrum* occurs in the southern extremity of South America and on some of the islands of the South Atlantic including the Falkland Islands where it is widespread. The latter may be regarded as almost sub-antarctic in character, although not in location. The species concerned is *Empetrum rubrum* which forms a dense bush, somewhat taller than that of *E. hermaphroditum*. The appearance of a stand of this plant is very heath-like, and a number of associated species such as *Gaultheria antarctica, Pernettya pumila* and *Myrteola Myrtus nummularia* have a more or less ericoid dwarf-shrub habit. On a number of sub-antarctic islands, communities dominated by *Azorella* spp. (Umbelliferae) or *Acaena adscendens* (Rosaceae) are sometimes described as 'heath', but although of about the same stature the communities are not, in fact, characterized by ericoid dwarf-shrubs.

(*iii*) Heaths occur at high altitudes in various parts of the world. They are conspicuous above the tree-limit on the mountains of western Europe, from Scandinavia and Britain to the Pyrenees, as might be expected from the prevalence of heath species in this region and the extent of heaths at lower altitudes. They range also to the mountains of central, eastern and southern Europe at appropriate altitudes.

In South Africa, another area of rich speciation of the genus *Erica*, heath-like mountain 'bush' occurs above about 900 m near the coast and also further inland at altitudes where rainfall and summer mists maintain high humidity in association with cool temperatures. As in other heaths under these conditions, raw humus accumulates at the soil surface. Vegetation described as 'moorland' occurs near the summits of many of the east African mountains, particularly those exposed to the more humid conditions. Hedberg (1951) refers to an 'ericaceous belt' on these mountains, the lower parts of which are occupied by 'tree-heath' or tall ericoid scrub which falls outside the definition of heath adopted here. However, this grades into open 'moorland' at the higher altitudes where in addition to considerable cover contributed by tussock-forming grasses, ericoid dwarf-shrubs such as species of *Erica* and *Philippia* are prominent. Like the European heaths, such communities are to some extent fire-resistant and the zone they occupy has in many instances been extended downwards by burning.

Similar communities occur under equivalent conditions in the mountains of New Guinea (Plate 2), eastern Australia and Tasmania. The species concerned are almost entirely different from those of heaths in other parts of the world, but structurally the resemblance is evident. Among the dominants concerned in Australia are species of *Epacris* and *Kunzea* which form a low heath, and a number of taller dwarf-shrubs which become low and prostrate in the more exposed conditions (Costin, 1959, 1967; McVean 1969).

Most of the heath-types referred to, whether lowland or upland, merge into other, taller types of sclerophyllous scrub or woodland on passing into conditions favourable for larger shrubs and a more complex stratification. In south-west Europe, for example, where oceanic cool-temperate conditions give place to the Mediterranean climate, maquis scrub appears instead of heath, while environmental gradients associated with descending altitude in numerous parts of the world cause parallel changes in community structure. Depending on the evolutionary history of the floras concerned, there is variation in the extent to which heaths stand out as distinct from adjacent vegetation-types. But the chief structural features of the heath formation are reproduced with some regularity even in widely separated regions, and in appearance the communities may show striking resemblances even when composed largely of unrelated species. Evidently the ericoid dwarf-shrub life-form, which contributes so much to the structure of heathland communities, is of adaptive significance under the particular environmental conditions obtaining in the regions described.

2 Environment and ecological history of European heathlands

As indicated at the close of Chapter 1, examples of communities having the physiognomic characteristics of dwarf-shrub heath occur in many parts of the world. However, they are for the most part limited in extent, except in western Europe where at least until recently they constituted a major element of the vegetation in lowland as well as upland parts of the region. This regional importance has earned the European heaths special attention, which they have received as much on account of their economic and social significance as of their scientific value. In addition, the dominant role of a few species of the family Ericaceae, together with other features of floristic composition, distinguishes them rather sharply from communities of similar structure in other regions. Thus they have a character and a unity, serving to establish them as a category deserving the exclusive treatment given in this book.

Heath communities occur at various altitudes in the mountains of southern and central Europe as well as those of Britain and Scandinavia, but the 'heath region' of Europe is best expressed by the extent of heathland as a lowland vegetation type (Fig. 2). Such heathland belongs essentially to the oceanic and sub-oceanic regions of west Europe, notably the broad west European coastal plain in countries bordering the North Sea and the English Channel. This lowland heath-region includes south-west Sweden (particularly the provinces of Halland – Plate 3 – and Skåne), Denmark (especially Jutland), the north German plain, parts of the Netherlands and Belgium, southern England (East Anglia and the southern and south-western counties) and north France (Normandy, Brittany). Fine examples of heathland country are also found on the lower slopes or foothills of mountainous territory, as in south-west Norway and in northern England and Scotland, where they have come to be associated with all that is most typical of the scenery of these areas. Northwards, heaths are well represented in the Atlantic islands of Orkney, Shetland and the Faroes, and occur also in northern Scandinavia, including Iceland. To the south, they extend along the Bay of Biscay, particularly in the 'Landes' region of south-west France, and into the narrow coastal belt between the mountains of north Spain and the sea. Both at the northern and southern extremes of their range, heaths give place to related types of vegetation, grading on the one hand towards tundra and on the other towards certain kinds of sclerophyllous

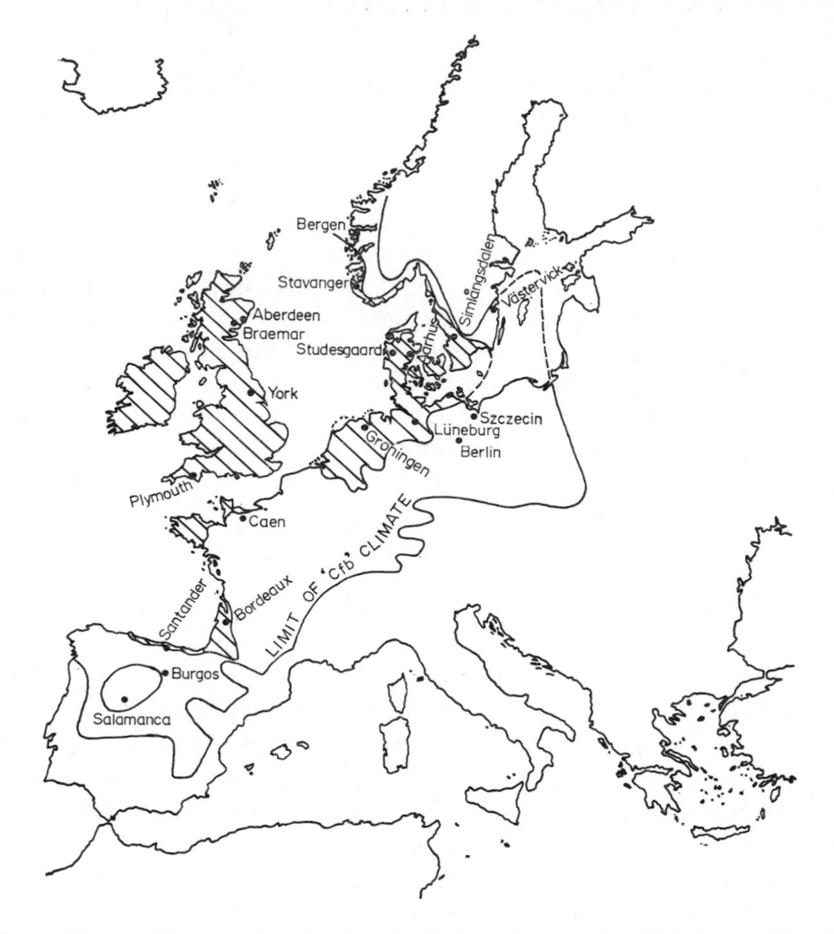

Fig. 2. Western Europe, showing (hatched) the main areas in which lowland heaths occur. Also shown are the localities from which the meteorological data given in Table 1 were obtained, and the approximate boundary of Köppen's 'Cfb' climatic zone.

scrub (Chapter 1). To the east, the representation of heath becomes increasingly restricted as the more continental zones are reached, for example in Poland and the Baltic states.

Climate

The climatic regime under which heathlands flourish may be judged from the geographical area just outlined. Fig. 2 shows that this area is completely contained within the zone of Köppen's 'Cfb' type of climate, i.e. a moist temperate climate with mild winters, in which the mean temperature of the warmest month is under 22°C and at least four months have mean temperatures above 10°C. This may be described as an oceanic type of climate, lacking temperature extremes, whether high or low, but with abundant and well-distributed rainfall and the maintenance of a generally high humidity. The limits of the heath region, however, fall well inside the boundaries of the 'Cfb' climatic zone. Conditions required for the development of heath may be judged from Table 1, in which meteorological data for localities within and beyond the main areas of heathland may be compared.

The heath region corresponds in general with the tracks of winter cyclones which pass along the north-west coast of Europe. For much of the winter, air-streams moving in over a warm underlying sea surface are drawn from the south-west northwards, and consequently maintain relatively mild conditions. In summer, the proximity of the sea has the indirect effect of limiting the rise of mean temperatures. Where the mean temperature for the hottest month exceeds *c.* 20°C, conditions are usually unsuitable for heath communities, and this factor may set a southern limit on their distribution. However, it is probably the associated summer evaporation stress which is important, rather than a particular temperature level. As a formation, heaths belong to a relatively humid belt where the duration of drought periods is very limited. This implies a relatively high rainfall and, more especially, one that is spread fairly evenly throughout the year. Rainfall varies within the approximate limits of 60 cm to 110 cm per annum in the heath region, but the mean number of rainy days in the year is seldom less than 115. Measurements of the average annual potential water deficit for stations in Britain (Green, 1964) give values of between 1·3 and 15·5 cm for areas in which heaths are most prevalent. Such a climate, in which the total amplitude of seasonal variation is not great, is characterized by relatively long spring and autumn periods, the latter being important in ensuring a gradual hardening off with the approach of winter.

On the north-west fringe of the heath region, for example in the west of Ireland, western Scotland and parts of the extreme west of Norway, conditions become 'hyper-oceanic'. The average annual potential water deficit falls to below 1·25 cm and under these conditions heaths are less well-developed. This is not to say that they are entirely lacking, but that they are restricted to slopes or particularly freely drained soils, while related types of peat-forming vegetation are more widespread (e.g. blanket bog). An 'index of oceanicity' (Poore and McVean, 1957 – a modification

TABLE 1

Climatic means from localities within and outside the heath region

Except where indicated, the data are derived from *Tables of Temperature, Relative Humidity and Precipitation for the World* (Meteorological Office (Air Ministry) 1968).

Locality (see Map Fig. 2)	Mean temperature (°C) January	Mean temperature (°C) July	Rainfall mean for year (cm)	Rain days: mean for year	Mean relative humidity at mid-day in the dryest month (%)
A. Localities in the heath region					
NORWAY					
Stavanger	1·4	15·3	108·5	153[1]	64 (May)
SWEDEN					
Simlångsdalen*	−1·4	17·0	103·6	–	–
DENMARK					
Studsgaard	−0·6	16·1	79·0	133[1]	62 (May)
Aarhus	−0·6	16·7	67·6	117[1]	59 (May)
UNITED KINGDOM					
Aberdeen	3·0	14·2	83·8	199[2]	70 (June)
Braemar	1·1	13·1	92·5	202[2]	
York	3·6	16·7	62·7	179[2]	76 (June)
Plymouth	6·4	15·8	96·0	183[2]	72 (April, May)
NETHERLANDS					
Groningen	1·4	17·5	74·7	203[3]	59 (May)
GERMANY					
Lüneburg	0·0	17·2	61·2	123[1]	55 (May)
FRANCE					
Caen	4·7	17·2	68·1	155[3]	
Bordeaux	5·3	20·6	82·1	164[3]	54 (August)
SPAIN					
Santander	9·2	19·2	112·0	177[3]	71 (March)
B. Localities outside the heath region					
(i) 'Hyper-oceanic'					
NORWAY					
Bergen	1·7	16·4	200·2	180[1]	65 (May)
(ii) To east and south of heath region					
SWEDEN					
Västervik	−1·9	17·2	53·6	97[1]	59 (May)
POLAND					
Szczecin	−1·1	18·1	56·1	108[1]	52 (May)
GERMANY					
Berlin	0·8	18·1	58·7	108[1]	50 (May)
SPAIN					
Burgos	2·2	18·3	51·2	113[3]	46 (July)
Salamanca	6·4	21·1	42·2	93[3]	40 (July, August)

* Data from Malmström (1937).

The definition of 'rain-days' unfortunately varies from country to country, as indicated by suffixes to the figures in column 4:

1. Average number of days per year with 0·04 in (0·10 cm) of rain or more.
2. Average number of days per year with 0·01 in (0·025 cm) of rain or more.
3. Average number of days per year with 0·004 in (0·01 cm) of rain or more.

This must be taken in account when comparing the figures.

of Kotilainen's index*) can be used to characterize such areas, where it reaches values from 330 to 400 or more.

Climatic conditions throughout much of the region are therefore well within the range required for forest vegetation, but to the north and with increasing altitude dwarf-shrub heaths extend beyond the tree-limit. Winter snow-lie is in general beneficial to heath communities and, particularly on mountains, enables them to occur in situations where conditions would otherwise be too severe. Nevertheless, there is a critical duration beyond which the period of freedom from snow cover becomes too short for the survival of dwarf-shrubs as community dominants. Hence, while there are well-defined chionophilous heath associations (e.g. 'Vaccinietum chionophilum, McVean and Ratcliffe, 1962), these give way to communities dominated often by *Nardus stricta* where snow-lie is most prolonged. The fact that dwarf-shrub heaths occur at higher altitudes in the central European mountains (up to about 2500 m) than in Britain (up to about 1100 m) is no doubt due in part to the relatively uninterrupted snow cover during winter on the former. However, the depression of the zone of alpine heaths on British mountains is caused not only by the variable nature of the protective snow blanket but also to persistent cloud and mist reducing the effective growing season, and to other factors.

Heaths are tolerant of very considerable exposure to high winds and this no doubt in part explains their prevalence in mountain and coastal habitats, including islands. Forest gives way to heath in such environments even under temperature conditions adequate for tree growth.

Soils

Throughout its numerous variations the heathland flora is essentially acidophilous; hence the communities are generally indicative of oligotrophic, acid soils. The geographical area outlined above is one which contains high proportions both of podsolic soils and brown earths. In a very general way, these soil types are associated respectively with coniferous and broad-leaved deciduous forest as the natural regional vegetational complexes. Like the coniferous forest, ericaceous dwarf-shrub vegetation influences soil development in the direction of podsolization. Hence, the presence of well-developed heath vegetation on a freely-drained substratum is generally associated with evidence of podsolization. Characteristic podsol profiles are therefore to be found, associated with heathlands, interspersed with other soil types in areas both within and considerably to the south of the main European podsol zone. In the latter case they may often have been derived

* $K = \frac{N.\mathrm{d}t}{100\Delta}$ where N = precipitation in mm; dt = number of vernal or autumnal days with mean temps. 0°–10°C; Δ = difference between mean temps. of warmest and coldest months.

by degeneration of former brown soils, a process which was intensified, though probably not initiated, by the spread of heaths (Dimbleby, 1962; Iversen, 1964).

The parent materials of heathland soils range from wind-blown sand, through fluvio-glacial and other sandy and gravelly deposits, to glacial tills and the products of various rocks weathering *in situ*. With certain exceptions discussed below, heaths are absent or poorly represented on soils rich in exchangeable calcium (or in mineral nutrient ions in general). Thus, while dune heaths form on stabilized siliceous sand (silicate syrosem), they are lacking on calcareous shell-sand; similarly, in upland areas heaths are characteristic of soils found on the older, harder igneous rocks such as granites and non-calcareous schists, and less well represented on those derived from the younger, sedimentary rocks rich in the clay fraction or in calcium. Where, however, high precipitation and low temperatures together cause the soils to lose cations faster than they are released, heath may be found, for example, over slates and basic igneous rocks (e.g. in the Cabrach area, north-east Scotland). The most extensive and vigorous heath communities occur on freely-drained soils in which the rooting region, while occasionally deep, may often be quite shallow.

In Kubiena's (1953) terminology, the soils concerned include several types of podsol and semi-podsol, oligotrophic braunerde (even occasionally eutrophic braunerde, Coombe and Frost, 1956) and ranker soils (tundra ranker, dystrophic rankers, podsol rankers and occasionally brown rankers). Heath vegetation may also colonize purely organic deposits (peat), while humic gley soils are associated with wet heath communities. A very characteristic form of the podsol which develops under heath on glacial till, for example in Sweden, is an iron-podsol in which the A_1 horizon (of leached mineral grains mixed with organic material) is well marked and dark in colour. Sometimes the pale-coloured A_2 horizon of leached mineral grains free of organic matter may be thin or even lacking. In the B horizon there may be evidence of iron deposition spread throughout a fairly deep zone. Sometimes, however, particularly in upland regions of north England and Scotland, iron is deposited in the form of a thin hard iron pan ('thin iron pan soils', FitzPatrick, 1964) which may in time be the cause of impeded drainage and surface waterlogging leading to a change in vegetation and consequent peat formation.

On the more sandy soils a little further south, for example in Jutland, north Germany and parts of the south of England, iron-humus or humus podsols are characteristic of heaths. Here, at the upper part of the B horizon, a striking dark and often hard deposit of humic materials is found, known as a humus pan or humus ortstein. Below this there may or may not be obvious signs of iron deposition; sometimes the pan is projected downwards at intervals in deep tongues. Some of these sands date from

Tertiary times and are composed largely of quartz: in these instances the A_2 horizon is generally very thick.

Certain types of heath community may establish on rather richer, less readily podsolized substrata. For example, heath soils over serpentine at The Lizard, Cornwall, have been described by Coombe and Frost (1956) as brown rankers and eutrophic braunerde. The terms 'limestone heath' and 'chalk heath' have been used by many authors to describe communities containing both heath plants and calcicolous species (Moss, 1913; Tansley and Adamson, 1926; McLean, 1935; Hope-Simpson, 1941; Thomas, 1957; Grubb, Green and Merrifield, 1969). In many instances, such communities are associated with shallow non-calcareous deposits such as loess or clay-with-flints overlying chalk or limestone, or with decalcified drifts over limestones as for example in the north of Ireland. Where the pH at the surface lies between 5 and 6, *Calluna* and other ericaceous species can establish even if their deeper roots penetrate into the calcareous material below. It has been suggested that conditions suitable for their establishment may sometimes develop even in the absence of any admixture of non-calcareous deposits, presumably as a result of strong surface leaching. Once the heath species are established in quantity, their litter has a marked acidifying influence (Grubb, Green and Merrifield, 1969). Pollen analysis of heath soils by Dimbleby (1962) has shown that although the onset of podsolization may in many instances have preceded the entry of *Calluna*, its rise to dominance is closely linked with increasing acidification, the disappearance of deep-burrowing earthworms and the accumulation of raw humus.

Dwarf-shrub heaths also occur widely on acid peat, although they do not represent the original vegetation of such areas and do not contribute very substantially to peat formation. 'Wet heath' communities are, however, closely related both floristically and ecologically to peat-building communities of acid mires, rich in *Sphagnum* species. Slight drying-out at the peat surface or improvement of drainage and aeration may lead to replacement of bog vegetation by wet heath, which is also characteristic of the sloping surfaces at the margins of bogs, of wet acid hillsides and of poorly drained hollows in mineral substrata. Further drying-out and aeration of acid peats converts them into particularly favourable habitats for the development of luxuriant *Calluna*-dominated heaths, floristically similar to those of podsolic soils. Such heaths are widespread on lowland and hill peat in the north of England and Scotland; elsewhere, as in Scandinavia, the heath phase in such habitats has generally been temporary owing to rapid colonization by trees. The history of many peat deposits in this region, as shown by their stratigraphy, has included several changes from bog to heath or woodland vegetation and back again, which may be correlated with fluctuations in the wetness of the climate. At the present time, many additional factors including artificial drainage, burning and cultivation of the surrounding territory have contributed to the establishment of heaths on peat.

Certain physical, chemical and biological properties may be taken as characteristic of the majority of heath soils. They are markedly acid, although seldom reaching the extremes of acidity exhibited by certain types of bog. pH in the great majority of habitats carrying well-developed heath falls in the range 3·4–6·5. In most parts of the heath region the water content of the soil is maintained at, or often well above, field capacity throughout the greater part of the year and periods of water stress are of relatively short duration. In some regions soil water may be frozen for a time during the winter. However, heath development appears to be best where oxidizing conditions prevail at least during the summer (Pearsall, 1938). The mor (raw humus) under ericaceous heath, however, is not a well-aerated material (Rennie, 1961) and is often readily waterlogged. If this condition should become permanent, heath will give place to peat-building communities.

It follows that nitrification is probably not very active in these soils, which are correspondingly generally deficient in available nitrogen and show high C/N ratios, often exceeding 15 (Duchaufour, 1948, 1950). Compared with other soils the fungal populations tend to be rich, bacterial ones relatively poor. Deep burrowing earthworms are usually lacking, but the upper horizons may contain populations of small worms. Acarina figure prominently in the soil fauna.

Heath soils are characteristically poor in exchangeable cations, and show low levels of cation saturation. In particular, the content of exchangeable calcium is low. Apart from this, it is difficult to generalize about the levels of particular mineral nutrients since so many different parent materials are concerned. However, it is frequently observed (in Britain, Scandinavia, etc.) that heath soils are deficient in available phosphorus; in some cases they are deficient in potassium (Chapters 11, 12).

Past history and theories of origins

The foregoing summary of the geographical, climatic and edaphic ranges within which heathlands occur in western Europe is sufficient to indicate that there is no simple solution to the problem of their origins. Although bearing definite relationships to a particular type of climatic regime and to a certain range of soil conditions, heaths cannot be treated as a 'regional vegetational complex', since their area lies within the zones of two major west European forest types which are represented side by side with the heaths. Nor can they be generally regarded as edaphically determined, since for the most part (though not exclusively) the habitats occupied by heath are equally capable of supporting the structurally more complex forest communities. Invasion of heathlands by trees is a common occurrence; yet areas also exist in which it is difficult to promote tree establishment.

Several different interpretations of the status of heath vegetation have been advanced. Most authors, however, are in agreement that in a certain limited range of habitats heath communities have for long been the characteristic vegetation. These are habitats from which trees are naturally excluded, such as the altitudinal zones above the tree limit on mountains, and certain coastal locations. McVean and Ratcliffe (1962) point out that in continental climates the dwarf-shrub is the characteristic life-form of the low alpine zone; as mentioned above, dwarf-shrub heaths occupy this zone in the central and north-west European mountains. In oceanic climates the corresponding zones occur at lower altitudes, but again mountain heaths are widespread. Beijerinck (1940) held the view that, despite many instances throughout the whole region of heaths coming and going as a result of human intervention, this aspect could be overemphasized. He preferred to recall that from 'times immemorial' heaths have developed entirely naturally on dunes, drained peat and mountain slopes. This viewpoint made him reluctant to think of heath as an 'artificial' vegetation-type, although its extent may have been greatly increased by the hand of man. Tansley (1939), employing terminology adapted from Clements, concluded that on exposed coasts and mountain slopes 'heath is certainly a climax'.

The presence of heath communities in mountain and coastal districts throughout much of the post-glacial period must therefore be taken into account in considering the origins of the widespread lowland heathlands. A further point of importance is that there is a strong floristic affinity between heath communities and the lower strata of certain forest communities belonging to the boreal west European pine (*Pinus sylvestris*) or spruce-pine (*Picea abies-Pinus sylvestris*) forests, and even with similar strata of oak-woods (*Quercus* spp.) on acid soils (Table 2). Hence, within the areas occupied by these forests many of the species characteristic of heaths have for very long periods been present in the dwarf-shrub and ground strata, constituting in open spaces and clearings fragments of communities closely resembling those of heathlands in structure and floristic composition. Moreover, in the drier climatic periods many peat bogs in a 'stillstand' condition, as regards peat growth, were covered with heath-type communities.

Difference of opinion has arisen, however, as to whether the majority of heathlands are to be regarded as seral and 'sub-climax' communities, or whether they have been derived by degradation of former forest. The former point of view was expressed by Tansley (1939) with regard to English heaths, arising from the suggestion that 'many of them have been subject to burning or pasturage, or both, from the earliest times, and perhaps have never borne forest at all'. Hence they were interpreted as sub-climax or 'deflected climax'. There can, perhaps, be little doubt that on fixed acidic sand dunes, heaths have originated as part of the prisere on

TABLE 2

List of species which belong both to Heathland and Woodland communities

Shrubs and Dwarf-shrubs		
	Arctostaphylos uva-ursi	*Juniperus communis*
	Calluna vulgaris	*Vaccinium myrtillus*
	Empetrum nigrum	*V. uliginosum*
	Erica cinerea	*V. vitis-idaea*
Pteridophyta		
	Blechnum spicant	*Polypodium vulgare*
	Lycopodium annotinum	*Pteridium aquilinum*
Monocotyledons		
	Agrostis canina	*Listera cordata*
	Agrostis tenuis	*Luzula multiflora*
	Carex binervis	*Luzula pilosa*
	Deschampsia flexuosa	*Luzula sylvatica*
	Festuca ovina	*Molinia caerulea*
Herbaceous dicotyledons		
	Anemone nemorosa	*Oxalis acetosella*
	Chamaepericlymenum suecicum	*Potentilla erecta*
	Galium saxatile	*Pyrola media*
	Hypericum pulchrum	*Trientalis europaea*
	Lathyrus montanus	*Veronica officinalis*
	Melampyrum pratense	*Viola riviniana*
Bryophyta		
	Dicranum scoparium	*Polytrichum formosum*
	D. rugosum	*P. commune*
	Hylocomium splendens	*Pseudoscleropodium purum*
	Hypnum cupressiforme	*Ptilium crista-castrensis*
	Leucobryum glaucum	*Rhytidiadelphus loreus*
	Mnium hornum	*R. triquetrus*
	Plagiothecium undulatum	*Lophocolea bidentata*
	Pleurozium schreberi	*Ptilidium ciliare*
Lichens		
	Cladonia arbuscula	

blown sand and are actually or potentially in a state of seral change. However, to regard most other English heaths 'as a stage in the succession to forest' seems now, with hindsight, to suggest that the interpretation has been forced to fit a particular framework derived from Clements' emphasis of seral relationships. Tansley himself drew attention to Graebner's (1901) much earlier evidence for the derivation of north German heaths from forest, a conclusion which was also accepted by Smith (in Tansley, 1911) for many Scottish areas.

In Denmark, P. E. Müller (1924) took the view that heathlands had followed tundra-type communities under which indurated horizons had formed in the substratum, creating conditions in which the forest could not invade. This, although a different interpretation suggesting edaphic determination, also implies that heaths were not generally preceded by a more complex vegetation-type. However, with the accumulation of further evidence the conclusion has become inescapable that over the greater part of their region, heathlands have been derived from pre-existing forest. None the less, it would be unwise to suggest that all heathlands conform to any simple historical pattern and the only sound conclusions are those based on evidence relating to particular areas.

Following the indications that, at least in many areas, territory now heathland was formerly forest-covered, discussions have centred on the possible cause of this change. Graebner (1901), working before detailed vegetational histories could be constructed with the aid of pollen analysis, accounted for the degeneration of oak forest to heath on the sandy soils of north Germany by suggesting that this was the outcome of progressive podsolization assisted by the removal of timber. Podsolization processes were regarded as the result of a natural trend towards acidification of the substratum inherent in oceanic climatic conditions, which has been postulated by several authors including Salisbury (1921) and Pearsall (1950). The relationship between deterioration of forest, followed by gradual encroachment of heath, and the accumulation of raw humus accompanied by leaching, has more recently been further investigated in France by Duchaufour (1948, 1956) and in Britain by Dimbleby (1962). The former, like Iversen (1964), supports the view that these were natural processes accelerated by the influence of man, but the latter ascribes a decisive role to human impact upon the original forest.

Great caution, however, is required in assigning the role of cause to any one factor concerned in these changes. The undoubted importance of vegetation-type in promoting podsolization has sometimes led to the assumption that this process must be a consequence, rather than a precursor of the shift from forest to heath, at least in the deciduous forest region (Jonassen, 1950). None the less, Waterbolk (1954), Dimbleby (1962) and Iversen (1964) have found evidence from buried soils that *Calluna* is not necessarily the initial instigator of podsolization.

Early stages in the development of heath: results of pollen analysis

Further important evidence concerning the nature of the vegetational changes has become available from the results of pollen analysis. Peat profiles from localities in the heath region yield a picture of the forest history of the surrounding district, and the ratios of non-arboreal to arboreal pollen provide indications of changes in the extent of forest cover. The non-tree pollen, after allowances for strong representation of species

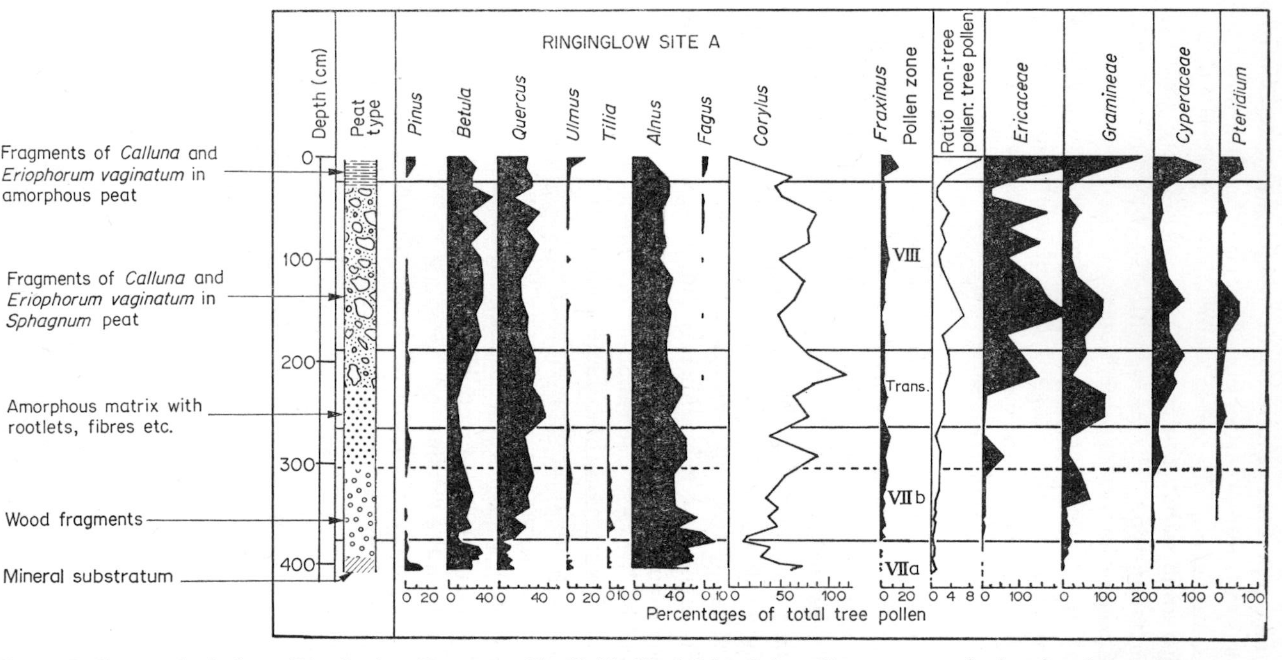

Fig. 3. Pollen analysis from Ringinglow Bog near Sheffield, Yorkshire (after Conway, 1947) showing increasing representation of Ericaceous pollen, and other changes, after the end of Zone VIIb (i.e. about the Sub-boreal–Sub-atlantic transition). The dotted line indicates the ending of woodland cover on the bog.

belonging to the local peat-bog surface, can be a valuable guide to the prevalence of heaths, grasslands or cultivated land in the surrounding countryside. The methods of pollen analysis have also been extended to deposits of mor and to soil profiles generally (Dimbleby, 1962; Iversen, 1964, 1969), contributing very significant data on the changes culminating in the establishment of heath communities.

Pollen diagrams from heathland localities provide convincing evidence that forest vegetation almost universally preceded the heaths. It is usually not until the upper parts of the profiles are reached that the proportion of tree pollen is seen to decline in relation to that of non-arboreal pollen. A marked expansion in the curve for Ericaceous pollen at about the same time indicates the development of open heath following reduction in forest cover (Figs. 3, 4). Although this pattern of change is repeated throughout much of the region, there was evidently great variation in the time of its occurrence. Indications of a shift from forest to heath as long ago as the

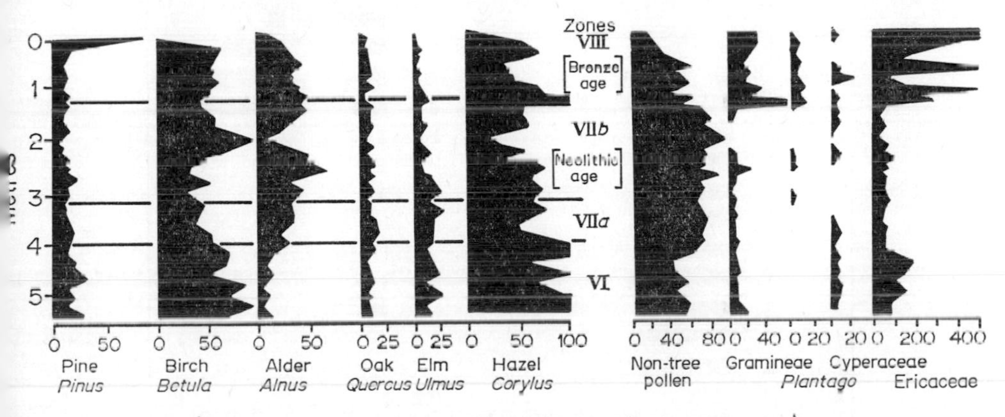

Fig. 4. Pollen analysis from Dalnaglar, Perthshire, central Scotland (after Durno, 1965). The fluctuations in the representation of Ericaceous pollen and the occurrence of *Plantago* in Neolithic and Bronze Age times provide evidence of 'landnam'.

Atlantic period (pollen zone VIIa, *c.* 4000 B.C.) have been found at a few localities (e.g. Mesolithic sites in Sussex and Hampshire, southern England: Rankine, Rankine and Dimbleby, 1960; Keef, Wymer and Dimbleby, 1965). Generally, however, it is not until the Sub-boreal (pollen zone VIIb, *c.* 3000 B.C.), that signs of forest deterioration first appear. This, together with much additional evidence, suggests that the process began to be effective in Neolithic times, gathering momentum in the Bronze Age (*c.* 1000 B.C.) and the Sub-boreal–Sub-atlantic transition which marked the beginning of the Iron Age (from *c.* 500 B.C. – Fig. 3). Sometimes there was a period of fluctuations of tree and Ericaceous pollen (pp. 24–5), indicating more than one sequence of retreat and recovery of forest, before its lasting replacement by heath. The expansion of heath at the expense of forest continued into Sub-atlantic and historical times (pollen zone VIII, Fig. 3).

The decline of forest and increase of heath is frequently associated with signs of human settlement and land cultivation, which include archaeological remains as well as the occurrence of the pollen of cereal crops and

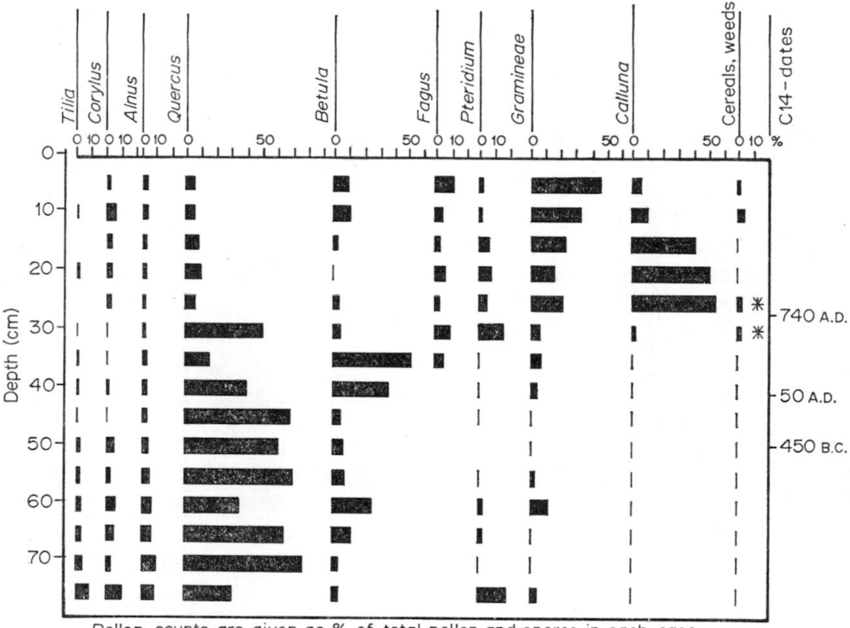

Fig. 5. Pollen analysis from a mor profile in Draved Forest, Jutland, Denmark (after Iversen, 1964). In this example, the sudden increase in *Calluna* pollen is dated at about A.D. 740 and coincides with the appearance of charcoal, and with pollen of cereals and weeds, indicating clearance of the forest by fire, followed by grazing and agricultural activity.

weeds such as *Plantago lanceolata* (Figs. 4, 5). The conclusions to be derived from these correlations are, however, by no means straightforward. On the one hand, it can be maintained that the role of man in destroying the woods was of prime importance in causing the change from forest to heath. On the other, it has been pointed out that the process generally started about the time of a widely recognized climatic change in the direction of a cooler, more oceanic regime (Sub-atlantic), favouring heath as against tree growth. Such an environmental shift may have been sufficient to alter the balance towards heathland development, with man playing a secondary part in accelerating a change for which the underlying cause had already been established.

No doubt numerous factors have contributed in various combinations to the final result, and on the basis of the available facts only speculation is possible as to the fundamental cause, which may differ from place to place. It is instructive, however, to compare the published data on the origins of heathland vegetation from different localities and to notice that according to local conditions it is possible to place widely differing interpretations (which may, indeed, be substantially correct) upon fundamentally similar sets of observations.

Faegri (1940) presents data from pollen analysis of peat deposits in the Jaeren peninsular of south-west Norway, where the environment today is not particularly favourable for tree growth. His findings led him to believe that, at least for the south-western coastal district of Norway, the disappearance of forest and its replacement by heath was the outcome of a climatically determined predisposition towards this change, which human cultural influences merely served to actualize. He discusses the almost universal modification of woodlands by man from early times, pointing out that these have by no means always caused their destruction. Severe encroachments have been made upon Finnish forests, for example, which none the less have survived in this more continental zone. From this it is argued that in south-west Norway, where conditions became colder and more oceanic early in the Sub-atlantic period, the climate was largely responsible for shifting the balance in favour of heath. There is some support for this interpretation from elsewhere in the oceanic northern parts of the heathland region. In north Scotland (east Sutherland and Caithness) pollen diagrams show pronounced and lasting increases in Ericaceous pollen, with corresponding decreases in tree pollen, before any significant interference by man is believed to have taken place in this area (Durno, 1958). Bøcher (1943) also takes the view that in the Faroes as well as in west Norway heaths may be regarded as 'natural', governed primarily by climatic factors.

On the other hand, Swedish authors agree in emphasizing the role of man in the origin of heathlands, particularly as regards southern Sweden. Romell (1951, 1952) attaches special importance to the fact that the relatively mild winters of the south-west permitted winter grazing, which he believes was a potent factor in preventing forest regeneration. However, Tamm (1948) and Malmström (1937, 1939) place more emphasis upon burning as the means by which forest clearance was generally initiated. The change from forest to heath took place over extensive areas in southern Sweden but apparently began rather later than in most of the countries bordering the North Sea, mainly in late medieval times and up to the sixteenth century in the south-west (Atlestam, 1942; Malmer, 1965), though perhaps somewhat earlier on certain islands off the south-east coast (mid-Sub-atlantic, *c.* A.D. 400–1000: Berglund, 1962).

Among the earlier detailed pollen analytical investigations bearing on this problem were those of Iversen (1941, 1949), in Denmark. Attention was drawn to increases in the quantity of Ericaceous pollen coincident with declines in the tree pollen, starting as early as the Neolithic period (pollen zone VIIb, *c.* 2500 B.C.). At first these were temporary fluctuations in quantities, preceding a more general decline in forest tree pollen. However, the temporary reductions in the predominance of forest were with great regularity associated with the appearance of pollen of cereal crops and weeds of cultivation such as *Plantago lanceolata*, and with an increase in that of Gramineae, Cyperaceae and Ericaceae. All the evidence pointed towards an explanation in terms of local and temporary reductions in tree density caused by human interference. Three stages were usually evident; a stage of clearance possibly by the use of fire which left charcoal layers, a stage of agriculture, and finally a stage in which the forest returned after the area had been abandoned. This sequence was described by the term 'landnam' (land-taking) since it probably represented a system of shifting cultivation in which, after a period of time, a site was abandoned in favour of new clearances. This naturally permitted the temporary spread of species of grassland and heath. Several such reversible encroachments into the dominance of the forest evidently took place before a more permanent decline set in at about the beginning of the Iron Age (*c.* 500 years B.C.). This was associated with further increase in Ericaceous pollen, and with renewed evidence of agricultural activity (Fig. 5).

Jonassen (1950) added to the documentation of Danish vegetational history, describing pollen diagrams from Jutland, an area of once extensive heath development. He supported the conclusions of Iversen, arguing that in the change from forest to heath the destructive role of man was of prime importance, although doubtless the changing climatic conditions encouraged the spread of heath on the cleared areas. Early clearances were shown in the Neolithic and Bronze ages, from about 3000 B.C. onwards, in connection with agriculture and cattle grazing. The concentration of numbers of Bronze Age tumuli on heathland areas both in Denmark and England is not without significance in this connection. New clearances were made as the older ones reverted to woodland, and it is in these cleared areas that *Calluna* and its associated species are presumed to have spread. At first this spread was restricted by the capacity of forest to return on abandoned areas, but with the arrival of the Iron Age (about the Sub-boreal to Sub-atlantic transition) and contemporary deterioration of the climate, the open heathland began to expand. Jonassen suggests that on the morainic and fluvio-glacial out-wash plains of Jutland the environment was never highly favourable for forest, and that some climatic deterioration might be sufficient to change the balance in the direction of stabilizing heath vegetation, once it had appeared as a result of forest destruction. With the resulting intensification of podsolization under stands of *Calluna*, conditions became such that today forest cannot be re-established on certain parts of the heath plains without relatively elaborate methods of sylviculture. In Jonassen's view, however, this cannot be taken as evidence that a deteriorating environment was adequate cause of replacement of forest by heath, since much of the edaphic change is regarded as consequent upon the vegetational change. Similar conclusions are reached by other Danish authors (see Bøcher, 1943).

In Britain, the 'landnam' phases have been recognized in England by Godwin (East Anglia, etc.: 1944*a* and *b*, 1948, 1956), in Ireland by Mitchell (1951, 1956), Morrison (1959) and Smith and Willis (1962) and in Scotland by Durno (1965) (Fig. 4). Godwin's data from the East Anglian breckland, an area densely settled in Neolithic times, gives striking evidence of forest clearance during that period. 'The pollen curves appear to indicate that the heaths once established remained open, thereafter probably extending . . . during the Iron Age' (Godwin, 1956). Dimbleby (1962, 1965), writing on the development of English heathlands and their soils, concludes that the impact on forest began in earnest in Neolithic or Bronze Age times (Sub-boreal) and supports this view with striking evidence from soil profiles buried under Bronze Age tumuli. In some instances, brown soils were covered and preserved where today heathland and heath soils extend around and over the mounds; in others the buried soils showed signs of podsolization. The latter, however, are 'patently immature versions of the present podsols of the area'. It is evident that the tumuli were built at a time when profound changes were occurring in the soils and vegetation. This is borne out by the results of pollen analysis which indicate that the earthworks were generally constructed in clearings or open country. 'There is abundant evidence . . . to give a picture of the degeneration of forest.' Concerning the question as to whether or not these changes were induced by a shift in climate, Dimbleby shows that, although in many cases they can be traced to the Sub-boreal period, there are also numerous instances of a later start at various times through to the present. He is therefore firmly of the opinion that they were the outcome of disturbances of the ecological equilibrium by man, and points to accompanying signs of fire (fragments of charcoal, especially in the uplands) and pastoral or agricultural activities, the latter particularly in the lowlands. The beginning of podsolization probably preceded the entry of *Calluna* and other heath species, but as soon as raw humus started to accumulate the record passes to one of *Calluna* dominance. Substantially similar evidence has been given for localities in Scotland by Durno (1957, 1965). In Perthshire and coastal Aberdeenshire, for example, the major expansion of heath and retreat of forest is dated at about 500 B.C. and associated with Bronze Age occupation (Fig. 4), although further inland in the north it was delayed until more recent times (pollen zone VIII 'modern').

Waterbolk (1957) has added the Netherlands to the list of areas for which there is evidence of the origin of some heathland in Neolithic times before climatic deterioration could be responsible. However, both Neolithic and Bronze Age earthworks are fairly numerous in Holland. Most of the former cover unleached soils, but those buried under Bronze Age mounds are generally markedly podsolized, suggesting that the tumuli were constructed on already well-developed heathland sites. In north-west Germany also the destruction of woods and their replacement by heath has been attributed to the activities of man by Overbeck and Schmitz (1931), Tüxen (1938) and others.

Although in lowland areas the pollen of cereals indicates that some of the land cleared of forest was used for cultivation, much of it must have been devoted to grazing. It is often difficult to distinguish between these two types of activity on the basis of the pollen record alone, for *Plantago lanceolata* and other weeds may be associated with either. Pearsall (1934) held that grazing was the greatest single factor involved in the destruction of the woodlands of north England and Ellenberg (1954) argued that the pasturing of animals in susceptible types of woodland ecosystem was responsible for the origin of many of the north-west European heaths, even as far back as the Neolithic. However, although grazing may often have been responsible for initial thinning of the forest, the frequency of charcoal layers (Fig. 5) suggests that further clearance was in many instances achieved with the aid of fire. Such clearances as were made in the Mesolithic before the introduction of any kind of agriculture presumably also depended on the use of fire.

In addition to heathland, forest clearance led to an extension of grassland on the better soils. Some of these did not escape soil deterioration as time went on, and this may offer an explanation of the origin of many of the heaths of northern France which appear to have arisen more recently than those of the countries hitherto considered. Lemée (1938) and Vanden Berghen (1959) refer to the derivation of heaths in this area from abandoned meadows and waste land.

To summarize, in many areas the early stages of the derivation of heathlands from former forest are becoming increasingly well documented. It remains impossible to determine with any certainty the chief causes of this change, but there is a suggestion that climatic influence is of prime importance in some highly oceanic northern parts of the region, while further south there seems little doubt that man has been largely responsible. Bøcher (1943) concludes that 'it is very possible that the role of the climate dominates as the climate becomes more oceanic', while Durno (1957) emphasizes that 'by destroying the forests to make way for grazing and cultivation the early inhabitants also cleared the way for an expansion of heath'.

Later development

It would be misleading to suggest that more than a small proportion of forest was permanently destroyed as early as the Neolithic, Bronze and Iron Ages. The evidence from numerous pollen profiles from west Europe suggests, indeed, that this was the beginning of a long process, sometimes proceeding rapidly for short periods, at others almost halted. Much of the forest destruction took place within the historical period: some of the cleared areas were destined for cultivation, but a large proportion remained free from intensive cultivation as grazing land. On the more fertile terrain this became grassland but elsewhere heath expanded, right up to the late nineteenth century.

Evidence that certain areas of heath originated during the Viking period in Denmark is given by Iversen (1964, 1969). These are interpreted as heath-covered glades derived from beech–oak forest by clearing, associated with burning. 'Fire-horizons' mark a rather abrupt transition in the pollen record contained in a mor deposit, indicative of a change from forest to heath. The C_{14} content of charcoal from these horizons dates them at about A.D. 740 or A.D. 830. The heath-glades were evidently maintained 'by repeated burning until this practice was abandoned probably about two hundred years ago'. Iversen regards this as confirming Romell's view of the former importance of heath in north-west Europe, especially as a source of winter fodder (p. 23). There was also a pronounced phase of forest clearance about A.D. 1200. This no doubt led to the expansion of heathland, which in some districts further extended around 1500, in others about 1600, while elsewhere the greatest expansion was about 1800. In detailed maps of the province of Halland, south-west Sweden, Malmström (1939) traces the decline of forest from around 1650 to 1850, and quotes evidence that the process continued at least until 1890. Much the same picture can be drawn for Britain, where forest clearances were pronounced in England about the year 1100 (e.g. in the Pennine areas where heaths are now widespread, Conway 1947). Further destruction followed the expansion of agriculture, particularly in the thirteenth century. From the Iron Age onwards charcoal had been used for iron-ore smelting, and this, together with timber requirements, led to depletion of the forests. This had already reached severe proportions by the sixteenth century, and, apart from insignificant fragments of surviving native woodlands, was virtually complete by the end of the seventeenth. Much of the resulting open country was devoted to pasture and arable cultivation, but on acid soils, in the absence of intensive farming, heath communities were established. This took place principally in Wales and northern England but also, for example, on the sandy soils of the London and Hampshire basins and the upland areas of the south-west.

Forests survived longest in the more remote districts. In the Highland area of Scotland, they began to suffer probably at the time of the Viking invasion in the first and second centuries A.D. (Darling, 1947), but more extensive clearances awaited the sixteenth and later centuries when even here timber was in demand as charcoal for iron-smelting. The latter continued in this region until late in the eighteenth century. The increasing acreage of open country attracted sheep farmers who built up their flocks on the hill grazings of Scotland in the eighteenth and nineteenth centuries. This set the seal on the destruction of what remained of the native pine forest and its replacement by grass-moors, sedge-moors and, particularly in the eastern half of the country, by *Calluna* heathland.

Present trends in use, management and reclamation

Detailed consideration of ecological aspects of the uses of heathlands is deferred until Chapters 9–12 but an account of the origins and development of heathlands is incomplete without reference to contemporary changes. Throughout most of the heath region there has, in recent years, been a marked decline in the extent of heathland vegetation, much of which has been planted, cultivated or otherwise reclaimed or rendered more productive. Prior to about 100 years ago the pattern of land use on heathlands had been rather uniform, at least from south Sweden, through Denmark and north Germany, to Holland and Belgium. This represented a marginal type of agriculture in which use was made of the heath vegetation for grazing, usually with a necessarily low density of animals. Frequently, both cattle and sheep were grazed on heaths, and burning was employed irregularly to prevent the spread of trees or shrubs and to promote the growth of new young shoots of *Calluna* which formed the main food of the animals. The greater the importance of sheep, the more necessary it became to burn. In some areas, at intervals in the region of 25 years, sods of the surface humus were removed, either for use as fuel or after trampling down in cattle sheds and incorporation of manure, as a fertilizer for arable fields.

The result of this form of land use was effectively to maintain and perpetuate the heath vegetation over considerable areas. But with increasing agricultural efficiency on more fertile soils, and in particular with increasing output of dairy produce, it became economically less viable. In south-west Sweden, for example, sheep have almost disappeared in the last 100 years, and with them has gone the practice of heather-burning. Cattle-grazing on heather has also largely died out, since it became increasingly apparent that the heathlands could become both more productive and economically rewarding if they were employed to augment the country's most valuable resource, forest timber. Correspondingly, by means of afforestation the heathlands once more support a vegetation more akin to its

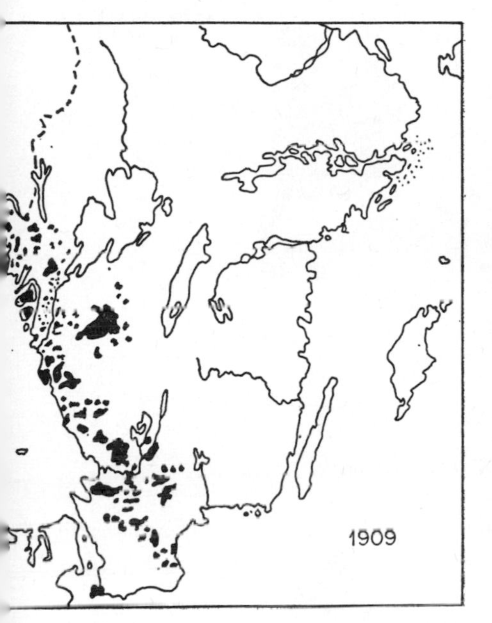

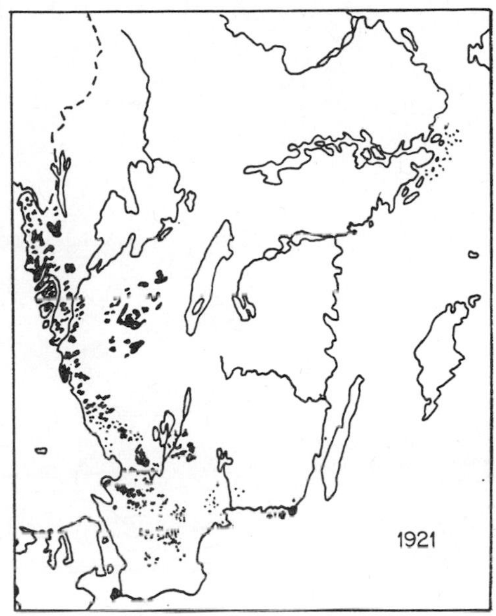

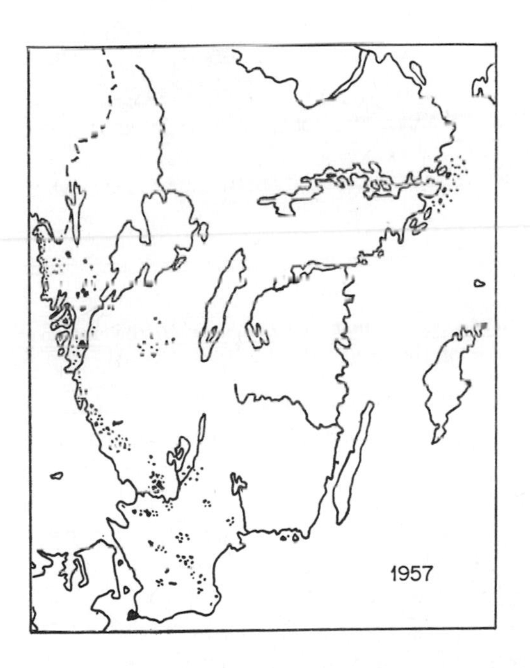

g. 6. Maps illustrating the decline in the extent of heathland in southern Sweden from 1909. (After Damman, 1957; the 1909 map from Schager and the 1921 map from Schotte.)

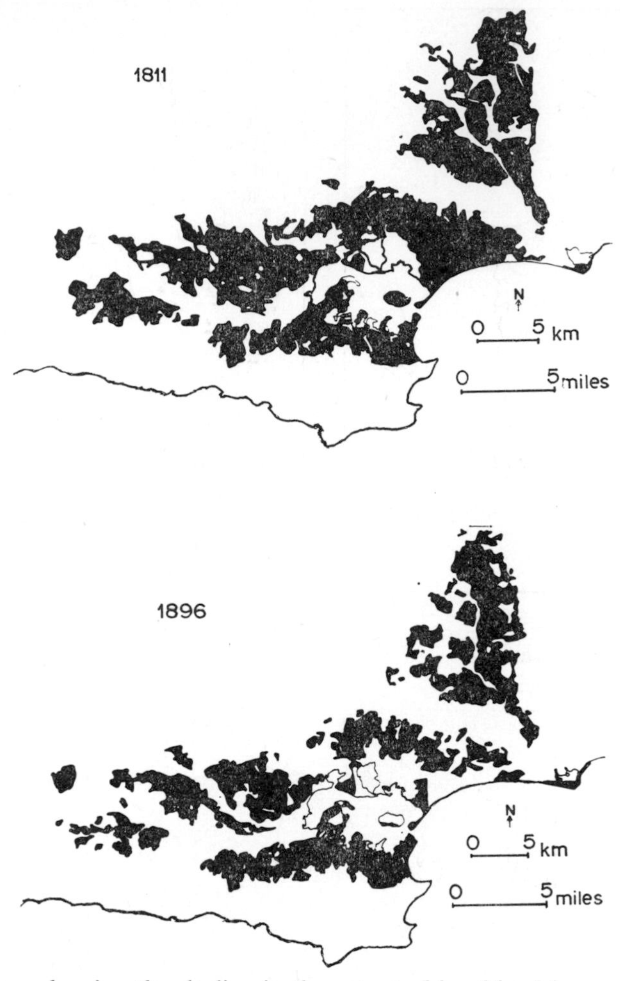

Fig. 7. Maps showing the decline in the extent of heathland in east Dorset and Hampshire west of the River Avon, from 1811. (From Moore, 1962.)

original cover. Clear illustration of the decline in the area of heath in south Sweden is given by three maps, dated 1909, 1921 and 1957 in Damman (1957) (Fig. 6). Malmström's (1939) map of the province of Halland in 1920 shows a great contrast to that of 1850, when heaths accounted for about 75% of the total non-cultivated area. Today they represent only about 5%, and their rate of disappearance has been such that it has become urgently necessary to establish some heath reserves in which this distinctive type of community will be preserved. This has been done in some parts of the province, where such areas provide a striking comparison with near-by 70-year-old forest, planted on former heathland.

A substantially similar history applies to the vast heathland area of

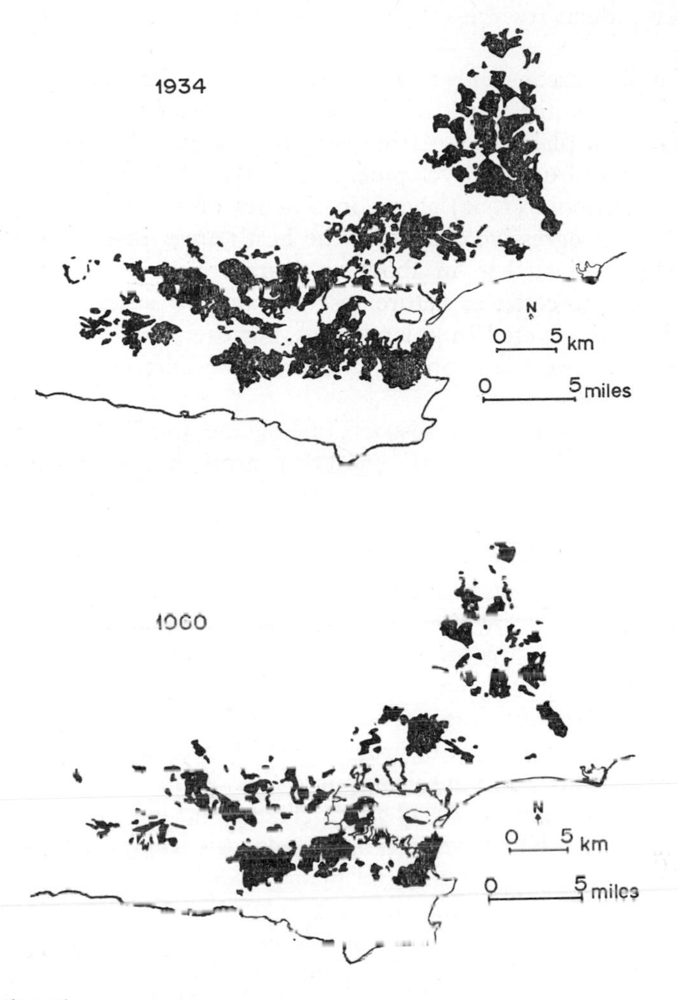

Fig 7 Continued

Jutland, in Denmark, with the exception that in this predominantly agricultural country the trend has been towards reclamation of heathlands for improved agricultural use as well as for forestry. Heathland has given place to arable land as a result of mechanized farming on the flat plains, together with considerable expenditure on soil improvement. Again the change was at its most rapid in the late nineteenth and early twentieth centuries. A remarkable private organization, the 'Danish Heath Society', was established in 1866 to encourage the depressed farming population of the Jutland heath plains to progress from their state of marginal to more productive farming and to establish tree plantations. The activities of the Society continue to have much influence and once again, apart from certain interesting areas set aside as reserves, the heathlands have been reduced to vanishing point.

For similar reasons, only relatively small areas of heath remain in Holland and Belgium, while much of the formerly more extensive heaths of the north German plains (apart from reserves) has given place to agricultural land, forest and urban development. In England the same is true of lowland areas. Moore (1962) shows in a series of maps from 1811 to 1960 (Fig. 7) the progressive decline of the heath areas in Dorset and Hampshire. He predicts that 'in about 30 years or less no heath will remain except that protected as nature reserves and for small areas of disused mineral workings, etc.' In upland areas, however, as for example Exmoor, Dartmoor, Wales, the Pennines, and the Lake District larger acreages of heathland remain.

In some of these upland districts of England and Wales, and more particularly in Scotland, there is a marked contrast in the present attitude to heathlands with that just described. In these areas, heaths are at present actively maintained and regarded as an integral part of the pattern of land use (although this attitude is subject to much current controversy). This is because sheep farming, although no longer quite so extensive or productive as in the eighteenth and nineteenth centuries, remains an important feature of the agricultural system. There is considerable demand for home-produced mutton and wool, and especially for hardy breeding stock, reared on the hills. At the same time, the potentialities of wide expanses of *Calluna*-covered hills for the sport of grouse-shooting were quickly recognized by land-owners who found in this a source of revenue which often more than made up for declining stocks of sheep. Today, this has perhaps lost some of the importance it had 50 years ago, but none the less it retains a very significant influence upon the use and management of heathland areas in the Highlands of Scotland. Since in these districts both hill sheep and grouse depend to a very large extent upon *Calluna* for food, a system of management involving regular burning to maintain uniform stands of young *Calluna* has evolved, and the interests of sheep farming and grouse rearing can be combined on one and the same area of heath. Although forestry has converted a significant and increasing amount of heathland, the result of these factors has been the maintenance of expanses of heathland not now paralleled elsewhere in Europe.

The main lowland examples of west European dwarf-shrub heathland therefore illustrate an unusual situation. They represent a largely man-made type of vegetation, now threatened with extinction as a result of changes in the pattern of land use. On the one hand, a thorough ecological understanding of heath communities is required if these changes are to lead to greater productivity with a minimum of habitat deterioration. On the other, conservation of adequate representative areas is a vital need both to avoid the loss of a richly varied type of vegetation and fauna of great ecological value, and to preserve the background of a way of life fast disappearing. In this instance, almost more than any other, conservation can be no passive procedure of mere protection, for the origins of heathlands – at least over much of their range – appear to have been a result of active intervention by man. Their maintenance demands continuation of at least some features of this intervention.

3 Community structure

The heath formation was identified in Chapter 1 as a type of vegetation dominated by ericoid dwarf-shrubs, with trees sparse or absent. This is a physiognomic description, based on those aspects of community structure which are responsible for determining the over-all appearance of the vegetation. It serves to differentiate heaths, in whatever part of the world they may occur, from other formations. However, considerable structural diversity is embraced within this broad definition, as illustrated by reference to recent systems of classifying vegetation according to its physiognomy and structure by Ellenberg and Mueller-Dombois (1966*a*) and Fosberg (1967). Extracts from their schemes, which were prepared as aids to survey and mapping, are given in Table 3.

The term formation is applied by these authors in a more restricted sense than that adopted above, while their 'formations' are linked in various ways to form categories of higher rank. These differences of procedure are of no concern here, but attention is drawn to the resulting formulation of several fairly distinct categories ('formations'), all of which fall within the compass of the definition of heath. In this way some indication is already given of extensive variation in the structural details of heath communities. Many important aspects of heathland ecology can be interpreted only in the light of a thorough understanding of this aspect of the organization of the vegetation, which must therefore be examined further.

The structure of a community is the product of an assortment of individual plants of differing form and mode of life, and is dependent upon the numerical proportions in which the several types are represented, their arrangement in space, and their mutual interactions. It may be analysed as follows:

(*a*) By determining the range of life-forms present. This contributes a summary of features of size, morphology and life-history, and may be elaborated by demonstrating which of the life-forms are predominant, and which of lesser significance. Detailed analysis along these lines establishes the extent of diversification into different synusiae, or groups of individuals of similar life-form making generally similar demands on the habitat.

TABLE 3

Extracts from classifications of plant formations by (1) Ellenberg and Mueller-Dombois (1966*a*) and (2) Fosberg (1967)

(1) Ellenberg and Mueller-Dombois (adapted)

Formation Class	IV DWARF-SCRUB AND RELATED COMMUNITIES, rarely exceeding 50 cm in height (sometimes called heaths or heathlike formations). According to the density of the dwarf-shrub cover are distinguished: *dwarf-shrub thicket* = branches interlocked; *dwarf-shrubland* = individual dwarf-shrubs more or less isolated or in clumps; *cryptogamic formations with dwarf-shrubs* = surface densely covered with mosses or lichens (thallo-chamaephytes); dwarf-shrubs occurring in small clumps or individually. In the case of bogs locally dominating graminoid communities may be included.
Formation Subclass	A. **Mainly evergreen dwarf-scrub**. Most dwarf-shrubs evergreen.
Formation Group	1. *Evergreen dwarf-shrub thickets*. Densely closed dwarf-shrub cover, dominating the landscape ('dwarf-shrub heath' in the proper sense).
Formations	(*a*) Evergreen caespitose dwarf-shrub thicket. Most of the branches standing in upright position, often occupied by foliose lichens. On the ground pulvinate mosses, fruticose lichens or herbaceous life-forms may play a role (e.g. *Calluna* heath.) (*b*) Evergreen creeping or matted dwarf-shrub thicket. Most branches creeping along the ground. Variously combined with thallo-chamaephytes in which the branches may be imbedded (e.g. *Loiseleuria* heath). (Subdivisions possible).
Formation Group	2. *Evergreen dwarf-shrublands*. Open or more loose cover of dwarf-shrubs.
Formation	(*b*) Evergreen mosaic dwarf shrubland. Colonies or clumps of dwarf-shrubs interrupted by other life-forms, bare soil or rocks (e.g. *Erica tetralix* swamp heath).
Formation Group	3. *Mixed evergreen dwarf-shrub and herbaceous formations*. More or less open stands of evergreen suffrutescent or herbaceous chamaephytes, various hemicryptophytes, geophytes etc.
Formations	(*a*) Truly evergreen dwarf-shrub and herb mixed formation (e.g. *Nardus-Calluna*-heath).
Formation Subclass	B. **Mainly deciduous dwarf-scrub**. Similar to A, but mostly consisting of deciduous species.
Formation Group	4. *Cold-deciduous dwarf-thickets* (or dwarf-shrublands). Shedding the leaves at the beginning of a cold season. Usually rich in cryptogamic chamaephytes.

(2) Fosberg (adapted)

Primary Structural Group	I CLOSED VEGETATION (Crownsor peripheries of plants touching or overlapping.)
Formation Class	1C DWARF-SCRUB (Closed predominantly woody vegetation less than 0·5 m tall.)
Formation Group	1C1 *Evergreen dwarf-scrub.*
Formation	2. Evergreen broad* sclerophyll dwarf-scrub.
Subformations	(*a*) Mesophyllous broad* sclerophyll dwarf-scrub (e.g. *Arctostaphylos uva-ursi* mat, Northern temperate region). (*b*) Microphyllous evergreen dwarf-scrub. Without significant peat accumulation (e.g. *Calluna* heath without peat; Western Europe). (*c*) Microphyllous evergreen dwarf heath. With peat accumulation (e.g. *Empetrum* heath, Arctic and Subarctic; *Loiseleuria* heath, Arctic).
Formation	3. Evergreen dwarf-shrub bog. Dwarf-shrub with significant peat accumulation, root systems of plants adapted to constant immersion (e.g. mountain bogs, more closed phases, Hawaii; *Chamaedaphne* bog, Eastern North America).
Formation Group	1C2 *Deciduous dwarf-scrub.*
Formation	1. Deciduous orthophyll dwarf-scrub.
Subformations	(*a*) Deciduous orthophyll dwarf-scrub. Without significant peat accumulation (e.g. low bush *Vaccinium* scrub, North temperate and Subarctic regions). (*b*) Deciduous orthophyll dwarf heath. With peat accumulation (e.g. *Vaccinium myrtillus* heath, Subarctic regions). 2. Deciduous dwarf scrub bog (e.g. *Vaccinium-Betula-Myrica* with *Sphagnum*, Sweden).
Formation Class	1F DWARF-SCRUB WITH SCATTERED TREES
Formation Group	1F1 *Evergreen dwarf-scrub with scattered trees.*
Formations	1. Microphyllous evergreen dwarf-scrub with trees. Without significant peat formation (e.g. *Calluna* heath with *Pinus*, England). 2. Microphyllous evergreen heath with trees. With peat accumulation (e.g. *Empetrum* phase of heath birch (*Betula*) forest, Lappland).
Formation Group	1F2 *Deciduous dwarf-scrub with trees.*
Formation	1. Deciduous heath with trees. With significant peat accumulation (e.g. *Vaccinium* phases of heath birch (*Betula*) forest, Lappland).

(*broad, in contradistinction to 'narrow' or needle-leaved.)

	Fosberg (adapted)–*continued*
Formation Class	1H OPEN DWARF-SCRUB WITH CLOSED GROUND COVER
Formation Group	1H1 *Open evergreen dwarf-scrub with closed ground cover.*
Formation	3. Open evergreen microphyllous dwarf-scrub (e.g. open *Calluna* and *Erica* heath lower phases, Western Europe).
Formation Group	1H2 *Open deciduous dwarf-scrub with closed ground cover.*
Formation	1. Open deciduous orthophyll dwarf-scrub. Without significant peat accumulation (e.g. open phases of low-bush *Vaccinium* scrub, Eastern U.S.). 2. Open deciduous orthophyll heath. With significant accumulation of peat (e.g. open phases of *Vaccinium myrtillus* heath, Subarctic)., 3. Open deciduous dwarf-scrub bog (e.g. *Vaccinium uliginosum-Betula-Myrica* with *Sphagnum*, Sweden).
Primary Structural Group	II OPEN VEGETATION (Plants or tufts of plants not touching but crowns not separated by more than their diameters; plants, not substratum, dominating landscape.)
Formation Class	2C DWARF-STEPPE-SCRUB (Open predominantly woody vegetation less than 0·5 m tall.)
Formation Group	2C1 *Evergreen dwarf-steppe-scrub*
Formation	4. Microphyllous evergreen dwarf-steppe-scrub (e.g. low open sand heath, Western Europe).

(*b*) By examining the ways in which components of the community concentrate their biomass in particular sectors of the habitat-space, both in the vertical and horizontal dimensions. Concentrations in the vertical dimension, above or below ground, constitute stratification; in the horizontal dimension they create pattern.

Life-forms

Raunkiaer's system (1907) is commonly used for initial synopsis of the life-forms represented in a sample of vegetation. Raunkiaer himself made use of his classification in characterizing formations, by calling them after the life-form to which their dominant species belong. Following this procedure he placed heaths among the 'formations of Chamaephytes', and throughout his analyses of Danish heaths, the dominant species of Ericaceae, notably *Calluna*, are invariably scored as Chamaephytes.

In this connection confusion has been introduced in subsequent literature. Most summaries of Raunkiaer's system by recent authors define Chamaephytes as perennial plants whose resting buds are situated up to

25 cm above the surface of the ground (Tansley, 1939; Clapham, Tutin and Warburg, 1962). According to this definition the life-form of *Calluna* is frequently given as either Nanophanerophyte or Chamaephyte depending upon its behaviour in different habitats (Gimingham, 1960; Clapham, Tutin and Warburg, 1962; Nicholson and Robertson, 1958). But it is certainly not the case that throughout the lowland heaths of Denmark surveyed by Raunkiaer, *Calluna* is universally shorter than 25 cm in height. It seems that Raunkiaer himself never intended the Chamaephyte life-form to be arbitrarily delimited in this way, for he described it in more general terms, with the comment that the surviving buds 'do not project more than 20 or 30 cm' above ground. Subsequent authors, desiring a brief and unambiguous designation, are apparently responsible for introducing the 25 cm limit.

This is recognized by Ellenberg and Mueller-Dombois (1966*b*), who, in their revision of Raunkiaer's system, describe Chamaephytes as plants whose mature branch system remains perennially within 25–30 cm above ground surface (or dies back periodically to that height). They 'have typically a shoot-crowding habit', and if in particularly favourable habitats plants generally behaving as Chamaephytes are taller, the height limit may be extended to 100 cm for purposes of classification. Interpreted in this way, *Calluna* and its chief associates may be regarded as almost always Chamaephytes, and we may return to Raunkiaer's view of heaths as dominated physiognomically by Chamaephytes.

While an essential character of heaths is the lack of trees, this is not to be taken in an absolute sense. The Phanerophyte life-form may be represented by scattered trees, and, particularly in certain heaths of the European mainland, *Juniperus communis* contributes a significant structural component referable usually to the Microphanerophytes (small trees, 2–5 m Plate 3). However, in a heath, the stand of such Phanerophytes as are present is very open, and it is the Chamaephyte life-form which produces a dense perennial thicket exerting a strong influence upon other features of community structure. Annuals (Therophytes) are almost, often completely, excluded; the remaining Angiosperm life-forms present are Hemicryptophytes and Geophytes. In addition, an important role is played by either or both of the Bryophyta and Lichens (Thallo-chamaephytes and Thallo-hemicryptophytes in the terminology of Ellenberg and Mueller-Dombois). The low stature of the vegetation implies few Epiphytes, but epiphytic lichens, and in a few cases facultatively epiphytic bryophytes, occur. Vascular parasites (e.g. *Cuscuta*) and semi-parasites (e.g. *Rhinanthus*) belong in some heath communities and, not unnaturally, various fungal parasites are associated with the autotrophic components.

'Biological spectra' for heaths, showing the number of species in each life-form class as a percentage of the total (vascular-plant) flora, usually show the Hemicryptophyte class in first place, more or less closely followed

TABLE 4

Representation of life-forms in selected examples of heath species
(From Gimingham, 1964*a*, recalculated)

Locality	Community-type	Proportion of species in each life-form class as % of total number of vascular plant species N	Ch*	H	G	No. of vascular plant species	No. of Bryophyte species	No. of Lichen species
BRITAIN								
Yesnaby, Orkney	Oceanic, *Calluna-Erica cinerea*	–	42·9	57·1	–	14	8	2
Daviot, Inverness-shire	*Calluna-Vaccinium*	–	33·3	55·6	11·1	18	7	2
Strathfinella Hill, Kincardineshire	*Calluna-Vaccinium*	3·8	23·1	65·4	7·7	26	16	4
Dinnet, Aberdeenshire	*Calluna-Arctostaphylos*	–	31·8	68·2	–	22	10	5
Forvie, Aberdeenshire	*Calluna-Empetrum nigrum* dune heath	–	16·7	83·3	–	12	7	11
SCANDINAVIA								
Karmøy, South-west Norway	Oceanic *Calluna-Erica cinera*	3·6	21·4	67·9	7·1	28	11	4
Hallandsåsen, South-west Sweden	*Calluna-Vaccinium*	5·9	23·5	58·8	11·8	17	11	3
Nörrevosborg, West Jutland, Denmark	*Calluna-Empetrum nigrum*	–	50·0	33·3	16·7	6	6	12

N – Nanophanerophytes H – Hemicryptophytes Ch – Chamaephytes G – Geophytes. Based on flora lists from areas of 40 m^2

(*Defined as on p. 38)

by the Chamaephytes. Phanerophytes and Geophytes have small percentages or may be absent (Table 4). Variations on this pattern are relatively slight and, in comparison with examples of other formations, the spectra for heaths are quite characteristic. With some exceptions (Chapter 4) the communities are not floristically rich, but lists of bryophytes and lichens may be long and their variations often give a valuable reflection of habitat differences.

TABLE 5

Representation of life-forms calculated from species frequencies (from Raunkiaer, 1910, 1913).
(% of the sum of the frequency figures for each stand)

	N	Ch	H	G	
Calluna-Erica tetralix wet heath	1·5	67·5	13	18	
Calluna-Empetrum nigrum	1	97	–	2	Aadum-Varde
Calluna-Empetrum-Arctostaphylos	–	91	5	4	Bakkeø,
Calluna-Arctostaphylos	–	95	4	1	Jutland, Denmark.
Calluna dominant	–	90	3	7	
Calluna-Erica tetralix-Salix repens heath	–	66	26	8	Skagens
Calluna heath, mostly with *Empetrum*	–	65	29	6	Odde, Jutland, Denmark.

(For key to column headings see Table 4, p. 39).

However, 'spectra' expressed in this way give little impression of the structural significance of the life-forms present. This was appreciated by Raunkiaer who, in an attempt to quantify the contribution of different life-forms to the composition of an example of vegetation, gave weightings to the species in each group according to their frequency in quadrat samples (Table 5). Despite the limitations of frequency as a measure of relative importance, this served to demonstrate in numerical terms the predominance of Chamaephytes and the usually small contribution from other life-forms, except sometimes Hemicryptophytes.

Synusiae

As societies of individuals having generally similar life-form, synusiae may be regarded as groups of plants occupying a particular niche in the community. In this context, the concept of life-form extends beyond that of Raunkiaer's main groups to incorporate further details of organization. The value of treatment in terms of synusiae is that, having established the range of synusiae present in a well-developed example of a formation, departures from this pattern may have general ecological significance irrespective of the particular species concerned.

The following is a suggested synopsis of the synusiae which might be represented in a well-developed heath:

A. PHANEROPHYTES

1. Synusia of trees and/or small trees and shrubs, e.g. *Pinus sylvestris, Quercus* spp., *Betula* spp., *Juniperus communis*. It is inherent in the definition of heath that trees and tall shrubs, although frequently present, are scattered and nowhere form a closed canopy. Accordingly, they may be treated for the present purpose as constituting one synusia whatever their size, since they occupy a part of the habitat-space not entered by the main constituents of heath communities.

B. CHAMAEPHYTES

2. Synusia of more or less caespitose, woody dwarf-shrubs, generally in the range 30–100 cm in height. This is normally the synusia which is physiognomically dominant. It may be subdivided as follows:
   (*a*) Microphyllous, evergreen, e.g. *Calluna vulgaris, Erica cinerea*, or aphyllous, e.g. *Ulex gallii, U. minor*.
   (*b*) Broad-leaved (*i*) Evergreen and sclerophyllous, e.g. *Vaccinium vitis-idaea*; (*ii*) Deciduous, e.g. *Vaccinium myrtillus*.
   Species such as *Vaccinium myrtillus* and *Genista anglica* differ from the others in combining a height often in excess of about 30 cm with a spreading branch-system rather than a shoot-crowding habit. This is particularly evident in the former when, in certain habitats, it becomes dominant in almost pure stands. Separate treatment as Nanophanerophytes would be justifiable, but since in most heaths plants of this group occur intricately mixed with the other Chamaephytes, reaching about the same height and occupying the same part of the habitat-space, they are better regarded as belonging to the same synusia.
3. Synusia of creeping dwarf-shrubs, stems often inter-twined to form extensive mats:
   (*a*) Microphyllous, evergreen, e.g. *Empetrum nigrum*.
   (*b*) Broad-leaved and sclerophyllous, e.g. *Arctostaphylos uva-ursi*.
   Where species of this synusia occur in dense stands of taller plants they frequently adopt a semi-scrambling habit, depending for support in part upon the dominants.

C. HEMICRYPTOPHYTES

4. Synusia of caespitose, densely tillering herbs with narrow erect leaves:
   (*a*) evergreen or partly evergreen, e.g. *Festuca ovina, Eriophorum vaginatum, Juncus squarrosus*.
   (*b*) not or very sparingly evergreen, e.g. *Nardus stricta, Molinia caerulea, Carex binervis*.
   This is a synusia of plants which habitually form sizeable tufts or tussocks.

5. Synusia of sparingly tufted, more or less rhizomatous herbs with narrow erect leaves: e.g. *Deschampsia flexuosa*, *Agrostis tenuis*.
   Whereas the tussocks of no. 4 very generally interrupt the canopy of Chamaephytes, the species of this synusia spread beneath it and occur usually as scattered groups of slender tillers. Where gaps in the canopy permit, they may form dense patches not unlike small versions of no. 4.
6. Synusia of deciduous forbs with erect or scrambling leafy stems in summer, e.g. *Potentilla erecta*, *Hypericum pulchrum*, *Lathyrus montanus*.
7. Synusia of rosette hemicryptophytes, e.g. *Succisa pratensis*, *Carex pilulifera*, *Blechnum spicant*, *Pyrola media*.
8. Synusia of creeping forbs, e.g. *Galium saxatile*.

D. GEOPHYTES

9. Synusia of summer-green geophytes, e.g. *Dactylorchis maculata* ssp. *ericetorum*, *Listera cordata*, *Trientalis europaea*.

E. THEROPHYTES

10. Synusia of summer-green annuals, e.g. *Linum catharticum*. A synusia of few members, frequently lacking.

F. THALLO-CHAMAEPHYTES

11. Synusia of hummock-forming mosses, e.g. *Sphagnum* spp., *Leucobryum glaucum*.
12. Synusia of robust weft mosses, e.g. *Hylocomium splendens*, *Pleurozium schreberi*.
13. Synusia of tall fruticose lichens, e.g. *Cladonia arbuscula*, *C1. gracilis*.

G. THALLO-HEMICRYPTOPHYTES

14. Synusia of mat-forming mosses and liverworts, e.g. *Hypnum cupressiforme*, *Plagiothecium undulatum*, *Lophocolea bidentata*.
15. Synusia of foliose lichens, e.g. *Peltigera canina*.
16. Synusia of crustose or gelatinous lichens, spreading over soil surface or litter, e.g.
    Cladonia spp. (including basal squamules of species with short podetia), *Lecidea uliginosa*.

H. THALIO-EPIPHYTES

17. Synusia of epiphytic lichens, e.g. *Parmelia physodes*. Sometimes certain bryophytes become partially epiphytic on the lower branches of the dwarf-shrubs, e.g. *Hypnum cupressiforme*, *Lophocolea bidentata*.

I. HETEROTROPHIC PLANTS

18. Synusia of vascular parasites, e.g. *Cuscuta epithymum*. Confined to southern heaths.

19. Synusia of fungal parasites, e.g. *Marasmius androsaceus.*
20. Synusia of saprophytic and mycorrhizal fungi.
    (In nos. 19 and 20 only fungi with macroscopic fructifications are considered.)

Despite the fact that heath vegetation is seldom rich in species, it is evident from this synopsis that it is as complex structurally as any similar community of similar stature. It is not suggested that every individual occurring in a heath community could be fitted into one of the above synusiae, but that they represent the biological categories of greatest significance structurally. Equally it is not necessarily implied that there is any actual community in which all 20 synusiae are present together, though this may perhaps be approached under favourable conditions towards the centre of the heathland region. On the other hand, towards its periphery certain synusiae are eliminated and others emphasized. For example, to the north and at high altitudes the Phanerophyte (1) and caespitose Chamaephyte (2) synusiae are eliminated, while the dominant synusia becomes that of the creeping dwarf-shrubs (3). At the other extreme, in the south of the heath region the Nanophanerophytes (shrubs) become increasingly important. Strong representation of a particular synusia may be associated with particular soil characteristics, for instance on the wet soils the tussock-forming herbaceous Hemicryptophytes (4) and the hummock-forming mosses (11) are conspicuous, while the Chamaephyte component is somewhat reduced. Other Hemicryptophyte synusiae (e.g. 6, 7, 8) may be particularly in evidence on the richer mineral soils.

Certain synusiae appear to be to some extent mutually exclusive, for example, those composed of tall weft mosses and of fruticose lichens. The environmental conditions favouring one or the other are not fully understood; each may contribute to communities which otherwise are very similar in structure. Finally, the representation of synusiae may reflect the past history of management of the heath; for example, complete absence of the tree synusia is indicative of an area effectively managed by regular burning; 16 (crustose or gelatinous lichens) and perhaps also 13 (fruticose lichens) of widespread provide evidence of recent burning (pp. 74; 200). Intensive grazing encourages the tussock-forming Hemicryptophytes (especially grasses) at the expense of the dwarf-shrubs.

Stratification

Where a community is composed of numerous synusiae there is usually a tendency for certain of these to associate in particular strata, to which they contribute the bulk of their biomass, especially foliage biomass. Owing to competition for canopy-space, individuals capable of greatest height-growth tend to equate about a particular level, having the effect of confining others which are generally shorter, though tolerant of conditions imposed by the dominant stratum, to competition amongst themselves at a more or

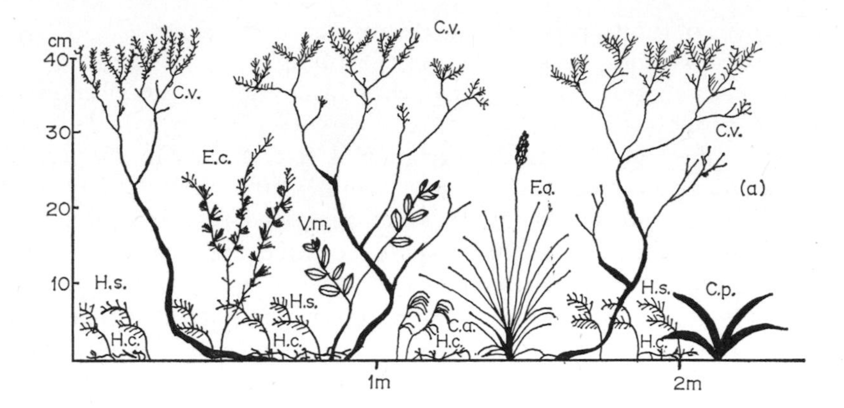
cm
40
30
20
10
C.v.
C.v.
E.c.
V.m.
F.a.
C.v.
(a)
H.s.
H.c.
H.s.
H.c.
C.a.
H.c.
H.s.
C.p.
H.c.
1m
2m

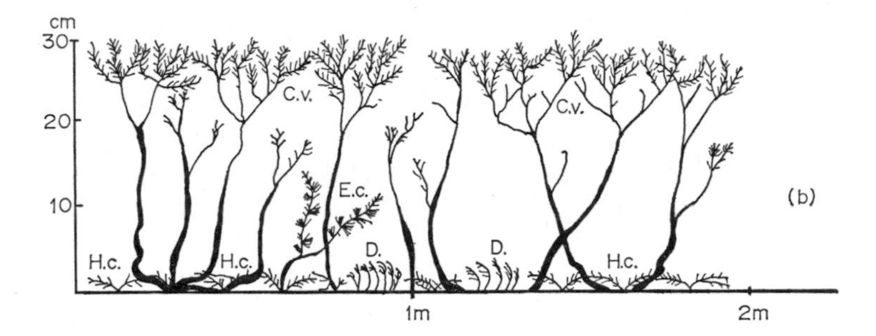
cm
30
20
10
C.v.
C.v.
E.c.
(b)
H.c.
H.c.
D.
D.
H.c.
1m
2m

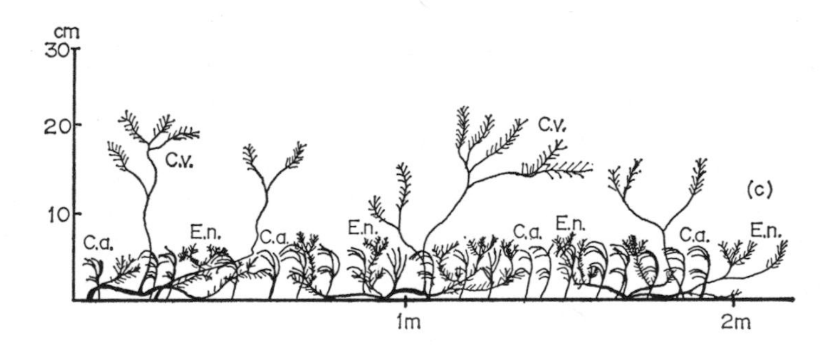
cm
30
20
10
C.v.
C.v.
(c)
C.a.
E.n.
C.a.
E.n.
C.a.
E.n.
C.a.
E.n.
1m
2m

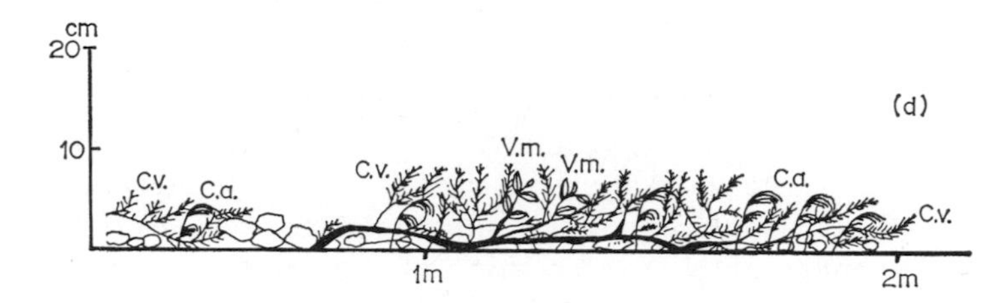
cm
20
10
(d)
C.v.
C.a.
C.v.
V.m.
V.m.
C.a.
C.v.
1m
2m

less distinct lower level. This, with repetition at progressively lower levels, gives rise to a series of strata which may be represented by photographs or diagrams of transect profiles, or by recording the quantitative contributions of the components of the community at different height intervals.

Diagrams of transect profiles in a series of heaths selected to demonstrate a range of types are given in Fig. 8. In all examples there is evidence of a stratified organization, suggesting in some cases at least four strata. However, illustrations alone provide only a visual impression of stratification and it may be investigated more objectively by measuring quantitatively the presence of plant components at every level in the community profile (Fig. 9). Where Phanerophytes are present, these contribute a discontinuous or localized uppermost stratum. For the rest, although stratification is not always distinct, there is good evidence in well-developed examples of heath (Fig. 8, a) of arrangement in four main strata (Gimingham, 1964*a*):

(*i*) Dominant stratum, consisting mainly of the dwarf-shrub canopy, e.g. *Calluna vulgaris*. This stratum, often at about 30–40 cm above ground, is composed largely of the caespitose dwarf-shrub synusia (2), but may incorporate foliage of some of the taller narrow-leaved or broad-leaved Hemicryptophyte synusia (4, 6).

(*ii*) Second stratum, discontinuous, formed beneath and in the gaps of the top stratum at about 10–20 cm. This is comprised partly of species such as *Erica cinerea* and *Vaccinium vitis-idaea* which, although referable to the same synusia as the dominant, are shorter and less densely branched; and partly of the creeping dwarf-shrub synusia (3), such as *Empetrum nigrum* and *Arctostaphylos uva-ursi*. The shorter grasses and Cyperaceae of Synusiae 4 and 5 may also contribute.

Fig. 8. Simplified diagrammatic profiles of selected heath communities to show variations in stratification.

(*a*) Uneven-aged stand of *Calluna* in a north-east Scottish *Calluna–Vaccinium* heath. Four more or less distinct strata.

(*b*) Even-staged stand of *Calluna* resulting from burning management. Two strata, with scattered plants at intermediate levels. Widespread in Britain.

(*c*) *Calluna–Empetrum nigrum* dry heath from a dune in north-east Scotland. Two strata, the lower occupied by *Empetrum* as well as lichens.

(*d*) Dwarf mountain *Calluna* heath at about 900 m, Cairngorm mountains. *Calluna*, other dwarf-shrubs and lichens all at the same level; scattered creeping bryophytes at ground level.

C.a.	*Cladonia arbuscula*	E.c.	*Erica cinerea*
C.p.	*Carex pilulifera*	F.	*Festuca ovina*
C.v.	*Calluna vulgaris*	H.c.	*Hypnum cupressiforme*
D.	*Dicranum scoparium*	H.s.	*Hylocomium splendens*
E.n.	*Empetrum nigrum*	V.m.	*Vaccinium myrtillus*

V.v. *Vaccinium vitis-idaea*

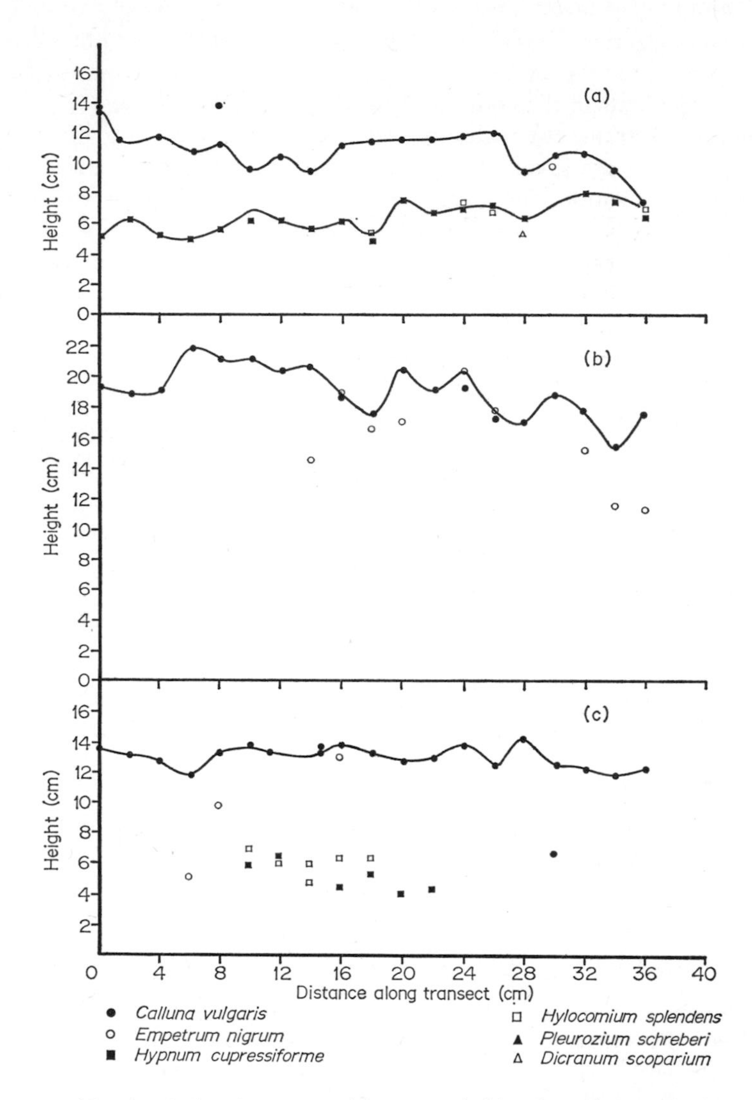

Fig. 9. Stratification in heath communities, revealed by the point method. (Needles were lowered into the vegetation at 2 cm intervals along 40 cm transects. Every contact with a plant is recorded.) (Data: Mrs E. L. Birse.)

(*a*) Two complete strata. Uppermost of *Calluna* with a little *Empetrum nigrum* at a height of 11 cm. Lower, of mosses, at *c.* 7 cm.

(*b*) One distinct stratum only, at about 19 cm: *Calluna* with *Empetrum nigrum*, the latter frequently at a slightly lower level.

(*c*) A continuous *Calluna* canopy (with a little *Empetrum nigrum*) at 14 cm, below which is a patch of mosses reaching 6 cm (see text, p. 48).

(*iii*) Third stratum, at about 5–10 cm, composed mainly of either or both of the robust weft mosses (synusia 12) or the tall fruticose lichens (13), together with rosette and creeping Hemicryptophytes, e.g. *Carex pilulifera*, *Galium saxatile*.

(*iv*) Fourth stratum, the ground stratum, of the mat-forming mosses and liverworts, and the foliose, crustose and gelatinous lichens (synusiae 14, 15, 16).

Other synusiae figure in the strata to which their height brings them.

The density and uniformity of stratum (*i*) are dependent partly upon the environment, but also very considerably on the age-structure of the stand. In a stand which has been free from disturbance long enough for a considerable number of individuals of the dominant to reach the degenerate phase (e.g. for more than 25 years, p. 127) the uppermost stratum becomes broken, and its canopy interrupted by gaps. With the further passage of time the whole stand may become uneven-aged and the top stratum distinctly discontinuous. A different modification is seen on hill slopes where many of the main branches of each plant become decumbent and follow the downslope in their orientation, producing an upper stratum which resembles a series of descending waves. Very wet soils inhibit the development of most of the dwarf-shrubs and result in a more open top stratum.

The density of the second stratum is generally inversely correlated with that of the one above, and this relationship continues down through the series. Hence, any practice which affects the density of one stratum will influence the rest. Moorland management by burning and grazing is designed to produce a dense, even canopy of *Calluna*, which amounts to a rather uniform smooth and uninterrupted upper stratum. Typical stands of the building phase (p. 126) in north-east Scotland show a concentration of the green shoots (leaf-bearing young long-shoots and short-shoots, p. 111) at a level usually between 30 cm and 40 cm, supported by the woody branch-system below (Fig. 10). Estimates of the leaf area index (Fig. 10) show that this reaches quite a high value at the 30–40 cm level, but very little foliage occurs at lower levels. Dense shade is cast by this canopy (Fig. 13) and subordinate strata are consequently very poorly represented or lacking. However, in the seral stages following a fire (Chapter 10) the third and fourth strata may be particularly well formed, and may for a time survive under the developing canopy of the higher strata.

Profiles of vigorous *Calluna* stands sometimes unexpectedly reveal a bryophyte stratum limited in its distribution to discontinuous patches (Fig. 9c). This may arise either from incomplete destruction of the bryophyte stratum when a stand is burnt, so that patches survive for a considerable time in the regenerating community, or by survival of bryophyte patches formed in the centres of old, degenerate *Calluna* bushes (pp. 128–31)

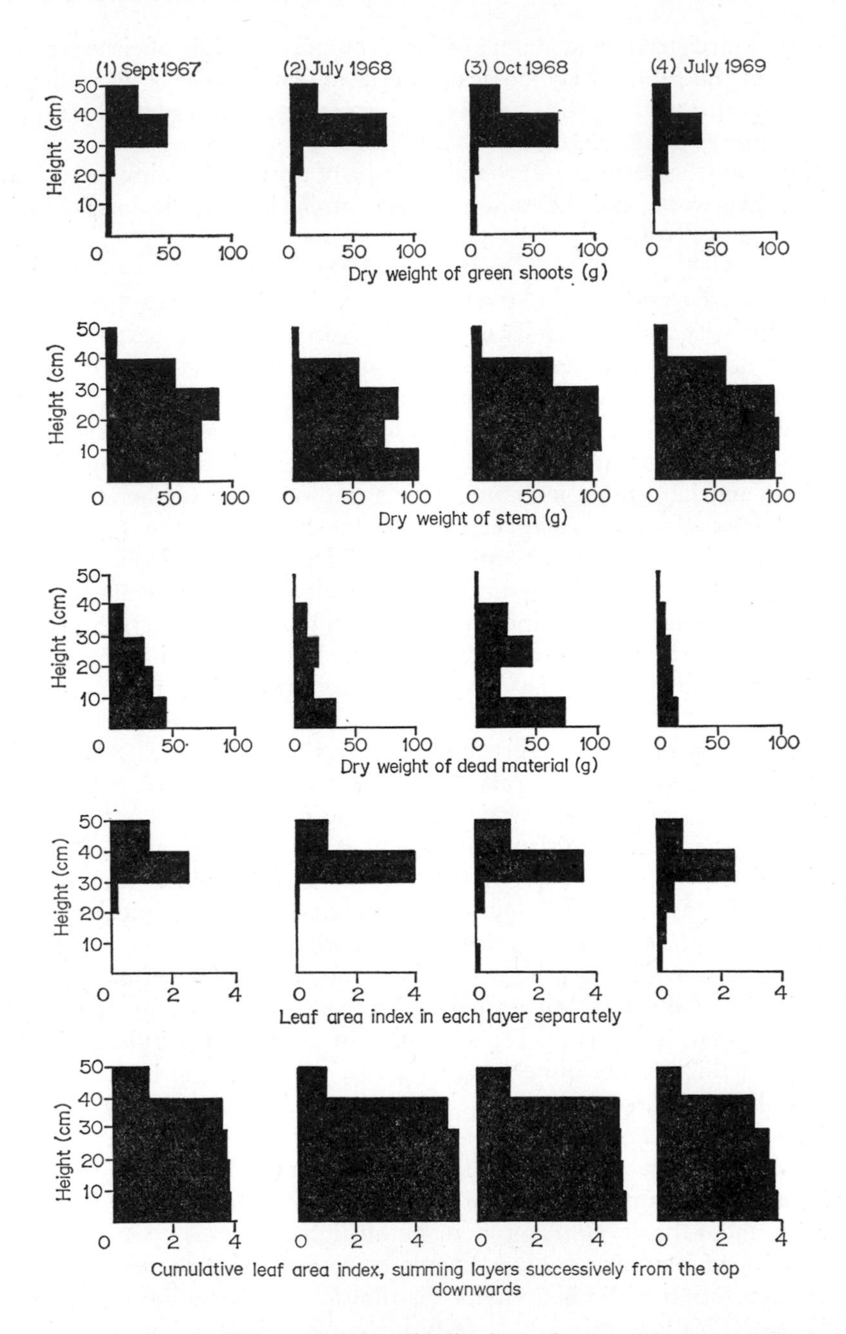

Fig. 10. Histograms illustrating the distribution of green shoots, stem, dead material, and leaf area index in successive layers of 10 cm each, above the ground surface in an even-aged stand of building *Calluna*. (Figure by D. K. L. MacKerron and Miss D. Goldie.)

after the canopy has been re-established as a new generation of *Calluna* plants colonize.

Exacting environments simplify the stratification by eliminating one or more strata. Dominance may then be exerted either by a species, such as *Calluna*, able in such habitats to form a canopy at a lower level, or by some other species normally occupying this level. For example, up to altitudes of about 970 m in Britain, *Calluna* may still be the dominant of a dwarf mountain heath, but beyond the limits of its ecological amplitude its place is taken by *Empetrum hermaphroditum* or *Loiseleuria procumbens*. According to conditions, the community structure may be reduced to three or only two strata (Figs. 8, 9); while the so-called 'lichen-heaths' or '*Rhacomitrium*-heath' of arctic-alpine or montane habitats may be regarded as related communities of a single stratum only, from which strata properly justifying the term 'heath' have been eliminated.

Micro-environment profiles

The relationship between synusiae and the development of stratification is of importance not merely as a description of community structure, but also because of their controlling influence on the micro-climate beneath the canopy. Furthermore, the structure of the vegetation governs the range of niches available to animals and largely determines the conditions of their environment. It was in connection with the ecology of heath invertebrates that Delany (1953) made a number of measurements of temperature and humidity profiles in *Calluna* heath in Devon. His object was to compare conditions under *Calluna* with those of neighbouring open patches, and for

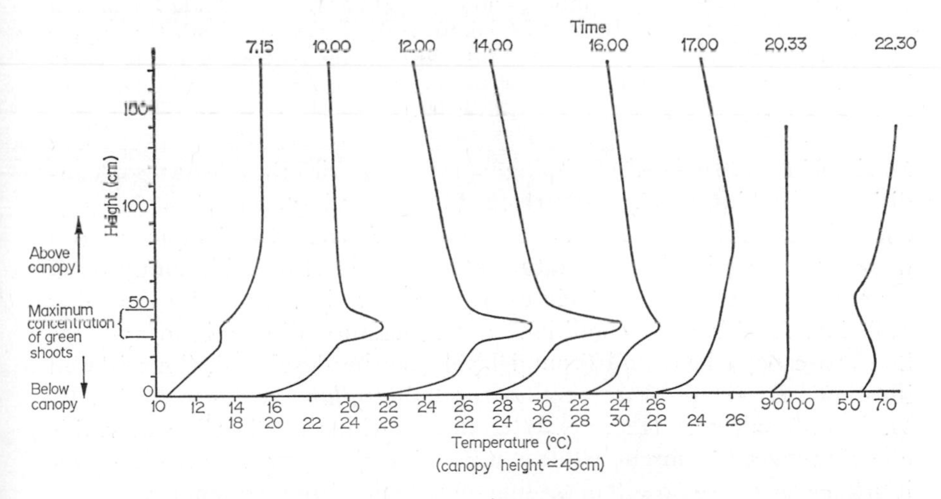

Fig. 11. Profiles of air temperature through an even-aged stand of building *Calluna*, at successive intervals during a fine day in June – north-east Scotland. (Figure by D. K. L. MacKerron.)

this reason small patches of pure *Calluna* were chosen rather than a well-developed stratified community. None the less, it was shown that during daylight the ground surface beneath the *Calluna* canopy is generally cooler and its temperature less variable than outside, while at night-time in cold weather it may remain somewhat warmer. Humidity at ground level within the stand is kept rather steadily at high values. Similar results were

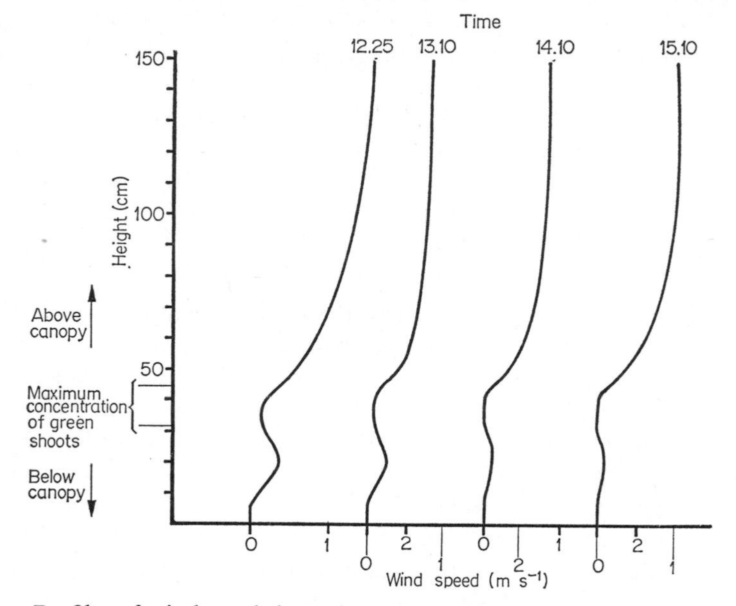

Fig. 12. Profiles of windspeed through an even-aged stand of building *Calluna*, at four times on a representative day in August – north-east Scotland. (Below the canopy there is evidence of a slight through-draught among stems, probably because the stand was of limited extent, unlike most heaths.) (Figure by D. K. L. MacKerron.)

obtained in Scottish heaths (Gimingham, 1964*a*), with the effects upon conditions at, and just above, the surface of the ground intensified by the presence of a bryophyte stratum. These measurements also demonstrated a marked increase in air temperature at canopy level on still, sunny days, when the upper stratum of close-spaced shoots is warmed by incoming radiation. Under such conditions a temperature difference of 5–10°C between canopy level and ground level may be developed. This is well illustrated by temperature profiles through *Calluna* stands recorded by D. K. L. MacKerron (Fig. 11). In cloudy or windy weather, however, this effect is reduced or absent, while at night-time there may be a tendency for it to be reversed, as a result of cooling by radiation from the canopy.

Stoutjestdijk (1959), working on Dutch heaths draws a similar picture, and gives in addition profiles showing gradients in vapour pressure deficit through the stand. The additive effects of the various strata begin to emerge in this work. The dense canopy of the dwarf-shrubs absorbs radiation and, while the resulting local increase in temperature may cause some increase in vapour pressure deficit, the overall effects at lower levels include reduced temperature, reduced incidence of radiation and reduced vapour pressure deficit. Further, except in the strongest wind the effects of air-movement are scarcely felt below the dense top stratum (Stocker, 1923; Gimingham,

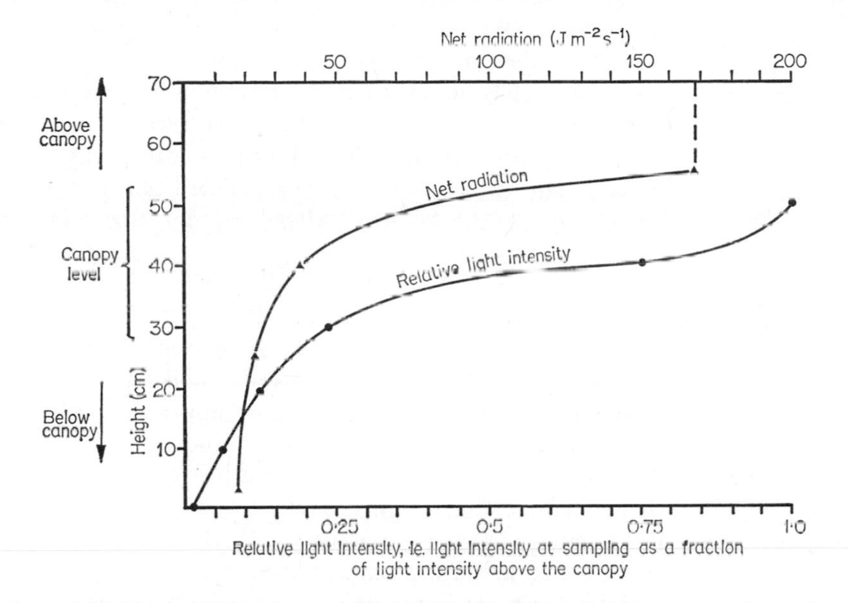

Fig. 13. Typical profiles of net radiation and relative light intensity through an even-aged stand of building *Calluna* – north-east Scotland. (Figure by D. K. L. MacKerron.)

1964*a*; Fig. 12). Where, however, a moss stratum is also present its properties of retaining moisture and further limiting temperature rise lead to the continuance of almost saturated conditions at ground level for much of the year. It has often been pointed out that heath vegetation is effective in conserving soil moisture.

In addition to gradients in temperature and moisture conditions, the penetration of light through the vegetation has been measured (Fig. 13). The dwarf-shrub strata may reduce illumination to anything from 20% of that in the open down to less than 1%, according to the density of the canopy. The extent to which bryophyte and lichen strata are developed is closely correlated with the degree of shading. A continuous weft-moss stratum or moss-lichen stratum will not develop where only about 2% or less of the photosynthetically active radiation reaches them (Stoutjesdijk, 1959).

Stratification of root systems

Where a number of species of various life-forms are associated their underground parts may also be stratified. Such vertical assortment of root systems does not necessarily bear any relationship to the aerial assortment of shoot systems, while, owing to variations in soil depth and other properties there is much variability in the disposition of roots. For example, in deep, freely drained soils including sand, podsols and peat, the roots of *Calluna vulgaris* may penetrate to as much as 1 m below the surface (Plate 5a). On the more highly leached podsol soils, however, very few *Calluna* roots are found in the A_2 horizon or below, and the existence of an indurated horizon or of a hard iron or humus pan sets a limit on the downward extension of roots and causes them to spread out laterally and form an inter-twined mat just above the barrier (Plate 5b). Rennie (1957) calculated that in a freely drained podsolic profile 80% of the roots of *Calluna* are concentrated in the upper 13 cm of the soil.

Practical difficulties have limited the investigation of rooting levels in heathland soils, but some information is available. By excavation of monoliths and washing to extract the root systems, great concentrations of *Calluna* roots in the upper 10 cm of soil have been demonstrated, almost irrespective of the soil-type (R. Boggie, personal communication). In addition to the finer laterals of the main root system, the older plants of *Calluna* produce abundant adventitious roots in the top 3 cm or more of litter and humus from stem-branches which have become procumbent and partially or completely buried at the base (Plate 5c, d). Other species whose roots are often largely contained in the same stratum include *Erica tetralix* and *Erica cinerea* (Fritsch and Salisbury, 1915; Heath and Luckwill, 1938; Rutter, 1955; Bannister, 1965, 1966; Sheikh and Rutter, 1969), while among plants frequently producing concentrations of roots at lower levels are *Deschampsia flexuosa, Eriophorum* spp., *Molinia caerulea* and, in dune heaths, *Carex arenaria*. In a heathland ecosystem on Bagshot sands in Dorset containing *Calluna vulgaris*, *Ulex minor*, *Erica cinerea*, *Agrostis setacea* and *Molinia caerulea*, Chapman (1970) demonstrated that 92% of all roots extracted from sample cores extending to a depth of 40 cm were contained in the upper 20 cm (Fig. 14).

Even these observations are of limited value in indicating the portions of available habitat-space functionally occupied by the roots of these species, as no distinction is made between roots which are active in absorption and those which are not. Boggie, Hunter and Knight (1958) have approached this problem by determining the uptake of radioactive tracers (e.g. P_{32}) placed at various depths in the soil under stands of heath vegetation. The results strikingly confirm the presence of a superficial stratum of about 10 cm in depth in which the absorptive activity of many species, including the dominant dwarf-shrubs, is concentrated. For those which

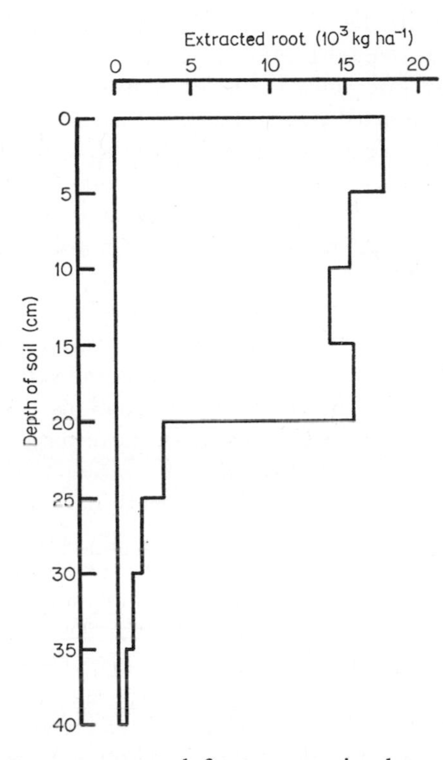

Fig. 14. Biomass of roots extracted from successive layers of a heathland soil profile: Dorset, south England. The roots are concentrated in the upper 20 cm of the profile. (From Chapman, 1970.)

range from sandy soils to peat, behaviour is, in numerous instances, unaffected by soil type. Lower strata, e.g. 10–30 cm, may be occupied by active roots of species such as *Agrostis tenuis*, *Holcus lanatus*, *Carex arenaria* (on the freely drained soils), but it would appear that, although as shown above a number of species have roots at greater depths, they can scarcely be said to constitute an active vegetational stratum, and may indeed largely be dead.

On deep peat the available evidence suggests greater development into the lower levels, e.g. by *Molinia caerulea* (particularly between 15 and 30 cm), *Trichophorum cespitosum*, *Eriophorum vaginatum* and *E. angustifolium* (particularly between 15 and 60 cm).

Pattern

Where the arrangement of components in the horizontal dimension departs from random dispersion and gives evidence of vegetational pattern, this also is an aspect of community structure. Analysis of heath communities by the methods of Grieg-Smith (1952) and Kershaw (1957) has provided ample evidence of pattern in the dispersion of some of the dominant species. Anderson (1961) has demonstrated patterns at two or three size-scales in *Calluna vulgaris* and *Vaccinium myrtillus* in an upland site in north Wales. On a dune heath in north-east Scotland, *Empetrum nigrum* also displays a marked pattern (Plate 4), while in moist heaths where *Calluna* is accompanied by species such as *Polytrichum commune*, *Eriophorum vaginatum*, etc., all these plants show patterns usually at two or more size-scales. Kershaw and Tallis (1958) report similar results for *Juncus squarrosus* at a high altitude site in Wales.

Vegetational pattern arises when the dispersion of any units, whether individual plants or parts of plants, is to some degree patchy or clustered. The evidence of pattern at several size-scales indicates that the primary clusters themselves occur in groups or aggregations, and these in turn in larger aggregates. In many species, a small scale of pattern will be displayed if sampling is sensitive enough to reflect the arrangement of foliage on the branches or tillers: such pattern is evident for example in *Vaccinium myrtillus*, *Calluna vulgaris* and *Empetrum nigrum* (Figs. 15, 16). From the standpoint of community structure, however, the aggregation of branches or tillers, determining the mean diameter of an individual plant, is more important. This is particularly true in heaths where many of the chief species are more or less caespitose in habit. The size-scale of this pattern naturally varies with environment and with the age-structure of the stand. In *Vaccinium myrtillus*, which is a species with a creeping rhizome, Anderson showed that in many stands there is pattern at a size-scale of about 80 cm in diameter, produced by the grouping of erect branches derived from a single rhizome system (Fig. 15). Evidence was also obtained of pattern in *Calluna* at about the same size-scale, believed to correspond to the size of individual bushes. This interpretation was verified by field measurements of bush diameter, the mean of which was about 70 cm. Analyses in north-east Scotland (Fig. 16) have shown that the size-scale of this element of pattern in *Calluna* varies from about 40 cm in a dune heath to 80 cm in a wet heath. In the former, *Empetrum nigrum* was intimately mixed with *Calluna* in the stand analysed and showed little sign of pattern at this scale (Fig. 16, *b*). However, in other parts of the same locality where *Calluna* is less vigorous and *E. nigrum* occurs in association with *Cladonia* spp., pattern analysis showed a marked peak for *Empetrum* corresponding with patches of about 64 cms in diameter, representing separate individuals occupying circular areas similar to those shown in Plate 4. In the wet heath, *Eriophorum vaginatum* was co-dominant with *Calluna* and the pattern analysis (Fig. 16, *e*) shows that its tussocks, interspersed with *Calluna* bushes, had approximately the same mean diameter as the latter. *Erica tetralix* (Fig. 16, *d*), however, produced rather smaller branch-clusters in this habitat, of the order of 20 cms in diameter.

In addition to the morphological behaviour of the plants, a further potent

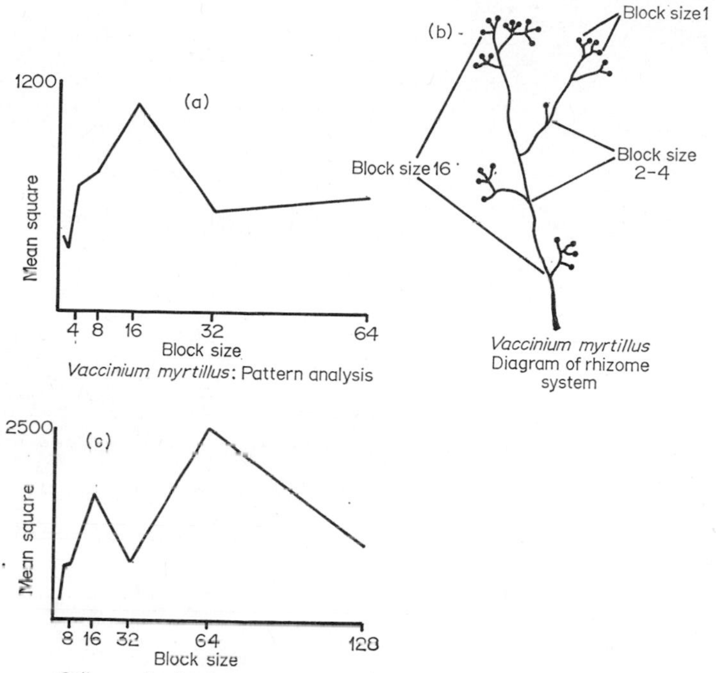

Fig. 15. Pattern in *Vaccinium myrtillus* and *Calluna vulgaris* on an upland heath in north Wales (after Anderson, 1961).

(*a*), (*b*) *Vaccinium myrtillus*

(*a*) Graph of mean square against block size, using point contacts along replicated transects. Interval between points: 1 cm. Basic unit or block = 5 points. Thus, block size 1 is equivalent to a distance of 5 cm, block size 2 to 10 cm, etc.

(*b*) Diagram of rhizome-system, explaining the size-scales of pattern revealed by the peaks in (*a*).

(*c*) *Calluna vulgaris*
The peak at block size 4 reflects clusters of shoots on a single main branch (diameter about 20 cm); that at block size 16 reflects the size of individual bushes (diameter about 80 cm); that at block size 64, clumps of bushes.

cause of pattern is non-random variation in factors of the environment. Anderson has shown how the environment may itself influence the size-scale of the morphological pattern. A reduction in the size of clumps of *Vaccinium myrtillus* belonging to individual rhizome systems was correlated with reduction in soil aeration due to trampling, or to a pattern of furrows persisting from former cultivation. Frequently, however, environmental variation leads to an additional large size-scale of pattern, which results from the occurrence of individual dwarf-shrubs or tussocks in groups. In the *Calluna* stand in upland north Wales, pattern on a scale of mean diameter nearly 300 cm was explained by the existence of a network of

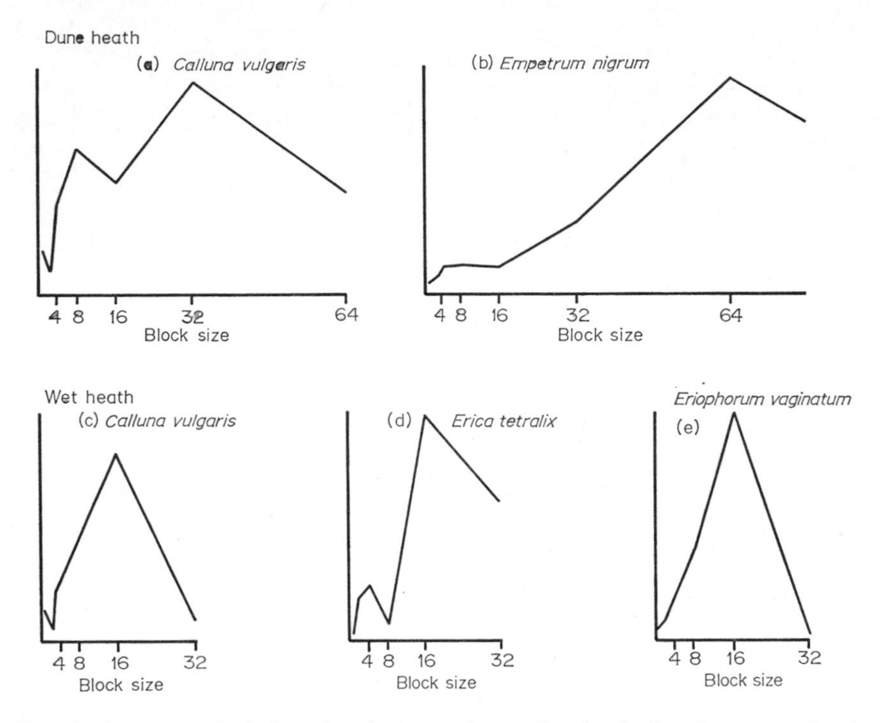

Fig. 16. Pattern analysis from heaths in north-east Scotland. Graphs prepared as in Fig. 15.

Dune heath (*a*) *Calluna vulgaris*
Peaks: Block size
1 – clusters of shoots on branches, diam. *c.* 5 cm.
8 – individual bushes, diam. *c.* 40 cm.
32 – clumps of individuals, diam. *c.* 160 cm.

(*b*) *Empetrum nigrum*
Peak at block size 64: unexplained clumping, diam. *c.* 320 cm.

Wet heath (*c*) *Calluna vulgaris*
Peaks: Block size
1 – clusters of shoots on branches, diam. *c.* 5 cm.
16 – individual bushes, diam. *c.* 80 cm.

(*d*) *Erica tetralix*
Peaks: Block size
4 – individual bushes, diam. *c.* 20 cm.
16 – clumps of bushes, diam. *c.* 80 cm.

(*e*) *Eriophorum vaginatum*
Peak at block size 16 – individual tussocks, diam. *c.* 80 cm.

sheep-tracks outlining the clumps (Fig. 15). No such pattern was found in very uniform even-aged stands of *Calluna* not dissected by sheep-tracks. *Vaccinium myrtillus* in a boulder-strewn area formed clumps measuring about 80 cm and 160 cm. These were always sited over boulders and were apparently confined to these positions by the sheep-tracks passing between.

In the dune heath from which graphs (*a*) and (*b*) of Fig. 16 were obtained there was evidence of aggregation of *Calluna* bushes at a mean diameter of about 160 cms and of pattern in *Empetrum nigrum* at a size-scale of about 320 cms, but in neither case has the cause been determined. In the wet heath, *Erica tetralix* revealed an additional scale of pattern with a diameter of about 80 cm. This perhaps can be attributed to clumping imposed by the two vigorous, caespitose dominants, *Calluna* and *Eriophorum vaginatum*. Further work in this community by T. Keatinge has revealed evidence of a very large ring-shaped pattern of arrangement of unit bushes of *Calluna*, sometimes approaching a diameter of 4 m (Fig. 17). Excavation has produced evidence of dead stems buried in the peat, connecting all these units to a single central point of origin. Although this interpretation is, as yet, tentative it suggests that in damp peaty areas, procumbent branches of old plants which root adventitiously (Plate 6d) may spread the plant vegetatively in a way which is unknown in drier sites, and may give rise to a further large-scale pattern.

Where control of community structure by management practices (burning and grazing) is lacking, processes of cyclical change become an important feature of heath ecosystems. Many of the dominant species mentioned,

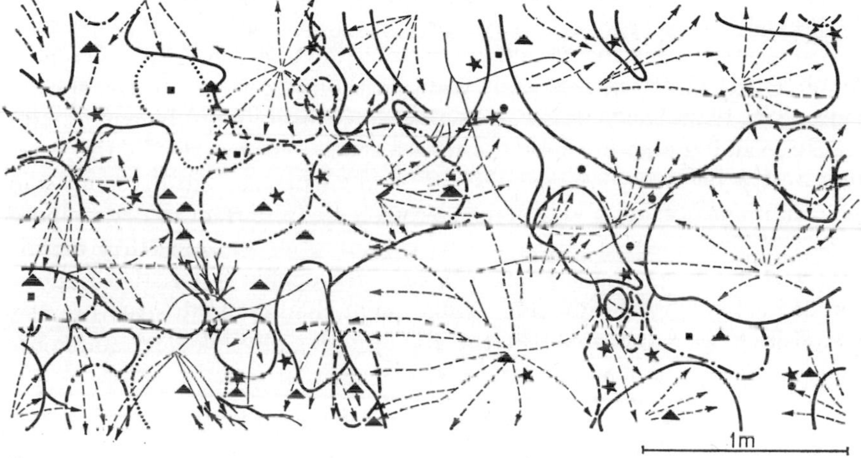

Key:
—— Perimeter of *Calluna vulgaris* bushes, --- Degenerate *C. vulgaris*, Visible dead stems of *C. vulgaris*,
-·-· Perimeter of *Eriophorum vaginatum* tussocks, ······ Degenerate *E. vaginatum* tussocks, ● *Sphagnum* spp.,
■ Other Bryophyta, ★ *Erica tetralix*, ▲ *Vaccinium myrtillus*, Buried stems of *C. vulgaris*

Fig. 17. Map of the distribution of bushes of *Calluna vulgaris* and *Erica tetralix*, and tussocks of *Eriophorum vaginatum*, in a sample of wet heath at Cammachmore, north-east Scotland. (Figure by T. Keatinge.)

The arrows show the position and direction of growth of procumbent, buried stems of *Calluna*, indicating how several of the bushes, now separate, may have been derived from a common origin.

notably *Calluna*, exhibit a morphological development which involves degeneration in the centre of the patch occupied by the individual (p. 127). A 'gap phase' is formed, frequently occupied for a time by other species of the community (Plates 22, 23). Where the stand is uneven-aged, adjacent patches may carry vegetation at different phases in this cycle, so constituting a very obvious and intense pattern. This source of pattern, in addition to those already discussed, is so important a feature of heath ecosystems that it is given separate consideration in Chapter 7.

In heath vegetation, a knowledge of community structure is more than a mere description of one aspect of the ecology of heathlands. It is fundamental to an understanding of the control of composition by the dominant species, and of the particular micro-environmental conditions obtaining. Hence, before the effects of management practices can be assessed and consideration given to improvement or conservation of heathlands, their impact upon community structure and, through this, on the micro-environment must be determined. A well-developed heath possesses a unique and highly characteristic structure conferred on it by the particular synusiae present, the type of stratification developed, and frequently by a relatively intense form of pattern. The latter is created by the inherited morphology of the chief species, most of which produce dense caespitose units with diameters about 60–80 cm and behave in such a way as to initiate cycles of change in the vegetation.

While variations in this structural organization due to habitat differences are obviously important, its basis rests on the morphology of the species concerned. Their characteristic growth-forms control both the community structure and the ability of the various species to co-exist in the ways described. Their morphological as well as their physiological responses to the conditions obtaining in different habitats are important in determining ecological amplitude and the range of communities to which they contribute. In Chapter 4 consideration is given to the details of floristic composition of heath communities and in subsequent chapters both structure and composition are interpreted in the light of the physiological and morphological characteristics of the species concerned.

4 The composition of heath communities, and their seral relationships

Much variation exists in the floristic composition of heath communities. Although the generally acid, oligotrophic habitats restrict the diversity of species in comparison, for example, with calcareous grassland, some heath communities can be quite rich in vascular plants and many contain a varied lichen or bryophyte flora. Floristic monotony is not a basic characteristic of heaths: where this obtains it probably stems in large part from the effects of management (Chapters 9, 10). Of particular interest, however, is the extent of variation in community composition in different parts of the heath region, and this has been the subject of numerous investigations. Studies have been directed on the one hand towards determining the nature and causes of these variations, and on the other, towards the formulation of acceptable means of classifying heath communities into recognizable and useful types. In this field, research on heathlands provides instructive examples of the advantages of applying several different techniques of analysis to the same subject matter, and offers an opportunity for assessing the compatibility of the results obtained.

As shown in Chapter 2, the geographical range of lowland heath vegetation in western Europe extends through parts of two major forest types, largely as a result of its derivation from former forest for the most part by human agency. Although associated with certain general features of climatic regime, it therefore spans a wide gradient of climatic and edaphic conditions. It is not surprising that this results in very striking variations in community composition, on to which are superimposed the effects of systems of management by burning and grazing. Floristic variation in European heathlands presents a complex situation for analysis.

Chronologically, attempts to build up systems of classification preceded the investigation of gradients in composition and the identification of 'directions of variation' (Sjörs, 1950). However, there is now much to be gained by examining the problems of classification in the light of prior consideration of the nature and causes of variation.

Ordination of heathland stands

In recent years the investigation of variation in community composition has been revolutionized by rapidly evolving techniques of stand ordination. Graphical representations of relationships between stands, based on measures of similarity in composition, offer a ready means of comparing gradients in floristic composition with gradients in environmental parameters. Stand ordination is also a valuable aid in detecting indicator species and identifying the principal factors responsible for variations in composition.

An elementary form of ordination can be carried out on the basis of

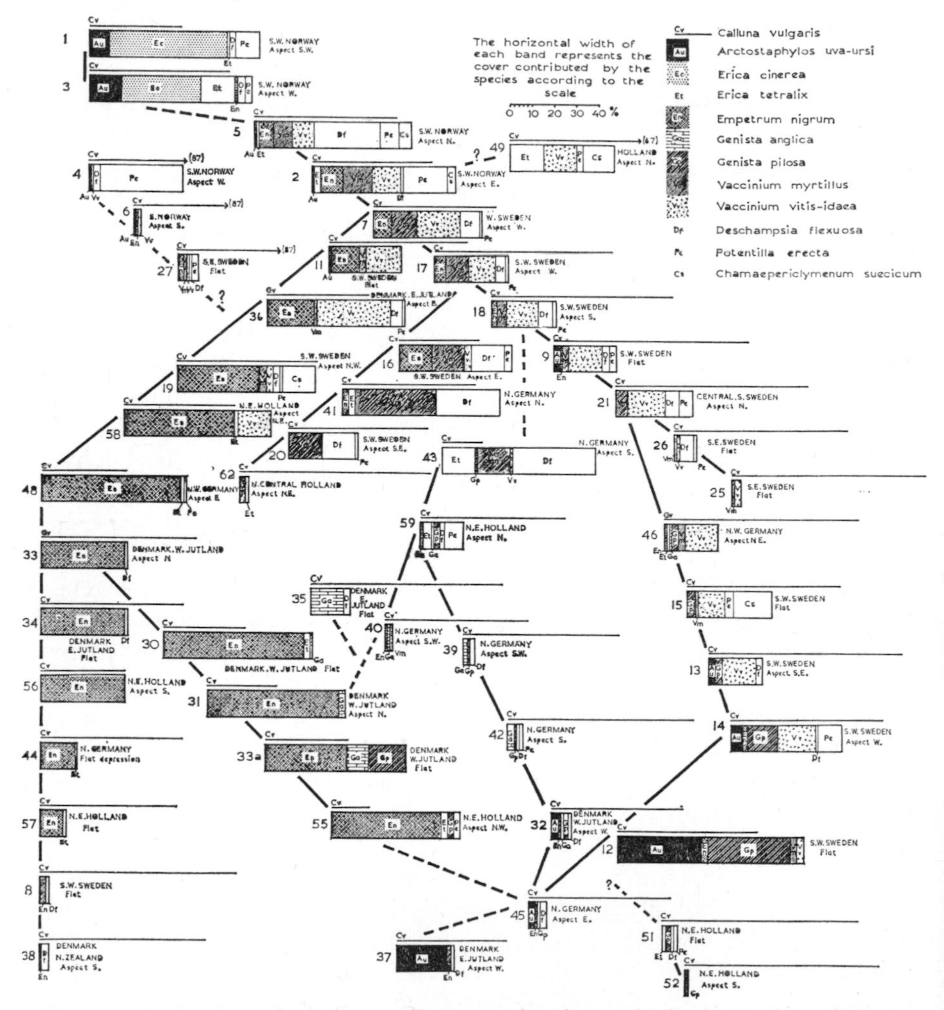

Fig. 18. Ordination of north-west European heath stands, by inspection of histograms showing cover contribution of dwarf-shrubs, etc. (From Gimingham, 1961.)

inspection of the results of quantitative community analysis of a number of stands. For example, data on the cover-contribution of dwarf-shrub species (and others of the same vegetational stratum) in 60 heathland stands from the northern sector of the west European heath region were compared by Gimingham (1961). Attention was confined to lowland heaths in south-west Norway, south Sweden, Denmark, north Germany and the Netherlands, where as many different examples were examined as time allowed. No other restrictions governed the choice of stands. With a few exceptions all the stands were alike in having *Calluna* as the dominant species and comparison was therefore based on the other species, whose cover values were plotted as histograms. These were arranged, by eye, into a number of interconnected series in which the representation of each of the main species varied progressively – rising or falling, or rising to a peak and then declining. In this way an arrangement was found which took the form of a network of variation, in effect a two-dimensional ordination of the stands (Fig. 18). It is not claimed that this subjective pattern is anything more than an approximate representation of stand relationships, but it succeeded in indicating two primary axes of variation, and other minor ones.

A major trend in composition is illustrated by the gradation from heaths rich in *Vaccinium* species (*V. vitis-idaea* and *V. myrtillus*) positioned in the upper part of the diagram, to heaths containing species of *Genista* (particularly *G. pilosa*) in the lower part. Comparison with the geographical location of stands shows that this gradient is generally, though not precisely, correlated with latitude. The obvious inference is that the floristic variation here expresses the effects of a climatic gradient, doubtless mainly those of temperature regime. This is borne out by reference to stands which appear to lie 'out of context' in the series, i.e. certain communities from the Netherlands, north Germany and Denmark, which appear amongst examples from more northerly situations. In these cases, the southern stands invariably occupied north-facing slopes where the local climate would resemble that obtaining more generally to the north, so producing floristic affinity.

A second axis of variation is shown by a gradient from right to left in the proportions of several species, particularly *Erica tetralix* and *Empetrum nigrum*, also *Vaccinium myrtillus* and *Erica cinerea* in the upper part of the diagram. The correlation here seems to be with a gradient of increasing oceanicity of climatic regime.

This subjective arrangement and assessment has many limitations and is open to errors. However, the same data have been used by Kershaw (1967) for a principal components analysis. The resulting ordination has identified three main axes of variation, the first two of which correlate with the same climatic gradients recognized above, namely a gradient of temperature regime and one of oceanicity. The third axis reveals variation which possibly relates to edaphic conditions since it derives from the contribution of species whose cover is known to depend on soil wetness. However, confirmatory measurements are lacking.

Although covering only a part of the heath region, an ordination such as this of widely scattered samples is clearly valuable in establishing some of the main trends in community composition governed by broad differences in climate. Ordination of stand data from a more restricted portion of the geographical or climatic range will not be expected to give the same prominence to climatic factors as causes of floristic variation. Hence, in the ordination by Ward (1968, again using principal components analysis) of 95 stands comprising 58 from north and north-east Scotland (*Calluna-Arctostaphylos* communities only) and 37 from the nearest parts of Scandinavia (south-west Norway, south-west Sweden and Jutland), the first axis of variation extracted is probably related to soil nutrient status. This is indicated by gradients in the representation of several indicator species (Fig. 19, a–d). The second axis, however, appears to express the influence of greater and lesser degrees of oceanicity in the climate (Fig. 19, e–h). These results fit into place when the geographical distribution of the stands, extensive in the west to east direction but limited from north to south, is considered. When the Scandinavian stands only were ordinated, the three axes of variation emerging were correlated (*i*) with soil nutrient status, (*ii*) with soil wetness and (*iii*) with past management practice, and no major climatic influence was shown. Similarly, ordination of even more restricted series of heath stands by Anderson (1963, north Wales) and Ward (1970, Dinnet Moor, north-east Scotland) have drawn attention particularly to soil factors and management by burning.

The results of ordination indicate the major directions of variation in community composition which must be taken into account in forming any valid system of classifying heath communities, and also suggest certain species which may serve as valuable guides for this purpose.

Classification of heath communities

While these ordinations clearly indicate that heathland vegetation must be regarded as a continuum of variation, lacking any very discrete groupings of stands, it does not necessarily follow that classification is impossible or useless. Even a continuous gradient can be broken up arbitrarily into categories, and in any case continuity is generally somewhat over-emphasized by the techniques involved in ordination. Furthermore, a practical basis for classification arises in the event of stands which belong to certain parts of the continuum occupying greater areas or recurring with greater frequency than those in other parts. This is revealed in an ordination if stand sampling is proportional to area occupied, or to frequency of stands, but in such extensive sampling as that of the studies discussed this is often impracticable. For example climatic gradients are seldom smooth, and relatively wide areas across which there is little change may be separated by narrow belts

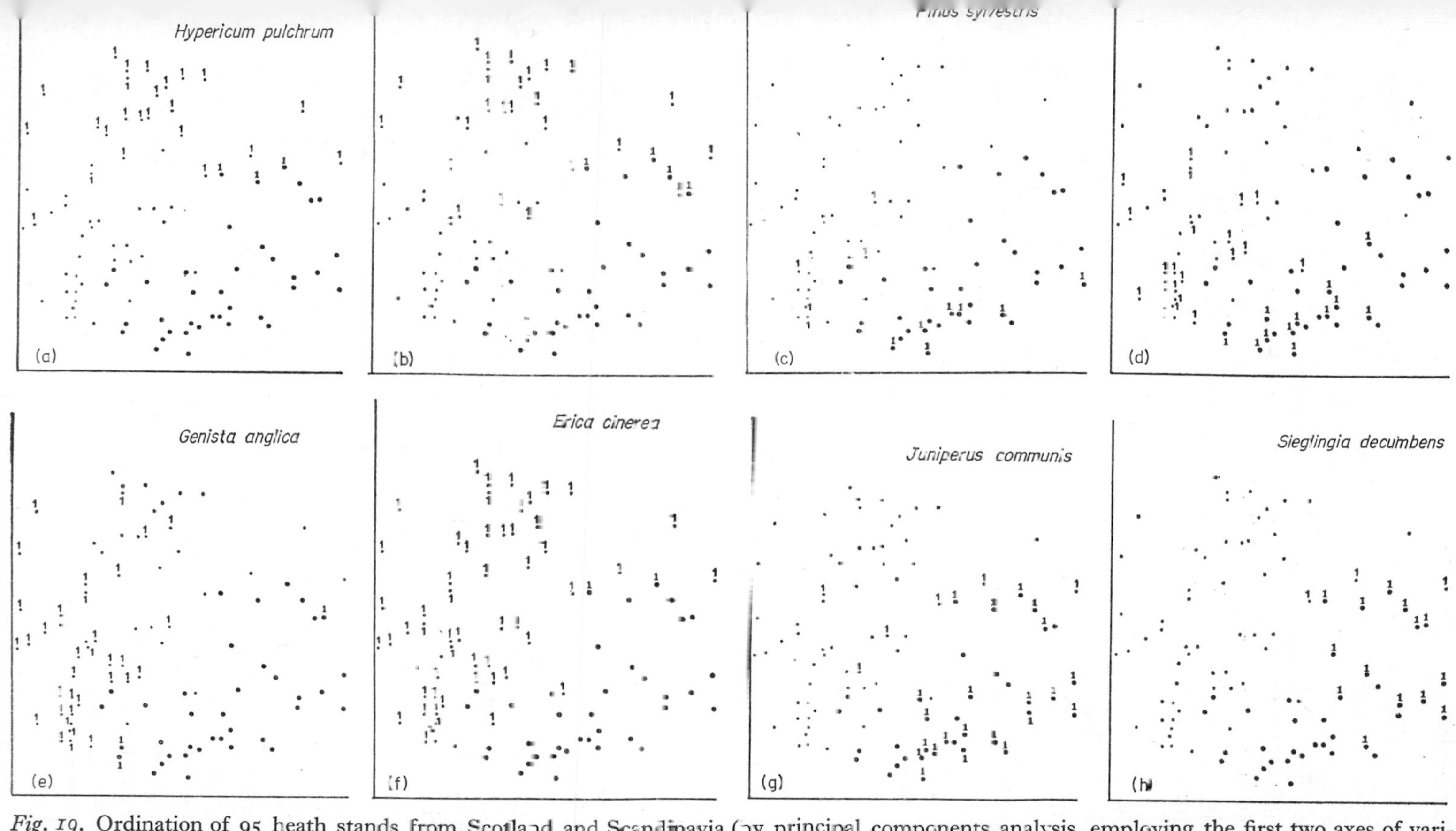

Fig. 19. Ordination of 95 heath stands from Scotland and Scandinavia (by principal components analysis, employing the first two axes of variation). (Figure by S. D. Ward.) Small dots: Scottish stands. Large dots: Scandinavian stands.

For eight selected species, occurrence in the stands is shown by the number 1. (*a*)–(*d*) Species correlated with the vertical axis of variation. (*a*) *Hypericum pulchrum*, (*b*) *Viola riviniana* are associated with soils of higher nutrient status; (*c*) *Pinus sylvestris*, (*d*) *Cladonia uncialis* are associated with the poorer soils. (*e*)–(*h*) Species correlated with the horizontal axis of variation. (*e*) *Genista anglica*, (*f*) *Erica cinerea* occur particularly in heaths of the more oceanic parts of the area represented; (*g*) *Juniperus communis*, (*h*) *Sieglingia decumbens* are (for various reasons) concentrated in the less highly oceanic parts.

across which the gradient is steep. Under these conditions heath vegetation may show floristic uniformity over quite wide areas, and rather rapid transitions to other 'types'. Phytosociological classifications may then express some degree of differing 'intensity' of variation, even in the absence of discontinuity. They may then be accepted as a further contribution to the investigation of community composition, without demanding the adoption of any particular 'community-type hypothesis' (McIntosh, 1967).

Although published well before any results of ordination were available, Bøcher's (1943) system of classification of communities of the North Atlantic heath formation recognizes precisely those main trends of variation shown up by the more comprehensive ordinations, i.e. the north to south gradient, and the oceanic to sub-continental gradient. Bøcher's aim was to classify rather than examine the nature of variation, but he chose deliberately to relate his classification to these primarily geographical aspects of community variation.* Accordingly he examined the geographical distribution of all the main species, and selected a number which showed representative types of area as guiding species. Naturally, a proportion of these are species for which ordination diagrams demonstrate correlation of occurrence with one of the main axes of stand variation. On the basis of the representation of these species Bøcher erected his groups, as shown in Table 6. The result is an effective classification giving expression to the main trends of regional variation in community composition. This permits the identification of localized communities which differ from the type characteristic of a region, so indicating the existence of a specialized local environment (cf. p. 61). Furthermore, areas in which several different groups are represented (e.g. parts of Denmark, south Sweden) merit special investigation.

The chief value of Bøcher's system is that it identifies types of heath community based on the main regional trends in composition. However, it may be argued that floristic variation may follow many other 'directions' and that the sole use of geographical guiding species obscures other important aspects of community composition. Although some types of community may be correlated with geographical (and hence climatic) regions, their distribution as already seen is seldom confined to the region of which they they are characteristic. Other types, determined by different factors, may be recognizable over much wider areas. Such arguments provide a basis for attempting classification on more general phytosociological criteria. An additional outcome of more general phytosociological systems is that floristic relationships between heath and other formations may also be examined.

Europe has seen the origin of more than one 'subjective' phytosociological system, and the dwarf-shrub heaths have appeared in numerous schemes of community classification both local and more general. A valuable opportunity is thus available to assess the measure of agreement between different workers, at least in so far as the classification of heaths is

*Phytogeographical aspects of the flora of north-west European heaths are also discussed in detail by De Smidt (1967).

TABLE 6

Arrangement of European heaths according to Bøcher (1943)

I. Dry heath communities

Arctic-alpine series

Phyllodoco-Myrtillion

Rhacomitrium group. Oceanic.

Phyllodoce coerulea group. Suboceanic-subcontinental and widely ranging arctic alpine species mixed.

Myrtillus group. Subcontinental and widely distributed, arctic alpine species (e.g. *Betula nana-Empetrum-Vaccinium-Hylocomium splendens* soc.).

Loiseleurieto-Arctostaphylion

Loiseleuria-Juncus trifidus group. Among the flowering plants, suboceanic and subcontinental species; ground layer frequently with continental lichens (*Alectoria ochroleuca*).

Arctostaphylos uva-ursi group. Subcontinental, widely distributed flowering plants, continental lichens. Suboceanic-arctic element much reduced.

Cassiope tetragona group. Continental, subcontinental (*Cassiope tetragona, Calamagrostis lapponica*), and widely distributed plants. Suboceanic element reduced.

Kobresieto-Dryadion

Continental element very prominent. Perhaps subtypes with many widely distributed and rather few continental plants or with a large number of frequent continental plants (*Cassiope tetragona-, Rhododendron lapponicum-, Arctostaphylos uva-ursi-,* and *Dryas*-sociations. The *Empetrum-Vaccinium* heath rich in *Dryas* from the Faroes is ecologically clearly related to *Dryadion*. It belongs, however, to another group or to another related oceanic alliance which may further occur in Ireland and Iceland.

Scano-Danish (Scotch) series

Myrtillion boreale

Chamaepericlymenum-Blechnum-Rhytidiadelphus loreus group.

Myrtillus-Hylocomium splendens-Rhytidiadelphus triquetrus group. Oceanic and northern oceanic element reduced.

Empetrion boreale

Empetrum-Vaccinium vitis-idaea group. Suboceanic-subcontinental.

Empetrum type.

Vaccinium vitis-idaea type.

Arctostaphylos uva-ursi group. Subcontinental.

Dutch-German series

Genistion

Genista anglica group. Oceanic.

Genista pilosa-Sarothamnus group. Suboceanic.

Genista germanica-tinctoria group. Subcontinental.

Genista sagittalis group. Suboceanic montane.

Baltic-submontane German series

Callunion balticum

Galium saxatile-Carex arenaria group. Suboceanic.

Lycopodium complanatum-Carex ericetorum group. Subcontinental northerly.

Filipendula hexapetala group. Subcontinental southerly.

I. Dry heath communities–*continued*

Eu-oceanic series

Ericion cinereae, a rather northerly alliance including heaths of the Faroes, west Norway, the British Islands, and northern France

Erica cinerea group. Northern-most main type.

Ulex gallii-europaeus group. More southerly type.

Ericion scopariae, the southern-most heath alliance occurring in France (Les Landes), Spain and Portugal. Transitions to Mediterranean *Maquis* are numerous.

Erica vagans group. Rather northerly type, extending to Cornwall and North France.

Erica scoparia group.

II. Wet heath and bog communities

Atlantic series

Ulicio-Ericion tetralicis

Myrica-Narthecium group. Northern oceanic.

Empetrum-Vaccinium uliginosum group. Northern.

Oxycoccus-Eriophorum vaginatum group. Northern-subcontinental species present.

Erica tetralix group. Southern, not extremely oceanic.

Ulex minor-Erica ciliaris group. Very oceanic and southerly.

Subatlantic-subcontinental series

Oxycocco-Andromedion

Betula nana group. Alpine-subalpine.

Trichophorum cespitosum group. Northerly-suboceanic.

Empetrum nigrum-Oxycoccus group. Subcontinental northerly element rather important.

Callunio-Juncion squarrosi. Suboceanic; rather southerly

Continental series

Ledio-Chamaedaphnion

concerned, and to benefit from the different insights into the relationships of these communities which each provides. It also becomes possible to test the compatibility of the findings of phytosociologists with the results of ordination, on the one hand, and of statistical methods of classification on the other. Among phytosociologists who follow the methods of Braun-Blanquet (1964) there has been a considerable measure of agreement concerning the basic units (i.e. 'associations') into which heath vegetation may be divided. The various associations have been described at different times by different authors, but in most cases they have been widely accepted in phytosociological literature. A good example is the Calluno-Genistetum*, first named by Tüxen in 1937 in his classification of the plant communities of north-west Germany. Numerous other authors have been able to

* Variously written as Calluneto-Genistetum or Calluno-Genistetum.

recognize stands clearly referable to this category in north Germany, Belgium, the Netherlands and elsewhere. Since associations are established on the basis of floristic similarity between stands (grouped in 'association tables') and then distinguished by 'differential' species, it is not surprising that in many instances they are closely comparable with Bøcher's types. In such cases the species which emerge as differentials are often also 'geographical guiding species'. However, the correspondence is not precise.

It is a further feature of the Braun-Blanquet system that associations may, on the one hand, be subdivided and, on the other, grouped in a hierarchy of assemblages of higher rank. Subdivisions generally reflect the presence or absence of sets of species having ecological requirements more specialized than those of the plants characterizing the whole assemblage. Hence, in many instances, subdivisions reflect floristic variation related to soil conditions (nutrient status, wetness, etc.) or the effects of management (e.g. burning). In general, therefore, phytosociological classification follows the chief directions of variation indicated by the results of ordination.

While there is substantial agreement about the constitution of the basic units (associations), much greater variety is shown in the way these have been arranged in hierarchies (Gimingham, 1969). This is perhaps not surprising, since a hierarchy is a linear or one-dimensional series of aggregations whereas floristic variation is multi-dimensional, as illustrated by the need to extract at least two or three axes in ordination. While no hierarchy can express fully the many relationships between associations, each may contribute some interesting suggestions. In Table 7, aspects of several different hierarchies are compared. That of Tüxen in 1937 drew attention to the relationships between lowland heaths of north-west Germany (Calluno-Genistetum) and a more montane type (*Calluna-Antennaria* association), while the alliance into which these are linked (Ulicion) is placed next to the Alpine Rhodoreto-Vaccinion. Other arrangements, such as those of Knapp (1942, referring to the Euro-Siberian Region) and Lebrun *et al.* (1949, referring to Belgium), are of interest in connection with the origins of heath from forest (Chapter 2), the former emphasizing affinity with coniferous and the latter with deciduous forests. Preising in 1949, and later, in a new arrangement, Tüxen (1955) take account of relationships between heaths and acid grasslands by placing Calluno-Ulicetalia next to Nardetalia and linking both in a class Nardo-Callunetea. This may give expression to the ecological affinity of these types, which frequently occupy similar soils and may in many cases have diverged as a result of management and biotic influences.

While each of these differing treatments has something to recommend it, there has in recent years been a tendency to settle on the type of hierarchy shown in Table 8 as about the most practical.

TABLE 7

Treatment of heath community units in selected phytosociological classifications.
(Details of Associations are given only for lowland heath-types on the drier soils.)
(From Gimingham 1969,)

Tüxen 1937 (Region: N. W. Germany).

Order	Ericeto-Ledetalia palustris			Calluneto-Ulicetalia		Rhodoreto-Vaccinietalia
Alliance	Ulicio-Ericion tetralicis	Oxycocco-Ericion	Oxycocco-Empetrion hermaphroditi	Ulicion		Rhodoreto-Vaccinion
Association				Calluneto-Genistetum	Calluno-Antennaria Assoc.	

Knapp 1942 (Eurosiberian Region)

Class	OXYCOCCO-SPHAGNETEA			BETULETO-PINETEA				
Order	Ericeto-Ledetalia palustris			Calluneto-Ulicetalia		Vaccinio-Piceietalia		Quercetalia roboris-sessiliflorae
Alliance	Ericion tetralicis	Oxycocco-Ericion	Oxycocco-Empetrion hermaphroditi	Ulicion		Pinion sylvestris	Vaccinio-Piceion	Quercion roboris-sessiliflorae
„Haupt-Assoziation"				Ulico-Ericetum cinereae	Calluneto-Genistetum			

Lebrun *et al.* 1949 (Region: Belgium)								
Class	OXYCOCCO-SPHAGNETEA		MOLINIO-JUNCETEA	ARRHENATHERETEA		QUERCETO-ULICETEA		
Order	Ericeto-Sphagnetalia					Calluno-Ulicetalia		Quercetalia roboris-sessiliflorae
Alliance	Sphagnion europaeum	Ericion tetralicis				Ulicion		
Association				Ass. à *Calluna* et *Genista anglica*	Ass. à *Calluna* et *Vaccinium vitis-idaea*	Ass. à *Calluna* et *Sieglingia decumbens*	Ass. à *Calluna* et *Antennaria dioica*	

Tüxen 1955 (Region: N.W. Germany)							
Class	OXYCOCCO-SPHAGNETEA			NARDO-CALLUNETEA			BETULETO-ADENOSTYLETEA
Order	Ericeto-Sphagnetalia	Sphagnetalia fusci	Nardetalia		Calluno-Ulicetalia		
Alliance	Ericion tetralicis	Sphagnion fusci	Nardo-Galion saxatilis	Empetrion boreale	Calluno-Genistion	Sarothamnion scopariae	
Association				Saliceto repentis-Empetretum	Calluno-Genistetum	*Calluna-Antennaria* assoc.	

TABLE 8
Phytosociological hierarchy for European heaths
(excluding mountain heaths)
Partly following Moore (1968) and Bridgewater (in preparation).

1. **Wet heaths**
   CLASS Oxycocco-Sphagnetea Br.-Bl. et Tx. 1943
   ORDER Ericetalia tetralicis Moore 1968
   ALLIANCE Ericion tetralicis Schwick. 1933
2. **Dry heaths**
   CLASS Nardo-Callunetea Prsg. 1949
   ORDER Calluno-Ulicetalia (Quantain 1935) Tx. 1937
   (north-west and Central Europe)
   ALLIANCES Ulicion*
   Calluno-Genistion Duvign. 1944
   Myrtillion†
   (In south-west Europe heaths grade into taller communities placed in the
   ORDER Erica-Ulicetalia Br.-Bl., Pinto da Silva, Rozeira 1964
   ALLIANCES Cistion hirsuti (Br.-Bl. et al. 1953) Br.-Bl., Pinto da Silva, Rozeira 1964
   Genista-Ericion aragonensis Rivas-Mar. 1962.

* Formed from the alliances Ulicion nanae Duvign. 1944 and Ulicion gallii des Abb. et Corillion 1949 (Bridgewater, in prep.).
†Formed from the alliances Empetrion boreale Bøcher 1943 and Myrtillion boreale Bøcher 1943 em. Schubert 1960 (Bridgewater, in prep.).

The works of Nordhagen (1928, 1936) and Dahl (1956) in Norway have also made an important contribution to the phytosociology of dwarf-shrub heaths, in particular those of the north European mountains. These authors differ from adherents to the Braun-Blanquet school of phytosociology by giving primary consideration to constancy, rather than fidelity of species occurrence, in diagnosing and naming associations. None the less, the procedure for classification is similar and the final groupings are of comparable status. It is therefore quite satisfactory to incorporate categories from both sources in attempting a synopsis of heath community-types (pp. 79–81).

British heath communities

An early attempt to classify heath vegetation within the British Isles was made by W. G. Smith (1905, 1911) as follows:

HEATHER ASSOCIATIONS

(*a*) Heath or dry heather moor (Heide) where *Calluna vulgaris* occurs with *Erica cinerea* and associates which prefer dry soils.

(*b*) Heather moor (Heide Moor) thoroughly dominated by *Calluna vulgaris* and associates preferring a slight depth of peat:
   (*i*) exclusively *Calluna*, often on sloping ground;
   (*ii*) *Calluna* and *Erica tetralix*, peat deeper and moister.

(A 'mixture of Grass Heath and Heather Associations' is also recognized.)

This classification depends partly on reference to the dominant species and partly on an aspect of the habitat – the nature of the substratum. Smith had, in fact, picked on one of the directions of variation in floristic composition, that associated with the increase in organic content of the substratum from podsol to peat. Tansley (1939) felt, however, that there was no good vegetational distinction between the '*Calluna* heath' and '*Calluna* moor' of Smith, and emphasized instead the altitudinal direction of variation,

TABLE 9

Classification of upland heaths in Scotland by Muir and Fraser (1940), as summarized by Zehetmayr (1960).

Dry heath types
Calluna-Erica cinerea *Calluna* with abundant *Erica cinerea* forming the ground cover, subsidiaries not very frequent and usually suppressed.
Calluna-Vaccinium myrtillus *Calluna* dominant, with abundant *V. myrtillus* co-dominant, *V. vitis-idaea* generally frequent.
Calluna-Arctostaphylos *Calluna* dominant but open. Undergrowth of *Arctostaphylos*. Many subsidiaries.
Calluna-Deschampsia-Vaccinium myrtillus *Calluna* dominant but open; sub-dominant or locally co-dominant *Deschampsia flexuosa* and *Vaccinium myrtillus*.
Moist heath types
Calluna-Deschampsia flexuosa *Calluna* dominant, with occasionally flowering, partly suppressed *D. flexuosa* abundant. Turf usually quite covered.
Calluna-Vaccinium *Calluna* dominant, with suppressed or poor growth of *V. myrtillus* and *V. vitis-idaea* abundant. Turf exposed and showing growth of encrusting lichens.
Calluna-Nardus *Calluna* normally dominant, with *Nardus* sub-dominant or co-dominant. *Pleurozium schreberi* the chief moss, along with patches of *Sphagnum*.
Submoorland types
Variable subtypes characterized by the presence of peat, subsidiary and co-dominant plants such as *Erica tetralix*, *Eriophorum vaginatum*, *Trichophorum cespitosum*, *Carex* and *Juncus* spp. abundant or locally abundant. *Sphagnum* spp. and *Pleurozium schreberi* usually the most frequent mosses.
Exposure types
Dry eroded *Calluna*. Wet *Calluna-Cladonia*.

contrasting 'upland heaths' with 'lowland heaths'. British vegetational classification relied for many years on using the dominant species to differentiate community types, and the term *Callunetum* describes heath vegetation in many publications. That there was much variation within *Callunetum* was, of course, recognized and subdivisions were made, sometimes on ecological grounds (Tansley, 1939) and sometimes by reference to other prominent species as in a valuable arrangement of Scottish heath communities by Muir and Fraser (1940) (Table 9).

It was evident, however, that more detailed treatments were required, and with the elaboration of systems of classification on the continent of Europe there was advantage to be gained by attempting to relate the British communities with those described in neighbouring countries. However, difficulties arise in relating the communities of near-by regions in a single phytosociological analysis. Even if closely similar, they are never identical, owing to responses to climatic gradients. The point at which the differences should be treated as of sufficient importance to warrant recognition of distinct associations inevitably depends upon subjective judgements. Some of the directions of variation in the composition of heath communities in Britain, for example those following the south to north climatic gradient, parallel similar variation on the European continent and may be expressed in similar terms. None the less, they are repeated under more oceanic conditions which introduce differences in community composition (for example, *Erica cinerea* is very generally present in British heaths on freely drained soils, but absent from their counterparts in Sweden, Denmark, Belgium, the Netherlands and Germany).

This problem may be partly responsible for a reluctance on the part of British workers to attempt the integration of British communities into systems of analysis of European heaths. A start was made, however, by Poore (1955) and Poore and McVean (1957) who compared communities from certain Scottish mountains with associations established by Nordhagen (1928, 1936) and Dahl (1956). More comprehensively, McVean and Ratcliffe (1962) and McVean (1964) classify the plant communities of the Scottish Highland Region, including the montane and alpine heath types, by methods based on those of Poore (1955) and Nordhagen (1928, 1936), indicating where possible their relationships or identity with European associations. Gimingham (1964*a*) lists a number of lowland heath community-types from Scotland, arranging them according to Bøcher's classification. The relationships between British and continental heath communities are discussed by Gimingham (1969), while more recently a classification of British lowland heath types on the European phytosociological pattern has been attempted by Bridgewater (in preparation).

Floristic variation in British heaths is evidently considerably greater than in their continental equivalents. Bridgewater's survey produced numerous associations, but for the purpose of linking them with related European types in a hierarchy he found the existing framework (Table 8) adequate. Of the dry heaths, eight associations were placed in the Ulicion, representing the southern and western oceanic communities; another eight of the more northerly types fell into the Myrtillion; while five, dominated by *Calluna* and lacking strongly northern or oceanic species, were allocated to the Genistion. All five associations of wet heath were adequately covered by the Ericion tetralicis. In this way the affinities of British communities with groups of associations in neighbouring countries were demonstrated, but no detailed study was made of the degree of similarity between British examples and the most closely related associations from other countries. Recently Harrison (1970) has discussed the affinities between communities on two Sussex heaths (*Calluna-Ulex minor* and wet heath communities), analysed quantitatively, and certain phytosociological units occurring in north France. However, the relationships between British and European heaths would repay much further study aimed at expressing their similarities and differences quantitatively, as attempted by Ward (1968) in his investigation of the British and Scandinavian *Calluna-Arctostaphylos* heaths.

The effects of the greater oceanicity of the British climate are probably most evident in lowland districts and on the drier soils. In the mountains, although the altitudinal limits of successive climatic belts are markedly lower in oceanic than in continental regions, the main features of the zonation are similar. Hence as shown by McVean and Ratcliffe (1962) it is often possible to equate the community-types of Scottish montane heaths with those of Norway and in some cases to use the categories established by Nordhagen (1928, 1936) and Dahl (1956).

In the lowlands, vegetation on wet soils is less susceptible to the differences between oceanic and continental climatic regimes than on dry ground. All the associations of wet heath vegetation shown by Moore (1968) as occurring in Britain are also represented in the neighbouring countries surrounding the North Sea or the English Channel, and have been described and named at various times by continental workers. On the more freely drained soils, however, particularly in the most strongly oceanic western parts of the country, there are heath communities probably unique to Britain, representing part of Bøcher's Eu-oceanic series of heath-types. An example is the *Calluna-Scilla verna* heath of maritime cliffs in south-west England (Malloch, (1971)). The *Calluna-Arctostaphylos uva-ursi* community of medium altitudes in east central Scotland has as its only immediate counterpart the very restricted west Norwegian *Calluna-Erica cinerea* heaths with *Arctostaphylos*. These communities are very unlike the more sub-continental heaths containing *Arctostaphylos* of Denmark and south-west Sweden.

On the other hand, heath stands of the eastern half of north Britain, generally containing *Vaccinium myrtillus* and in some cases *V. vitis-idaea*, must be closely related to widespread similar communities in south-west Norway, south-west Sweden and parts of Denmark (with outlying stations in the Netherlands and north Germany), although the continental examples lack *Erica cinerea*. In the same way, certain heaths in south-east England (e.g. East Anglia) compare with those of north Germany in containing *Genista anglica*, *G. pilosa* (occasionally) and *Cuscuta epithymum*; while others in south-west England agree with those of north France, having *Erica vagans*, *Ulex gallii*, *Agrostis setacea* and other species in common.

There is one aspect of the composition of British heaths which is more difficult to reconcile with those of the Continent. This is the widespread occurrence of communities in which species other than *Calluna* are very poorly represented. Other dwarf-shrubs, except sometimes *Erica cinerea*, are generally lacking and often there are extensive pure stands of *Calluna*. These communities are described as 'dry *Calluna* moor' (Birse, 1968), '*Calluna-Erica cinerea* heaths with reduced oceanic floristic element' (Gimingham, 1964*a*), or simply Callunetum vulgaris (McVean and Ratcliffe, 1962), having as their constants only certain mosses (*Dicranum scoparium*, *Hylocomium splendens*, *Hypnum cupressiforme* and *Pleurozium schreberi*) in addition to *Calluna*. The only obvious determining factor common to these communities is that they occur generally where burning for sheep grazing or grouse rearing is regular, tending towards the production of a monoculture of *Calluna*. Extensive species-poor stands of *Calluna* also occur on the Lüneburg heath, north Germany, one of the few areas on the Continent in which sheep grazing is maintained.

Statistical methods of classification

Objections to the phytosociological systems so far discussed are based on the strongly subjective element both in the selection of stands and in the process of grouping them. The latter difficulty may be eliminated by the use of statistical methods which generate classifications from the data. At present, these methods have been applied only to restricted sets of samples but the results provide an interesting commentary on the more comprehensive subjective schemes. There are several instances of the application of association-analysis to data obtained by sampling within a limited area of heathland (Williams and Lambert, 1959; Harrison, 1970; Ward, 1970). The groups of quadrats created can usually be given a satisfactory ecological interpretation: in most cases they illustrate soil differences (wetness, nutrient status) and the effects of management by burning. Ward's (1970) classification of samples from Dinnet Moor, Aberdeenshire, north-east Scotland, distinguished between lichen-rich and moss-rich groups (Fig. 20). The former he related to stages in recovery after burning, while further division of the latter produced herb-rich groups indicative of the better soils.

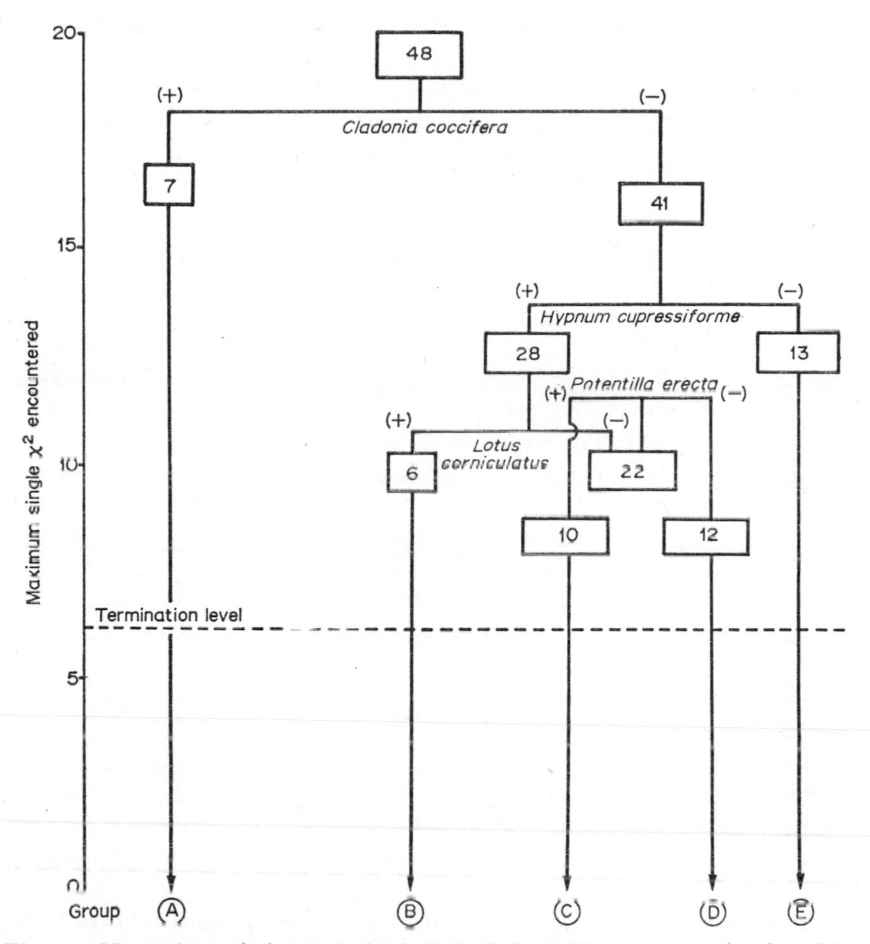

Fig. 20. Normal association analysis of 48 stands from Dinnet Moor, Aberdeenshire. (From Ward, 1970.) All species.

Group A – Lichen-rich
B – Moss- and herb-rich
C – Moss-rich, intermediate in herb-richness
D – Moss-rich, herb-poor
E – apparently not a natural group

(Stands are successively divided into two groups on the basis of the presence or absence of the species with the largest sum of χ^2 for association (+ or —) with other species. Each of the resulting groups is re-examined and again divided in the same way, the process being repeated until an arbitrarily determined limit is reached. The species concerned in the divisions are shown in the diagram on the line marking each dichotomy.)

A wider geographical range is represented by Gimingham's (1961) stands, chosen for the preliminary ordination described on pp. 60–61, have also been classified by the normal association-analysis of Williams and Lambert (1959, 1960). In the first place, data on the occurrence of all species (including bryophytes and lichens) were used by Cormack (reported in Gimingham, 1969), to produce the dendrogram shown in Fig. 21; while secondly, a similar analysis, using data for the dwarf-shrub species only, (Fig. 22) was carried out by Bannister (also reported in Gimingham, 1969). The latter classification produced groups corresponding more or less closely with those of the general phytosociological systems, e.g. groups of stands based on the presence of *Vaccinium vitis-idaea*, *Erica cinerea*, and *Genista pilosa* respectively. The former dendrogram, however, first identified herb-rich stands, secondly lichen-rich stands and thirdly a group of dune-heath communities, and only after separating these were the remaining stands divided according to the occurrence of northern or southern floristic components. Thus when the whole flora was considered, distinctions due largely to edaphic factors were the first to emerge.

Only by considering these two rather different classifications together can their bearing on the comprehensive subjective system of classification be assessed. In arrangements following the Braun-Blanquet procedure it is usually the sub-divisions of associations which display the effects of soil differences on community composition. Parallel variants may be recognized in several related associations on the basis of the occurrence of a particular group of indicator species. For example, plants typical of soils of relatively high nutrient status such as *Agrostis tenuis*, *Lotus corniculatus*, *Viola riviniana*, *Veronica officinalis* are often used to differentiate herb-rich variants in heath communities belonging to several different associations. Naturally these recurrent groupings will generate high measures of association between their component species when subject to association analysis, stands, chosen for the preliminary ordination described on pp. 60–61, which and these will then produce the first assemblages of stands. Once these are set aside, or if the species concerned are eliminated from the data being analysed (as in the second dendrogram), the classification proceeds on the basis of species which are more restricted regionally and climatically. These will also be the ones which are 'geographical guiding species' and, very often, differential species in the Braun-Blanquet terminology. A good example is provided by Ward's (1968) classifications of *Arctostaphylos* heaths and related communities from Scotland and Scandinavia. Once the influence of burning had been eliminated from the investigation by restricting the choice of stands to areas which had not been burnt for 15 years, the main division in the hierarchy took place on the presence or absence of *Hypericum pulchrum* which proved to be an important differential species with regard to a group of oceanic *Calluna-Erica cinerea-Arctostaphylos* communities (Fig. 23).

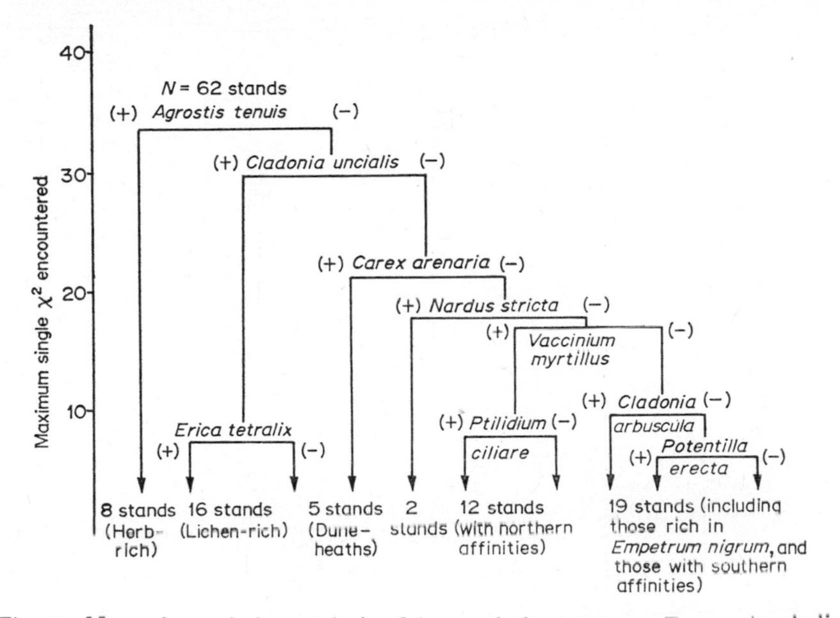

Fig. 21. Normal association analysis of 62 stands from western Europe (excluding Britain), using all species. (From Gimingham, 1969 – analysis by R. M. Cormack).

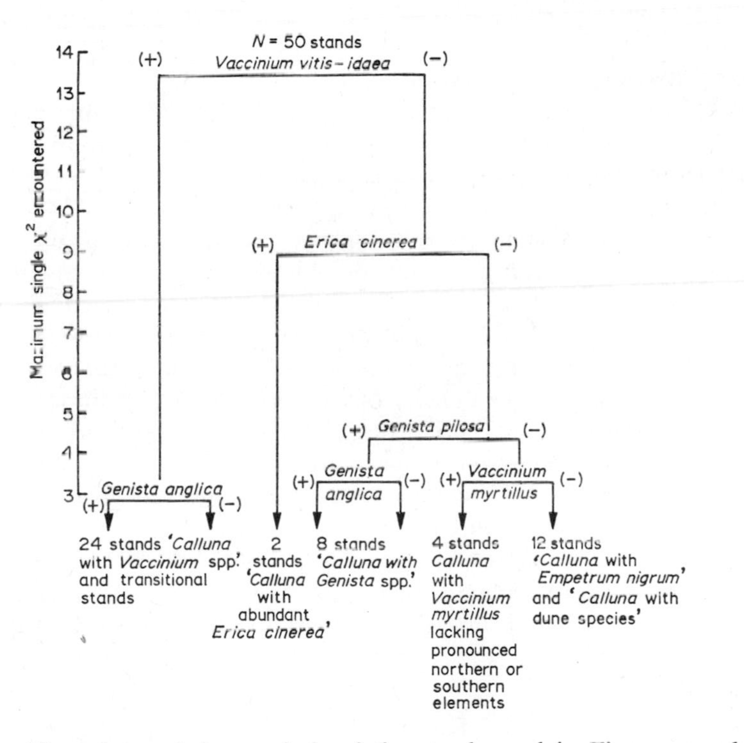

Fig. 22. Normal association analysis of the stands used in Fig. 19, employing dwarf-shrub species only. (From Gimingham, 1969 – analysis by P. Bannister.)

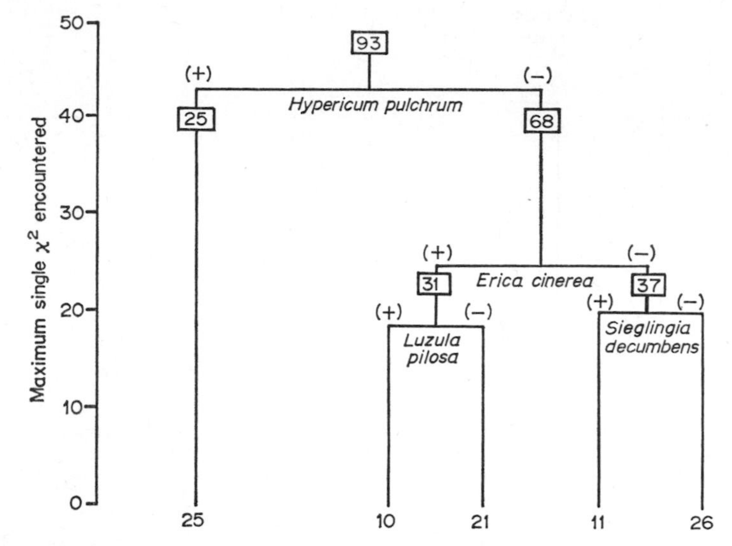

Fig. 23. Normal association analysis of *Arctostaphylos* – containing heaths and related communities from Scotland and Scandinavia. (Figure by S. D. Ward.) (Figures in boxes show numbers of quadrats in each group.)

Classification and ordination

Evidently ordination, comprehensive subjective classifications and more limited statistical classifications all contribute to an understanding of community composition in heathland vegetation, particularly when the results from all sources are compared. Classification provides more or less agreed categories of some value in practice, but can deal with only one direction of variation at a time, leading to considerable diversity in treatment at the upper levels of hierarchy. Ordination deals with several directions at once and hence provides valuable insight into the nature of variation. Using all the information, a list of important points on the continuum of variation can be drawn up (pp. 79–81). In many cases the communities referred to correspond with community-types established by one or other of the main phytosociological systems. Where this is so the correspondence will be indicated, but it is not intended to offer a definitive scheme according to any one set of principles of classification. Hence, the headings are descriptive only and do not adopt standard phytosociological nomenclature. The purpose is to provide a brief synopsis of the main kinds of community composition encountered throughout the heath region of western Europe.

List of main categories of heath communities

A. Mountain heaths

1. **Dwarf mountain heaths.** Exposed habitats, little snow-cover ('chionophobous'). Acid substrata. At medium altitudes *Calluna* with one or more of *Empetrum hermaphroditum, Loiseleuria procumbens, Arctostaphylos uva-ursi, Arctous alpinus*; at higher altitudes combinations of these, lacking *Calluna*. Mountains of Scandinavia, Britain, central and southern Europe. Also communities with *Cassiope tetragona* and species having sub-continental and continental affinities: not in Britain.

2. **Dwarf juniper heath.** Chionophobous. *Juniperus communis* subspecies *nana* associated usually with various species listed in 1 above. High altitudes on continental mountains; also apparently similar communities at lower altitudes on the oceanic seaboard of western Europe, e.g. in north-west Scotland and Ireland.

3. **Mountain *Vaccinium* heaths.** Habitats in which snow lies, including snow-patch areas ('chionophilous'). *Vaccinium myrtillus* and *V. uliginosum*, generally with *Empetrum hermaphroditum*. Often rich in *Cladonia* spp. Mountains of Scandinavia, Britain, central and southern Europe. Also communities with *Phyllodoce caerulea* and species of suboceanic to subcontinental affinities: not in Britain.

4. ***Rhododendron* heaths.** Taller heaths on the lower slopes. Usually rich in mountain species with subcontinental to continental affinities. Mountains of central and southern Europe. *Calluna* plays little part.

5. **Submontane *Calluna-Antennaria dioica* heaths.** Herb-rich heaths in which grasses are prominent on soils of relatively high nutrient status in upland areas of Britain and north and central Europe.

6. **Dryas heaths.** Chionophobous, on calcareous soils. Characterized by *Dryas octopetala* in intricate mixture with *Empetrum* spp. At high altitudes on mountains of Scandinavia, Britain and Europe; also similar communities at low levels in the Faroes and north Scotland.

B. 'Dry' heaths: lowland and upland, but not extending to high altitudes

(Three broad categories, with complex intergradations can be recognized; (*i*) heaths of the highly oceanic western seaboard; (*ii*) a group with predominantly northern affinities; (*iii*) a group with predominantly southern affinities. There is a general but not precise correlation with Bøcher's (*i*) Eu-oceanic Series, (*ii*) Scano-Danish (Scotch) Series, (*iii*) Dutch-German Series (Table 6); and with (*i*) Ulicion, (*ii*) Myrtillion, (*iii*) Calluna-Genistion (Table 8).)

(*i*) OCEANIC HEATHS

1. ***Calluna-Empetrum hermaphroditum* heaths.** Both northern and oceanic occurring at low altitudes in the Faroes and Shetland Islands. May contain *Juniperus communis* ssp. *nana*: probably closely related to A2.

2. ***Calluna-Erica cinerea* heaths,** often with *Arctostaphylos uva-ursi*. Generally in localities with a strong maritime influence, or on soils of relatively high nutrient status. Often herb-rich. Examples in coastal south-west Norway and north Scotland have affinities with the *Calluna-Arctostaphylos* community of upland habitats in the east-central Highlands of Scotland. Species with northern affinities are strongly represented, e.g. *Vaccinium vitis-idaea*.

3. ***Calluna-Ulex gallii* heaths** (Plate 6). Lowland heaths of western regions from Ireland and Wales to south-west England (Cornwall, Devon, Dorset and stations in Hampshire) and north-west France.

4. ***Calluna-Ulex minor* heaths** (Plate 7). Lowland heaths generally to the east of B3 and hence less strongly oceanic. The two types interdigitate in the Dorset–Hampshire region. Also in the corresponding parts of north France.

5. ***Erica vagans* heath.** A highly oceanic type belonging to the extreme south-west of England, Brittany and other parts of western France.

6. ***Calluna-Scilla verna* maritime heath,** with *Festuca ovina*. Largely confined to cliffs in the extreme south-west of England. (Malloch, 1970.)

(*ii*) 'NORTHERN' HEATHS.

7. ***Calluna-Vaccinium* heaths** (Plate 8). Generally on peaty podsols or peat. *Calluna* with one or more of *Vaccinium myrtillus*, *V. vitis-idaea*, *Empetrum nigrum* and abundant mosses, notably *Hylocomium splendens*, *Pleurozium schreberi* and *Hypnum cupressiforme*. Widespread in south Norway, south-west Sweden and north Britain.

8. ***Calluna-Empetrum nigrum* dry heath** (Plate 5). Usually on coastal sand dunes; in Denmark, south-west Sweden, Scotland. Usually rich in lichens.

9. ***Calluna-Arctostaphylos uva-ursi* heath with sub-continental affinities.** Despite the occurrence of *Arctostaphylos* in this type of community as well as in B2 they are rather dissimilar floristically. This community contains species with continental distribution patterns as well as some with southern affinities and therefore occupies a transitional position. Denmark and south-west Sweden (Halland).

10. ***Calluna-Sieglingia decumbens* heaths.** Generally herb-rich and grass-rich, lacking most of the northern species and so, like B8, transitional in character. Netherlands and Belgium.

(*iii*) 'SOUTHERN' HEATHS

11. ***Calluna-Genista* heaths**. Generally on deep podsols on sandy soils. *Genista pilosa* is the chief differential species: *Genista anglica* is also well represented (but extends into other heath types). Also other spp. of *Genista*. Generally lichen-rich and poor in mosses. Mainly in north Germany and eastwards into adjacent territories.

12. ***Erica scoparia* heaths.** Strongly oceanic as well as southern. Contains many species which extend southwards into related Mediterranean types of vegetation, e.g. *Erica arborea*, *E. mediterranea*. South-west France and north Spain; but communities showing some affinities occur in the west of Ireland.
(Note: Bøcher also distinguishes some communities with subcontinental affinities in a Baltic-submontane German series (Table 6) which may not be fully covered by the above groups.)

C. 'Humid' and 'wet' heaths

---

1. **Humid heaths with *Calluna* and *Erica tetralix*.** In habitats of intermediate moisture status. No strong development of *Sphagnum* spp. Widespread in the more oceanic parts of the heath region.

2. **Humid heaths with *Calluna* and *Erica ciliaris*.** Closely related to C1 but occupying suitable habitats in the western part of southern England from Dorset to Cornwall, and corresponding parts of north-west France.

3. **Wet heaths with *Calluna* and *Erica tetralix*.** Permanently wet soils – gleys and peats. *Sphagnum compactum* and *S. tenellum* and other *Sphagnum* species prominent. Other typical species include *Juncus squarrosus, J. acutiflorus, Molinia caerulea, Trichophorum cespitosum, Narthecium ossifragum.* Widespread throughout the more oceanic parts of the heath region.

4. ***Calluna-Eriophorum vaginatum* wet heath.** On peat: alternating tussocks of *E. vaginatum* and bushes of *Calluna*, sometimes with *Empetrum nigrum, Trichophorum cespitosum* and *Polytrichum commune.* Widespread, *Empetrum nigrum* only in the northern parts of the region.

5. **Upland *Calluna-Eriophorum vaginatum* wet heath.** On hill peat in the northern parts of the region. *Rubus chamaemorus* is highly characteristic.

Seral relationships

The investigation of community composition by means of ordination and classification is concerned only with communities as they are found at the time of analysis. No account is taken of the extent to which a single analysis represents not only the community as it is at a certain time, but also what it has been for a number of years and will be in the future. This does not necessarily assume a static view of vegetation, but merely confines attention to the study of variation of composition in relation to the conditions obtaining at a particular time. None the less, dynamic changes in composition are an important feature of many types of vegetation, and no full interpretation of community composition in heaths is possible without consideration of the extent to which seral change may be taking place.

Much has been written about heath as a seral vegetation-type and it is therefore important to examine this interpretation. As mentioned in Chapter 2, Tansley (1939) regarded lowland heaths as 'a stage in the succession to forest,' visualized as 'subclimax' or 'deflected climax.' This view implies assumptions about their earlier history which in most cases have not been supported by more recent studies of quarternary vegetational developments. However, undoubtedly some heathland communities owe their origin to seral changes. Noteworthy in this connection are, first, the reversible changes towards heath which have occurred in the vegetation of peat bogs. The evidence for these lies in the accumulation of *Calluna* remains at certain levels in peat profiles, notably at the Sub-boreal – Sub-atlantic transition. The explanation is presumed to lie in long-term climatic trends, resulting in a gradual drying-out of bog surfaces. On the return of wetter conditions, *Calluna* and associated species gave place, for example, to *Sphagnum*, and active peat growth resumed. Such sequences may have been repeated several times in the same peat bog.

Secondly, it is widely accepted that dune heaths represent a stage in the course of priseres on certain types of sand dune. As these have received considerable attention they merit a rather more extended treatment here. In most instances it has to be admitted that the directions of vegetational change have been guessed at rather than substantiated. Few attempts have been made to obtain direct evidence of invasion by *Calluna* and other heath species or of replacement of dune communities by heath.

This problem, in so far as it is exemplified by Scottish sand dune vegetation, was discussed by Gimingham (1964*a*). Heath commonly develops on dune systems where the composition of the sand is such that at the stage of surface fixation the pH falls to about 6·5 or less and the concentration of calcium is low. Its position in the general zonation of the system is normally consistent with the hypothesis that it repesents a stage in succession following the formation of a closed community and the beginning of decline in *Ammophila arenaria*. Both *Ammophila* and *Carex arenaria* are regularly represented in the heath community, although usually with low vitality. However, firm evidence for the dynamic nature of this relationship is difficult to obtain. Seedlings of *Calluna* are very rarely encountered as colonists in the *Carex arenaria-Ammophila* fixed dune or even in the transition zones between the communities, which are generally characterized by scattered well-established plants of *Calluna vulgaris* or sometimes (on the west coast of Britain) *Erica cinerea*. None the less, the quantities of rhizomes of *Carex arenaria* which are often found buried under dune heath provide direct evidence that the *Calluna* stands have developed on sites formerly occupied by a *Carex arenaria* fixed dune community. It is probable that replacement of the one community by the other is not a steady process but is confined to infrequent periods when conditions are favourable for the establishment of *Calluna* or *Erica* in the preceding community. In addition to the effects of physical factors of the environment, rabbit grazing may be responsible for preventing the entry of heath species. In certain localities a complete absence of dune heath may be attributed to rabbits; in others a very occasional specimen of *Calluna* may be found in dune pasture communities. These presumably gained entry when rabbit pressure was temporarily reduced, but are prevented from multiplying. At St Cyrus on the north-east coast of Scotland, for example, the appearance of a single individual of *Calluna* was dated at about 1952–3 when the numbers of rabbits were much reduced by myxomatosis. Any subsequent spread has presumably been prevented by the recovery of the rabbit population.

Further direct evidence for seral colonization of certain habitats in dune systems by heath plants is available from the Sands of Forvie, also in north-east Scotland. Here a number of flat sandy plains have been exposed by the northward advance of a series of moving dune ridges (Gimingham, 1946*a*). One example of these 'deflation-plains' contained damp grassland with only one small plant of *Salix repens* and a single patch of *Empetrum nigrum* when surveyed by Mrs. S. Y. Landsberg in 1954, but ten years later included numerous young plants of *Calluna vulgaris* as well as *Empetrum nigrum* and *Salix repens*.

In some areas heath plants may establish in parts of the dune where the surface sand is not already fixed by the usual community of *Ammophila arenaria, Festuca rubra, Senecio jacobaea* and other turf-forming and rosette species, mosses, etc. On dunes at the Bay of Luce, south-west Scotland, both *Calluna* and *Erica cinerea* are represented as scattered individuals in the bare sand amongst clumps of *Ammophila arenaria*. Here they are clearly colonists, for on passing inland their density increases until their canopy coalesces into a dune-heath stand. At the time of investigation in 1959, young plants could not be found but both species had evidently behaved as invaders at various times in the past when, presumably, favourable conditions obtained.

In a similar way, *Empetrum nigrum* is even more clearly an invader on certain dune systems in Denmark, west Sweden, and north Scotland (Plate 4). Its requirements for germination and establishment (Chapter 5) indicate that an open situation with a moist mineral substratum are favourable. These conditions may be expected to occur with sufficient frequency on dunes in the northern oceanic parts of the European coastline. In the areas mentioned, young plants are frequently observed colonising amongst mosses or lichens in the early stages of dune fixation, before conditions are favourable for the entry of *Calluna*. The long shoots adopt a prostrate habit, spreading rapidly from the centre and forming expanding circular patches which may eventually coalesce to form a dense carpet effectively covering and fixing the sand surface. Under these conditions, *Ammophila arenaria* very quickly disappears. *Salix repens* and *Calluna vulgaris* may then enter the community, the latter like the pioneer *Empetrum* forming low, asymmetrical mounds elongated in the direction of the prevailing wind, gathering hummocks of sand and flowering mainly on the lee side of the bush. In so far as zonation on passing inland reflects succession, *Calluna* in time increases in quantity until it becomes dominant with 80–90% cover. The canopy remains low and dense, seldom exceeding 30 cm in height, probably as a result of wind-exposure, but in some areas (e.g. north-east Scotland) rabbit-grazing contributes to this effect. Frequently, robust mosses such as *Pleurozium schreberi* and *Rhytidiadelphus triquetrus* extend throughout the bushes of the ericaceous species, keeping pace with their rather slow growth and contributing to the canopy by occupying the spaces between the peripheral shoots of the dwarf-shrubs. *Cladonia arbuscula* or *C. impexa* may also become prominent in such communities, with the development on raised, dry areas of an almost pure *Empetrum*-lichen heath.

Further seral trends in dune heaths are so generally obscured by grazing, burning or trampling that it is virtually impossible to reconstruct them. Where seed-parents are available it is reasonable to suppose that *Betula* scrub might colonize, to be followed later by some form of woodland, but clear examples are lacking. Podsolization can become quite marked below dune heaths and this will presumably affect further developments. On moist dune heaths with abundant *Erica tetralix*, examples of colonization by *Myrica gale* and willow scrub (*Salix* spp.) are quite plentiful.

Heath communities which owe their existence to the drying-out of bog surfaces, or to primary successions on non-calcareous dune sand, can clearly be regarded as stages in natural seral processes. The majority of the communities referred to earlier in this chapter, however, have not originated this way but have been derived from structurally more complex vegetation, often as a result of the influence of management. The chief means of management, grazing and burning, are themselves the causes of many changes in the composition of heath communities which are successional in nature. These are of particular importance in relation to the use and productivity of heathland and are further considered in Chapters 9–12.

It remains to draw attention to the fact that in most cases the perpetuation of heath depends upon the continuance of management. This implies that the majority of heath communities, whatever their origins, are unstable and that successional change will normally follow the cessation of management. Typically, this change is in the direction of colonization by trees and development towards woodland. The rate of change varies greatly and depends upon the availability of sources of seed, the nature of the soil and the structure of the heath community. Where the latter is a dense, even-aged stand, trees are slow to colonize, but where numerous gaps are present in an uneven-aged stand these provide sites for the establishment of tree seedlings. Generally *Betula* spp. are the first to invade, sometimes appearing after a few years of freedom from grazing and burning. In time, thickets of *Betula* dense enough to cause local elimination of *Calluna* are established, within which grasses become dominant. Dimbleby (1952) has shown that in Yorkshire stands of *Betula* developing on heath podsols gradually modify the soil profile. With the entry of deep-burrowing earthworms a mull type of humus eventually replaces the mor. By this time profound differences from the original heathland have developed both in the plant community and the habitat. In the northern part of the heath region *Pinus sylvestris* may frequently be associated with *Betula* in the colonization of heath. Further south it is common to see *Quercus* as well as *Betula* spp. invading heaths where management has been discontinued, and even on occasion *Fagus sylvatica*.

Probably not all heaths are unstable and subject to seral change. If, as suggested by Faegri, a chief cause of the decline of forest in south-west Norway and its replacement by heath was climatic change, there would be no return of trees to the heathland unless the climate improved. Elsewhere long-continued monoculture of *Calluna* may have so affected the soil as to inhibit the natural re-entry of trees. These, however, are the exceptions and as a general rule heathland ecosystems, especially in the lowlands, are unstable and liable to undergo successional change if controlling factors are removed. Hence the many variations discussed earlier which are revealed by ordination and classification cannot be regarded as static entities. Their composition is determined, in most cases, not only by environment but also by past history and present management, and is liable to change in response to any alteration of the latter.

5 The physiological ecology of *Calluna vulgaris* and some associated species

The heath formation is unusual, possibly unique, in the pre-eminence throughout the entire region, at least on the drier soils, of a single species – *Calluna vulgaris*. Apart from the moist or wet heaths, *Calluna* is present in almost all the communities and is clearly dominant in the majority. Other species may be prominently represented as associates and in certain of the more specialized habitats may achieve dominance, even occasionally to the exclusion of *Calluna*. However, heaths provide as good an example as can be had of the widespread dominance of one species throughout a considerable range of habitat conditions and community composition. An explanation of this fact is of prime importance as the background to an understanding of heathland ecology, and of dynamic processes operating within the vegetation.

The area within which climatic conditions approach their optimum for *Calluna* coincides generally with the area in which heathlands are best developed (Figs. 2 and 24). Historical factors have played their part, while the responses of *Calluna* to such practices as burning and grazing are also significant. Furthermore, partly because of the prevailing oligotrophy of the habitats and partly because of management activities, the communities are generally far from rich in number and diversity of species. These are precisely the conditions under which one species, favoured by a certain combination of conditions and meeting few effective competitors, can become widely dominant. If, as with *Calluna*, it is also a species which profoundly modifies its own habitat, the tendency may be reinforced and extended. A thorough knowledge of the autecology of *Calluna* is therefore required, both to determine those aspects of its biology which equip it particularly well for the heathland environment, and to serve as a foundation for the general ecological analysis of heath communities. The following account considers establishment, development, persistence and reproduction in *Calluna*, briefly comparing its behaviour in these respects with that of certain other species with which it is commonly associated. This is not intended to provide an exhaustive treatment of the autecology of these plants since much of the relevant information is more appropriately contained in other chapters, while reference can also be made to the series of contributions to the Biological Flora of the British Isles for the following species: *Calluna vulgaris* (Gimingham, 1960); *Erica cinerea* (Bannister, 1965); *Erica tetralix* (Bannister, 1966), *Vaccinium myrtillus* (Ritchie, 1956) and *Vaccinium vitis-idaea* (Ritchie, 1955).

Fig. 24. Map of the main area of distribution of *Calluna vulgaris* in Europe (outlying stations omitted).
– – – – – approximate limits of distribution
–·–·–·–·– area within which *Calluna* is commonly a community dominant.
———— approximate extent of ecologically optimal habitats for *Calluna*.
(After Beijerinck, 1940; modified.)

Seed germination

The ability of any species to invade a habitat open to it depends in the first instance upon the arrival of its seeds, and thereafter upon their capacity to germinate in the conditions obtaining, and that of the seedlings to survive. Continued occupancy requires tolerance, throughout all stages of its life history, of the environmental conditions as modified by the plant's own influences upon the habitat, and renewal either by vegetative means or from seed. To this must be added, in the case of *Calluna*, the ability to respond favourably to the effects of management practices including burning and grazing.

Seeds of *Calluna* are very small, measuring only about 0·6 × 0·35 mm. Whenever the plant is abundant they are shed on to the soil surface in very large numbers. Germination takes place readily under suitable conditions, but while it begins quickly in some seeds, in others it is delayed for varying periods. In experimental sowings, the first signs of germination usually appear after 8 to 14 days, and normally within the following two weeks about half the viable seeds have germinated. By six weeks to two months after sowing, a percentage germination of more than 70 is usually reached, but intermittent further germination may bring this figure up to 95 within periods not usually exceeding six months. This prolongation of the germination period is observed in the field, for numerous seedlings may appear in the autumn after seed shedding, while a proportion of seeds remain dormant until conditions again become favourable in the following spring and summer, some perhaps for an even longer period. This may play an important part in maintaining potentiality for regeneration if, for example, the first batches of seedlings were eliminated by climatic extremes, fire, etc. A similar effect is produced by retention in the capsule of a few seeds which fail to get shaken out in the autumn. These may be shed later, or may remain enclosed in the capsule which later falls to the ground where it may blow about for some time before the trapped seeds meet with conditions suitable for germination. Table 10 shows the number of seeds germinating from within fallen capsules gathered in floral debris from beneath *Calluna* plants in autumn.

TABLE 10

Calluna vulgaris: development of seedling clusters from seed trapped in fallen capsules. (Data: E. Whittaker)
(Counts of seedlings growing from remains of capsules lodged in moist moss.)

No. of seedlings in cluster	% of capsules producing this number
1	47·0
2	19·7
3	9·3
4	4·9
5	5·5
6	6·0
7	0·5
8	1·6
9	0·5
10	0·5
<10	2·7

(Total number of capsules from which seedlings had emerged: 183.
Total number of seedlings: 481.)

As distinct from a tendency to spread out the germination period when conditions are favourable, ungerminated *Calluna* seeds are also capable of retaining their viability for a considerable time. Little diminution of germination is observed after three years of storage, and a case has been reported of a seed germinating after a period of 11 years in a dry soil sample (D. E. Coombe). Seeds buried for some years at depths of 1 cm or more under litter and humus may occasionally be exposed by removal of this layer in a severe heath fire, and can then produce a dense crop of seedlings.

Temperature and light both affect the number of seeds germinating and the rate of germination. Under the artificial conditions of a constant, unvarying environment, temperatures within the range of 17°C to 25°C seem to be the most favourable, while at 30°C there is a considerable fall both in the rate of germination and in the final percentage of seeds germinating. However, several workers have shown that better results can be obtained when the seeds are subjected to varying temperatures during the period of germination, such as alternation between eight hour periods at 30°C and 15 hour periods at 20°C. Thus the experiments indicate that germination is favoured by fluctuating temperatures which are particularly characteristic of the sites open to colonization by *Calluna* seedlings (p. 195).

It has been reported that germination of *Calluna* is almost completely inhibited by darkness. This, in fact, seems to apply mainly when temperature is held constant. If the more favourable fluctuating temperatures are provided, germination will occur, though not as readily as in the light. Good germination has also been obtained when seeds sown in the dark have been given relatively small 'doses' of light, by bringing the plates out for inspection for about half an hour every few days. It is uncertain, therefore, what ecological importance should be attached to the effects of light upon germination. However, seeds are seldom, if ever, found germinating under the dense shade of a vigorous *Calluna* canopy, although other factors in addition to the lack of light may be operative here. Young seedlings sometimes appear in the developing gap in the centre of an old *Calluna* bush, and in the spaces between bushes where these are not very closely crowded. Otherwise, they are generally confined to open patches of soil or peat, often on ground which has been disturbed, cleared or burnt.

Germination depends upon a supply of water for imbibition by the seed. Some adverse effect upon germination is evident when soil water tensions exceed about two atmospheres, equivalent to a pF of 3·3–3·6. However, on any soil suitable for *Calluna* ranging from purely mineral to organic the water supply is generally adequate for free germination if it is kept at or above field capacity (Bannister, 1964*b*). Although peat and raw humus may not be so favourable chemically for germination as certain mineral substrata (see below, pp. 90–91), slightly better total germination is sometimes obtained on organic than on mineral soils, presumably because of more effective retention of water in the immediate vicinity of the seeds at the surface. This accords with the observation that, after heath fires, regeneration from seed is frequently quicker and more uniform on peat, humus, etc., than on mineral surfaces. Provided the atmosphere is not excessively desiccating these organic substrata, once moistened, may retain sufficient water in the surface for germination without further wetting. This is shown in figures obtained by Miss S. M. Anderson for the number of seeds germinating from 0·1 g of floral litter of *Calluna* spread out on the following surfaces, moistened once at the start of the experiment and kept in a saturated atmosphere:

Peat	166
Calluna litter	85
Slightly humic sand	8
Pure sand	3

For similar reasons, a consolidated substratum is better than a loose one (Wallace, 1917; King, 1960), and it has repeatedly been observed that seeds of *Calluna* germinate poorly on its own litter while this is fresh and loose, but more vigorously when partly decomposed and compacted into a raw humus layer.

Although successful germination of *Calluna* seed will take place over quite a wide range of conditions of water-supply, the observations and tests reviewed here clearly show that it is dependent upon the maintenance of relatively high levels of available water at the soil surface over periods of at least several weeks in the growing season. While this requirement may be more readily met on an organic than on a mineral substratum, it also implies a relatively humid climate. The ability of seed to germinate in the natural habitat at almost any time from spring to autumn is therefore dependent upon an oceanic climatic regime.

Some experiments have shown reduced germination if a water table is maintained above the surface of the substratum on which seeds were sown, but none the less quite a high proportion of seeds will germinate even when completely submerged, if the temperature is adequate. If not, seeds will survive soaking or storage in a saturated atmosphere for at least two to three months without suffering adverse effects on subsequent germination. This perhaps has some significance in an oceanic climate, for if seeds have not germinated before the onset of winter they are likely to lie in a saturated environment for long periods. With the return of favourable temperatures, germination can evidently take place without delay, whereas after only two weeks of exposure to a desiccating atmosphere subsequent germination is delayed.

There is also strong evidence that the chemical environment of the seed has an effect upon germination. For example, so long as water supply is not a limiting factor better germination can be obtained on a humic sand than on peat. Experiments using a range of natural substrata have shown effective germination between pH 3·2 and 7·0, corresponding closely to the observed range of *Calluna* in the field, but above pH 7·0 germination was markedly reduced. Using nutrient agar cultures, Poel (1949) showed that, while a very little germination took place at pH values as low as 2 and as high as 10, higher percentages were obtained between pH 4 and 8 with an optimum in the region of pH 4–5. However, little is known of the identity of the soil properties or substances which directly affect germination.

Pretreatment of the seed by heating before germination may contribute to the widespread dominance of *Calluna* in heathlands, most of which owe their existence to management by man. It was noticed by Rayner in 1913 and by Grevellius and Kirchner in 1923 that periods of heat treatment (e.g. up to a few hours at 60°–80°) could accelerate and increase germination. This was further investigated by Whittaker and Gimingham (1962) whose work yielded some indication that while temperatures of 200°C and over are always lethal, short periods at temperatures in the range 40°–160°C could increase and accelerate germination. The effect is related both to the temperature and its duration. Periods from half a minute upwards at 40°C and 80°C produced some evidence of improved germination and, judging from other work, extension of the time up to several hours causes no reversal of this result. However, at 120°C only periods of less than 30 seconds were effective and longer exposure led to a depression of germination. A similar stimulus resulted from up to 20 seconds at 160°C, but any increase in the time was accompanied by progressive reductions in germination and periods in excess of one minute were generally lethal.

Seeds lying on the soil surface will sometimes be killed during a heath fire, since temperatures at ground level may exceed 300°C and charring of the seed coat is in any case lethal. However, in a normal fire temperatures at the surface generally remain below 200°C, while in addition seeds are often protected by thin layers of litter, humus, etc. They will therefore frequently receive heat treatments such as those which have been shown to increase and accelerate germination. This has been verified by placing seed samples in containers at appropriate levels in the soil surface prior to a heath fire, after which they have shown the expected improved germination. The ability of *Calluna* to establish a uniform stand of seedlings rapidly after burning is doubtless one of the characteristics which have promoted its widespread dominance in heathlands which are managed by systems of regular burning.

Seedling establishment

The requirements of the developing seedling are in some respects more exacting than those of germination, and hence further limit the range of conditions in which the plant will regenerate. Again the water supply is critical and the saturation deficit of the atmosphere exerts a strong influence in addition to that of the availability of water to the developing root system. For example, young seedlings placed with their roots in glass wool adequately supplied with water will die within a month if kept in containers at 60% relative humidity or less (room temperature), while controls at 80% or more remain alive and healthy. As far as water supply in the substratum is concerned, if this falls below field capacity the number of germinated seeds which survive as established seedlings begins to decline (Bannister, 1964 *b*).

On the other hand, very wet soil regimes have an adverse effect upon seedling development. Any tendency towards waterlogging of the substratum leads to reduced growth, as indicated by height and dry weight, and to reduced root penetration. This has been shown in numerous experiments, the results of one of which are summarized in Table 11 as an

TABLE 11

Dry-matter production, height, and root penetration of *Calluna* seedlings grown in three water regimes and on three soil types (harvested after ten months.) (Data from Bannister, 1964 *b*)

	Water regime			*Soil type*		
	Wet	Moist	Dry	Mineral	Humus	Peat
Total dry weight per pot (mg)	120·0	309·1	199·9	222·2	272·2	123·6
Mean dry weight per plant (mg)	1·71	5·10	3·87	3·49	5·43	1·77
Mean height per plant (mm)	14·0	26·8	21·9	22·5	26·7	13·4
Root length as percentage of soil depth (angular scale)	50·5	90·0	90·0	70·8	83·5	76·2

Seedlings were growing on three soil types, viz. mineral, humus, and peat.
'Wet' = wetter than field capacity
'Moist' = around field capacity
'Dry' = drier than field capacity

example. In general, growth is poorer on peat than on humic or mainly mineral soils. The root system rapidly becomes more extensive and richly branched on mineral than on organic substrata.

Calluna will fail to establish seedlings if high saturation deficits are experienced in the air-layer close to the ground during the early phases of growth and development, and if the surface soil dries out beyond its field capacity. Again the association of the species with oceanic climatic regimes is linked to these requirements. Where they are satisfied performance in the early stages, as with germination, seems to be rather better on the more freely drained substrata than on peat, but the causes of this are obscure and probably complex.

In the field, the abundance of seedlings varies greatly. With its high reproductive capacity and very small seeds requiring light for both germination and seedling establishment, *Calluna* is a species equipped to colonize bare areas rather than closed communities. Seedlings are in fact very scarce or absent where *Calluna* forms dense stands or contributes to a closed community, but numerous where, for example, gaps have formed in an uneven-aged canopy, or cleared areas occur as a result of fire, peat-cutting, forest felling, etc. The availability of light at ground level is undoubtedly of prime importance in controlling the occurrence of seedlings. However, protection from desiccation is essential and in numerous measurements of relative humidity in the neighbourhood of vigorous seedlings, none lower than 65 % has been recorded, except on burned sites where

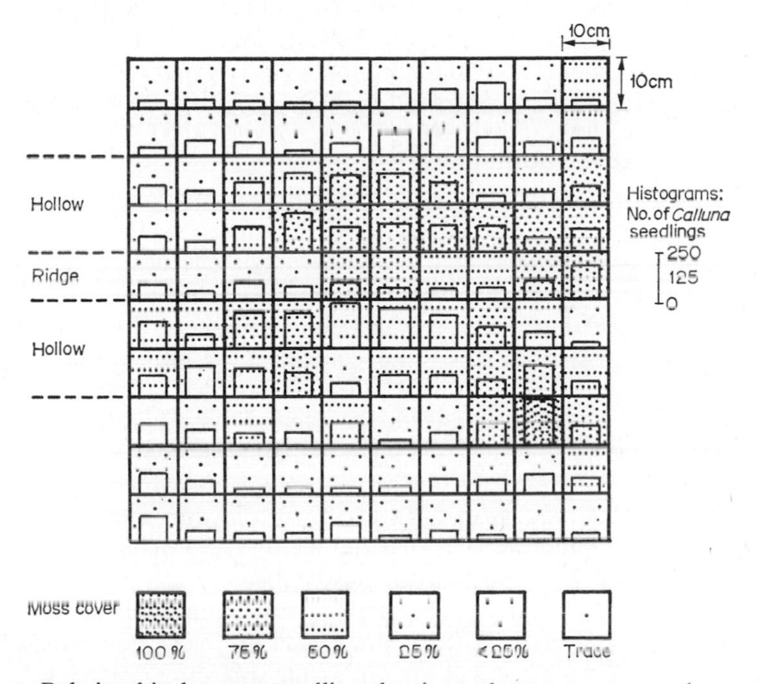

Fig. 25. Relationship between seedling density and moss cover on a burnt area: north-east Scotland. (Figure by Mrs. S. Anderson.) Each square represents an area 10 cm × 10 cm; the density of dots represents cover of *Campylopus flexuosus*, and the height of the histogram indicates the number of *Calluna* seedlings in the square.

atmospheric conditions are liable to wider fluctuations. In cleared areas such as these the moisture factor may be critical and is perhaps responsible for great variations in seedling density. In the same summer, a mean seedling density of 2·73 per 100 cm^2 was recorded on a recently burned area on a mainly mineral soil at Dinnet Moor, 12·9 per 100 cm^2 on the raw humus horizon of a podsol on Brimmond Hill, and 87·5 per 100 cm^2 on a peat at Cruden Moss, all in Aberdeenshire, north-east Scotland. In the latter site, the seedlings were markedly aggregated (Fig. 25), the clumps coinciding with small depressions in which the moss *Campylopus flexuosus* had formed a thin turf. Near by, similar depressions caused by vehicle wheels were densely colonized by the moss and by *Calluna* seedlings. Both the depression itself and the presence of the moss may have contributed to the maintenance of favourable moisture conditions in the atmosphere and in the substratum.

In these instances the features of organic substrata which are less favourable for germination and establishment are apparently outweighed by the better maintenance of water supply. The converse of this may explain the frequent scarcity of seedlings in dune heaths which, apart from their liability to desiccation, might be expected to provide a most suitable medium for regeneration. The dense cover which frequently develops in communities of stable dunes may constitute an additional unfavourable factor, and in general conditions suitable for regeneration by seedling establishment may here be restricted to infrequent intervals (Chapter 4).

Physiology of the adult plant

Experiments by Bannister (1964 *c*, *d*) brought to light some aspects of the water-physiology of *Calluna* which contribute to an understanding of its geographical and ecological range. He showed that various degrees of water-deficit can be detected in samples of the terminal shoots taken at different times of the year from plants in the field. Expressing these in terms of relative water content,* i.e. the water content at the time of sampling as a percentage of that at full turgidity, the values recorded ranged from almost 100% down to 52%. Hence it may be concluded that *Calluna* is tolerant of the development of rather low relative water contents in its shoots, and this was borne out by experiments with potted plants which sustained no damage at values of between 70% and 60%, at least during short periods of treatment. It was found, in fact, that the relative water content had to fall nearly to 30% before a plant was obviously killed.

As in many ericoid dwarf-shrubs, the total water content in relation to the dry weight of the plant is relatively low (Stocker, 1924). Bannister's (1964 *c*) data, however, confirm that under favourable circumstances *Calluna* shoots can show high rates of transpiration in relation to unit fresh weight of the transpiring organ. Furthermore, his graphs illustrating the reduction in relative water content of cut shoots over a period of time (Fig. 26) show no obvious check in the rate of decline until the value falls to about 70% (Bannister, 1964 *a*). It is suggested that the check which is then evident may be due to stomatal closure. If this is a correct interpretation, it indicates that control does not begin to become effective until a fairly low relative water content has been reached. It follows that when atmospheric deficits are rather high there will be a fast 'turn-over' of the free water in the plant. Any tendency for uptake to fail in counter-balancing loss will cause the relative water content to fall quickly to a value between 70% and 60%,

* = 'relative turgidity'.

and thereafter more slowly as the control mechanism becomes operative. The sclerophyllous construction of the plant may be interpreted in the light of the observed ability of *Calluna* to withstand low relative water contents in its tissues, and this in turn no doubt relates to the rather wide range of the plant with regard to environmental water regimes. Its vigour, however, is reduced in habitats regularly subject to drought.

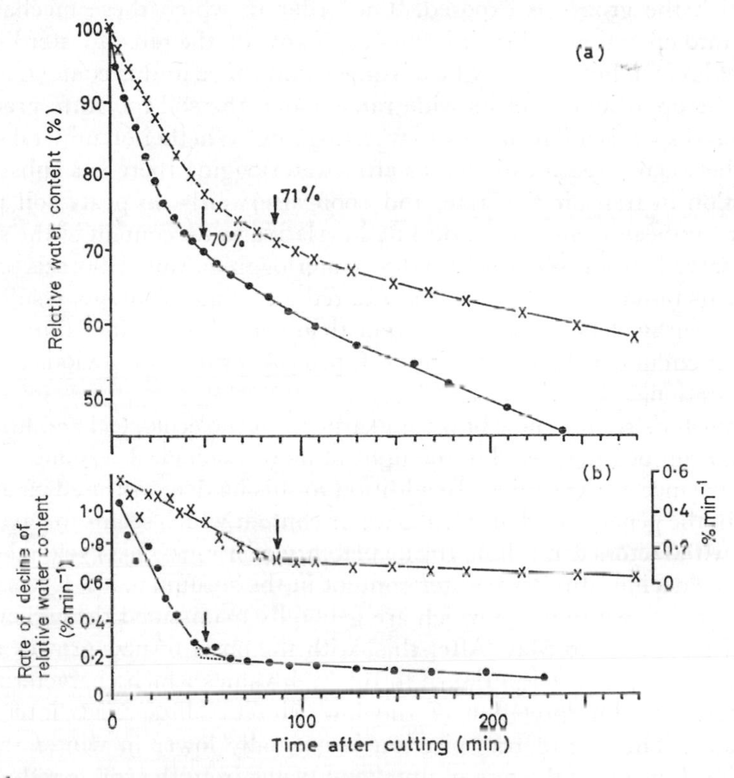

Fig. 26.

(*a*) The decline with time of relative water content in cut shoots of *Calluna vulgaris*, exposed to different evaporating conditions.

(*b*) The same data expressed as rate of loss of relative water content, in relation to time after cutting.

× — × low evaporation
• — • high evaporation
(From Bannister, 1964.)

Bannister has demonstrated a daily rhythm of transpiration, which normally reaches its peak about 1130 hrs, and then declines even if the atmospheric saturation deficit continues to increase as is often the case in summer. Some evidence has recently been obtained that a degree of control of water loss may begin to operate before the relative water content falls to between 70% and 60% (H. Hinshiri, personal communication). Examination of the shoots and leaves suggests mechanisms, additional to the stomata, which may have a function in controlling transpiration. The abaxial (lower) leaf surface to which the stomata are confined is restricted to a narrow groove containing hairs (Plate 9), the opening of which is adjustable. Further, movements of the leaves take place, increasing or decreasing the degree of imbrication on the short-shoots and modifying the extent to which the groove is exposed. The order in which these mechanisms come into operation and their influence, if any, on the rate of water loss are not yet known, but they may be of some significance in this context.

At the opposite end of its wide range along the soil moisture gradient, *Calluna* responds unfavourably to waterlogging. Whether on mineral soil or peat, between three and six weeks after waterlogging there is a substantial reduction in transpiration rate, and soon afterwards on peaty soil plants become unhealthy and die. A decline in relative water content of the shoots is observed from six weeks after waterlogging, but chlorosis occurs before its onset and it appears that the reduced water contents result from the unhealthy state of the plant rather than contribute to it. Failure of the root system under these conditions (cf. p. 116) is the most obvious cause of deterioration.

Bannister's results show how some aspects of the ecological amplitude of *Calluna* can be interpreted in the light of its physiological responses to the water balance of the habitat. In addition to this he demonstrated an annual cycle in the general level of relative water content which seems to have little to do with factors controlling the availability of water. This cycle takes the form of a decline of relative water content in the autumn to low levels (from about 85% down to 60%) which are generally maintained throughout the period December to May. After this, with the onset of new growth in the terminal shoots, there is a return to the high values which characterize the summer. The interpretation of the low winter values raises interesting problems. The rate of transpiration is normally lower in winter than in summer, but unless the rate of uptake of water from the soil is reduced to a greater extent there should be no fall in relative water content. Measurements in the rooting region of the plants show no rise in soil water tensions, since soil water is normally abundant in winter under the climatic conditions obtaining and, with the insulating effect of the vegetation, is rarely frozen. Bannister suggests that an internal resistance develops at this time of year, progressively reducing the rate of uptake by the roots, and that even the restricted transpiration is sufficient to reduce relative water contents gradually, towards levels at which stomatal closure occurs. Observations by Thren (1934) of rising osmotic pressures of the cell contents in winter correlate well with this finding, and evidently the plants enter a phase of physiological inactivity or dormancy.

None the less, temporary relief of the water-deficit may be observed

TABLE 12

Relative water content (%) of samples of *Calluna* from three altitudes, north-east Scotland.
(Means of 10 samples) (Data, H. Hinshiri)

Altitude	7 November 1969	20 January 1970	13 May 1970	16 June 1970	14 August 1970	4 November 1970	16 December 1970	16 February 1971	23 March 1971	20 April 1971	13 May 1971
870 m	48·4	90·0*	50·5	51·9	66·6	34·7	82·5	70·0*	91·5*	58·1	62·4
425–460 m	70·7	87·7	70·6	71·4	72·6	36·5	66·5	50·7	63·5	59·4	64·2
90 m	80·0	90·3	78·8	80·7	76·1	39·9	80·0	70·7	64·5	68·0	72·3

(*Snow-cover at highest altitude)

Plants at the intermediate and low altitudes show only slight variations in relative water content except in late winter and early spring when values fall. In contrast, at the high altitudes there are fluctuations of considerable amplitude related on the one hand to periods of exposure to desiccating conditions when values fall and to periods of snow cover (*) when they rise during the periods of wet weather or snow cover (Table 12). In the latter event, transpiration is doubtless completely stopped and even a very slow rate of uptake would cause an increase in relative turgidity, but with regard to the former there is some evidence that deficits may be made up by direct absorption of water through the aerial parts. On the other hand, values for relative water content down to 40% are sometimes found in shoots projecting through a snow cover, and in these conditions browning and death frequently occur. Winter browning has widely been termed 'frosting', but it is by no means always associated with freezing temperatures, while experimental application of low temperatures to potted plants has failed to produce the symptoms (Braid and Tervet, 1937). It seems much more probable that this localized browning and death is associated with the development of very low relative water contents as a result of desiccating conditions (particularly dry winds, especially when accompanied by bright sun) during the late winter when root uptake appears to be at a minimum. Under these conditions relative water contents as low as 34%, which experiments have shown is a potentially lethal level, have been recorded in the field. Watson, Miller and Green (1966) investigated the winter browning of *Calluna* in Glen Esk, north-east Scotland, finding that in April 1959 there was about twice as much dead foliage on the plants as was recorded at the same time in 1960 and 1961. During the 1958–9 winter (December–February) there were nearly twice as many hours when the relative humidity was below 70% as in same periods of 1959–60 and 1960–61. They concluded that, although winter browning can occur even when the soil is not frozen, it is most frequently produced by desiccation during periods of high evaporation potential when the soil is frozen and not thickly covered by snow. Consequently, most damage occurs in January and February, but the amount varies considerably from year to year and some has been recorded in other winter months and early spring. These findings suggest some of the influences which may be at work in determining the predominantly oceanic distribution-pattern of *Calluna*, for in oceanic regions desiccating conditions are seldom prolonged, particularly in winter. Where *Calluna* spreads to more continental territories it becomes increasingly confined to woodland.

Another characteristic of the oceanic type of climate to which *Calluna* and other ericoid species appear to be adapted is the prevalence of cloud cover. For much of the time incoming radiation is scattered and received by a plant from many directions, rather than predominantly from above. Grace and Woolhouse (1970) suggest that the microphyllous habit, in which the receiving surfaces of the numerous small leaves are variously orientated, may be adaptive in these conditions, and this effect is no doubt enhanced by the radiating branch system. They showed that, under experimental conditions with overhead illumination, young shoots of *Calluna* did not become light-saturated even when exposed to light intensities approaching those of a summer's day at noon, whereas this might be expected to occur at lower intensities if a greater proportion of the light were received from lower angles.

These authors calculated that the optimum temperature for net photosynthesis in plants taken from Moor House, at an altitude of about 560 m in the Pennines, was 18°C. Among other factors influencing the net photosynthetic rate, temperatures experienced by the plants during the days prior to testing were found to be important. Plants which had been exposed to low temperatures during winter showed reduced rates of photosynthesis. Compensation points were shown to be relatively high, reflecting the inability of *Calluna* to survive under substantial shading (Björkman, 1945; see also Chapter 6). Measurements of net photosynthetic rate in year-old short-shoots gave values around 60% of those for newly formed shoots, a point of some significance in regard both to seasonal variation in dry matter production and to changes with age, in view of evidence for a decline in the production of new shoots in older plants (Chapter 7). In general, *Calluna vulgaris* is a relatively slow-growing species with maximum photosynthetic rates below those of many other woody plants.

Relationships between the physiological characteristics of *Calluna* and its ecological range are also evident in regard to its acidophilous, calcifuge habit. In terms of pH its range extends from about 3·2 to just over 7·0, but specimens found, for example, on serpentine soils at pH 7·4 are generally chlorotic. Where it is vigorous enough to exert dominance, the pH lies usually between 3·5 and 6·5 (Chapter 2). Rather little work has been done on the mineral requirements of *Calluna* and the physiological reasons for its soil preferences. It is still occasionally stated that the endotrophic mycorrhizal fungus frequently present in the roots and other parts of the plant is capable of fixing atmospheric nitrogen, contributing to the success of the plant in soils poor in nitrates. This view arose from the work of Rayner (1915, 1921, 1922 *a* and *b*, 1923), Jones and Smith (1928) and Rayner and Smith (1929), but must be discarded in the light of experiments on the nitrogen fixing properties of symbiotic systems by Bond and Scott (1955) in which entirely negative results were obtained with *Calluna*.

Rayner further claimed that the mycorrhizal fungus was essential for healthy growth in *Calluna*, and that the reason for failure in calcium rich soils lay in the restricted development of the fungus (possibly owing to bacterial antagonism). Again, much doubt has been cast upon this conclusion, and upon the whole interpretation of the mycorrhizal association as an obligate one (Christoph, 1921; Knudson, 1928, 1929, 1933; Freisleben, 1935). Numerous observations of active roots descending into highly calcareous materials including limestone and chalk have been recorded. The possibility remains that the calcifuge behaviour of *Calluna* may be determined largely by the requirements for seed germination, seedling establishment, and vegetative growth.

Reproductive capacity, seed dispersal and vegetative propagation

As demonstrated by Salisbury (1942), there is a close relationship between reproductive behaviour and the type of habitat occupied by a given species. In the neighbourhood of Bergen, Norway, the capsules of *Calluna* usually contain between 20 and 32 seeds (Nordhagen, 1937). Reproductive capacity varies greatly with size of plant, age and growth phase, habitat, etc., and the number of seeds produced probably fluctuates from year to year. A very general indication may be obtained from Beijerinck's (1940) estimate that a 'robust plant' bore about 7900 flowers. This would give a yield of about 158 000 seeds in any year, and assuming 80% germination the reproductive capacity would be of the order of 126 400 viable seeds per plant, per year. This is in line with other estimates which have suggested the production of about one million seeds per square metre in a season (Nordhagen, 1937).

These seeds may be shed in the autumn when the capsule opens but the persistent calyx, which splits along the upper side only, restricts the shedding to windy periods. Beijerinck (1940), using Praeger's (1911) observations on the rate of fall in still air, calculated a dispersal radius of about 100 m in winds of velocity 10 m sec^{-1} and Nordhagen (1937) suggested distances of $\frac{1}{4}$ km in winds of 30–40 m sec^{-1}. Some seeds, however, fail to escape from the capsule which itself may be blown about after falling from the plant. These may eventually be released or germinate within the remains of the capsule (p. 88).

Calluna, then, is a species showing high reproductive capacity and relatively effective and rapid dispersal. These, according to Salisbury (1942), are characteristics of species which normally are colonists of open habitats. The light seeds are spread in large numbers and because food reserves are limited the seedlings will survive only where they receive little shade or root competition from established vegetation. In this light, *Calluna* appears as a temporary or cyclical dominant rather than one of 'climax' communities. This conclusion is in keeping with the interpretation of the status of heath vegetation in Chapters 2 and 4, and with the evident success of this species as a colonist of cleared and burnt areas. As shown earlier in this chapter, its life-span is normally limited to around 30 years and, unless burning, grazing or wind-pruning cause repeated renewal of young shoots from the stem bases, individuals seldom persist beyond this age-limit. There is some evidence in wet, peaty habitats of vegetative propagation by means of straggling branches which root adventitiously and give rise to new bushes as the original one dies away (p. 57), but this seems to be the exception rather than the rule. Regeneration from seed under its own canopy and on its own litter also is restricted, and so in general *Calluna* would be unlikely to feature as a very persistent species in entirely natural ecosystems, except in exposed habitats. Once again the evidence emphasizes the conclusion that the extensive and continued dominance of *Calluna* arises from artificial causes which either repeatedly rejuvenate the individual plants or perpetuate conditions favourable for establishment.

The interactions between *Calluna* and other heathland species

Erica cinerea and *Erica tetralix*

The ecological range of *Calluna* is determined not only by its own physiological requirements, but also by competition with species of overlapping requirements. Among the most frequent of these in the more oceanic parts of its range are *Erica cinerea*, generally on the drier or more freely drained mineral soils, and *Erica tetralix*, generally on the wet or waterlogged and often largely organic soils. There is considerable resemblance in the structure and growth-form of all three species, although neither of the *Ericas* attain as great a height as *Calluna* in optimal conditions. *Erica tetralix*, however, spreads vegetatively by means of rhizomes.

In Britain and other parts of the heath region where all three of these species are represented, their relative frequency is closely correlated with soil hydrology. Where the topography is undulating, resulting in repeated hydrologic sequences of soil profiles, this is reflected by a corresponding vegetational pattern. *Calluna* shows the widest range, intermediate between and overlapping with the others, while *Erica cinerea* increases in prominence on the drier, more freely drained soils and *E. tetralix* on those which are wet and poorly drained. Under extreme conditions at either end of the range, especially at the wet end, *Calluna* may disappear as the *Erica* species become dominant. This gradient in habitat and plant representation recurs so frequently that it may be designated a 'soil-vegetation catena'.

To interpret the relationship between these species a comparison is required between what is known of the physiological requirements of the two *Ericas*, and those of *Calluna*. A partial interpretation of the distribution pattern outlined above can be obtained on the basis of requirements for germination and establishment alone. As shown on pp. 87–91, *Calluna* performs well in these respects over a wide range of soil conditions. Whether the soil is wet, moist or relatively dry, mineral or organic, germination and establishment are in general numerically higher, and a higher proportion of the germinated seeds become established in *Calluna* than in either of the other two species. This finding, from a series of comparative tests by Bannister (1964 *b*), accords with the fact that the habitat-range of *Calluna* is wider than those of the other species. However, the tests demonstrated that at either end of the moisture gradient both germination and establishment in *Calluna* could be depressed, and for comparison the requirements of the other two species are summarized below.

Under experimental conditions, the percentage germination of *Erica cinerea* is always lower than in *Calluna*, 48% being the maximum obtained. As with *Calluna*, germination is best when temperatures fluctuate, and light is also beneficial. In *Erica cinerea* particularly, percentage germination remains very low in steady temperatures, especially in the dark. Again there is some evidence of stimulation by heat pre-treatment. However, in view of the abundant and rapid germination in the field (e.g. after a heath fire) it seems that the conditions required for germination have not yet been achieved in experiments.

Seeds begin to suffer after about two weeks if stored wet, sooner than in *Calluna*. The biggest contrast, however, lies in the conditions required for seedling establishment which is adversely affected by a high water-table and is distinctly better on mineral than organic soils. Even on the former, the advantage of good drainage is shown by an experiment in which seeds were sown in pots having water-tables at the surface and at 5 cm, 10 cm and 15 cm below it. Going down this series, the dry weight of the crop of seedlings after 12 months from the date of sowing showed a linear increase. The combination of a high water-table with an organic soil is generally lethal, whereas on the drier mineral soils *Erica cinerea* establishes as effectively as *Calluna* and outstrips the latter in height.

Erica tetralix seems to be less dependent upon fluctuating temperatures for good germination, but in the dark the percentage is again reduced. The optimum pH is probably rather lower than in *Calluna*. Wet storage seems to do little harm to the seeds, at least for periods of 12 weeks or more. On wet peat the total germination may be as high as in *Calluna* (e.g. up to 80%), although it is generally less on mineral soils, particularly where moisture content at the surface falls below field capacity. Establishment of seedlings is also poor on the drier soils and their performance, in terms of dry weight increment and height growth, is best in intermediate rather than the wettest regimes, and on mineral soils rather than peat. On wet soils, the seedlings may outstrip those of *Calluna*, but on peat the general performance is only about as good as, and not better than, that of *Calluna*.

These findings regarding establishment and seedling performance refer only to the respective species growing alone and take no account of interactions between them. However, they suggest some reasons for the observed distributions in the field, indicating that as far as young plants are concerned *Erica cinerea* may be in a position to compete with *Calluna* on mineral soils, while the same applies to *E. tetralix* on very wet soils. The latter, however, do not necessarily represent either the optimum or the mid-point of the edaphic range of *E. tetralix*, which sometimes figures on more mesic, mineral soils; whereas *E. cinerea* is very seldom represented in wet regimes.

The root systems of all three species tend to be concentrated in the upper 20 cm of the soil, with a high proportion just below the surface (Chapter 3). Hence, as pointed out by Sheikh (1970) for *Erica tetralix*, the plants may to some extent evade the full effects of reduced aeration in wet habitats, but in this respect there is little difference between the species.

Bannister's (1964 *c*, *d*) measurements of relative water contents and related aspects of water balance, however, show important differences between the two *Erica* species and *Calluna*. Under field conditions both of the former maintain higher relative water contents, which never fall to the low levels sometimes recorded for *Calluna*.

The relative water content in *E. cinerea* seldom falls below about 84%, except in a dune heath where values down to 72% were occasionally recorded. In cut shoots, the check in rate of decline of relative water content occurs usually when values between 95% and 85% are reached, suggesting that stomatal closure may begin to operate well before they drop to levels found in *Calluna*. Annual variations, although of smaller amplitude than in *Calluna*, follow a similar pattern, with a decline in December to May when from time to time relative water content falls below the 'closure' value.

The pattern of daily transpiration is, however, different, since the morning rate is generally maintained throughout the day till early evening. Furthermore, *Erica cinerea* appears to remain rather more active, physiologically, during the winter. Although gross transpiration is not much altered, the decline in relative water content is rather slight. Hence, it is unlikely that inhibition of water absorption by the roots, postulated for *Calluna*, plays much part in this species. This may be correlated with its restriction to more oceanic districts, since the absence of a period of winter dormancy may render *E. cinerea* rather less resistant to winter cold than *Calluna* (as noted by Nordhagen, 1920).

An additional feature of *E. cinerea* is its ability to maintain relative water contents of 80% or more even when soils dry out to the wilting point or beyond. Lethal reductions in turgidity only occur in very dry conditions. Furthermore, progressive increase in soil water tension leads to no substantial reduction in transpiration until conditions become severe. In these ways, this species differs from the other two and shows itself better equipped to remain active, without adverse effects, when soils become rather dry. Conversely, it is very susceptible to waterlogging, which results in reduced transpiration after a few days and death often within two weeks.

The usual habitat of *Erica tetralix* is one in which soil water is normally abundant throughout the year. It is therefore to be expected that relative water contents are on the whole uniformly high, seldom falling below about 82%. However, despite the ample supply of water available in the soil, there is evidence that transpiration is reduced, presumably by stomatal control, during the middle part of each day, suggesting that the plant is sensitive to water loss. It is, indeed, the least resistant of the three species to soil dryness. Transpiration is less effectively controlled and relative water content begins to fall rather earlier than in *Calluna* as a soil dries out, even before the wilting point is reached, and for *E. tetralix* levels between 64% and 48% may be lethal. On the other hand, *E. tetralix* remains healthy during long periods of waterlogging, showing little inhibition of transpiration and continuing active growth. In natural communities, this species can be dominant in the presence of a constantly high water-table, whereas fluctuations in level are frequently associated with increased prevalence of *Calluna* (Rutter, 1955).

The winter decline in relative water content, evident in the other two species, is present although less marked in *E. tetralix*. At the same time, there is little modification of transpiration behaviour and the suggestion is that physiological activity continues to some degree throughout the winter. As with *E. cinerea*, some connection is possible between this and a strongly oceanic type of distribution.

Further light has been shed on the differential response of the two species of *Erica* to waterlogging by Jones and Etherington (1970) and Jones (1971 *a* and *b*). In their experiments, waterlogged plants of both species took up significantly more iron than plants on drier soils, and there was a strong indication that the resulting concentrations of iron in the plants were considerably more toxic to *Erica cinerea* than to *E. tetralix*. Furthermore, comparisons of *E. cinerea* and *E. tetralix* suggested the possibility of a link between higher rates of transpiration in the former and its tendency to accumulate iron when growing on waterlogged substrata. Where the supply of available iron was limited, as on ombrogenous peat, the survival of *E. cinerea* in waterlogged conditions was improved. It is also evident from other work (Loach, 1966; Sheikh, 1969) that the performance of *Erica tetralix* is best where concentrations of the major nutrients (e.g. N, P, K, Ca) are low. Increases can readily become lethal. This too may have a considerable bearing on the habitat-range of this species and its competitive balance with the others in various situations.

While the conditions required for germination and establishment of seedlings set certain limits to the range of habitats occupied by each of the three species, their ecological amplitudes are also determined in part by the water relations and ion uptake systems of mature plants. At the seedling stage, the *Erica* species are in a position to equal the performance of *Calluna* only on the drier, mineral soils or in wet areas. The susceptibility of *E. cinerea* to waterlogging excludes it from the latter, while *E. tetralix* is eliminated at the drier end of the series by sensitivity to soil drought.

The relatively wide range of *Calluna* with respect to soil water regimes is apparently associated with its ability to maintain physiological activity despite the development of relatively high water deficits. However, *Erica cinerea* is of all three species clearly the least inhibited by dry soils and in these conditions can compete effectively, while on the wetter sites *E. tetralix*, less sensitive to waterlogging, can hold its own. It is probable that the two *Ericas* are examples of species whose normal habitat range is determined partly by competition. Conditions for their optimal representation in mixed communities are not necessarily the same as those for optimal performance in a controlled environment. For example, seedlings of *E. tetralix* show better performance on moist rather than on very wet soils, and growth of *E. cinerea* appears to be best also in fairly moist conditions. Sheikh (1970) reported that *E. tetralix* is rather sensitive to poor aeration in the root region. It may be that *Calluna* is the most efficient colonizer of the intermediate regimes and the other two species are restricted to the more extreme parts of the gradient where, because of their own physiological advantages, they can compete successfully with *Calluna* (Bannister, 1964 *d*). As far as *E. tetralix* is concerned, this is borne out in recent experiments by D. Smart (Fig. 27). On a wet soil, this species showed no depression of its

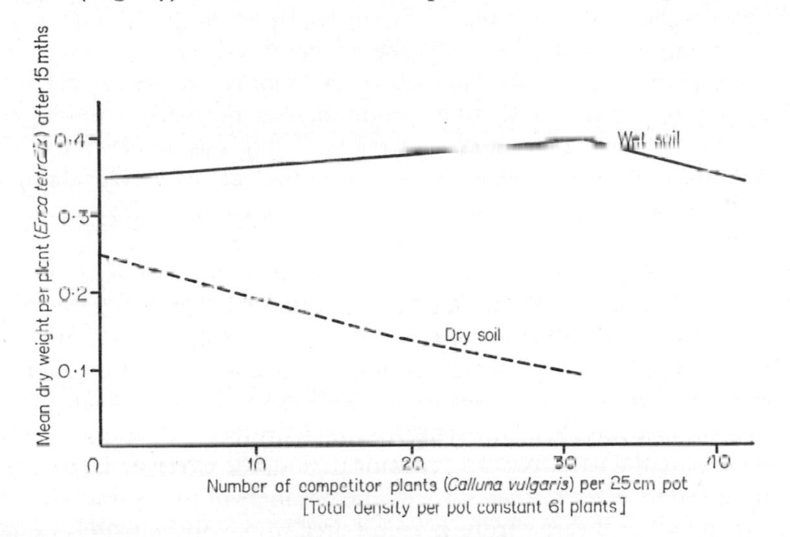

Fig. 27. Mean dry weight per plant of *Erica tetralix* grown with varying numbers of *Calluna* plants on two soil types. (Figure by D. Smart.) Plants harvested after 15 months.

performance when grown in mixtures with *Calluna*, whereas on a relatively dry soil there was a marked reduction in vigour under these conditions.

In the light of these conclusions it is not surprising to find, on the one hand, habitats in which all three species are represented, and on the other, communities dominated, for example, by *Erica tetralix* on relatively dry, nutrient-poor substrata if *Calluna* is for any reason lacking. A case in point is an *Erica tetralix-Cladonia* heath in Denmark and in Skanör, south Sweden on sandy soil (Bøcher, 1943; Gimingham, 1961).

In addition, the investigation of water relations has suggested a possible line on which an explanation may be sought for the more strictly oceanic distribution of the *Erica* species. The tendency for *Calluna* to be tolerant of low relative water contents, doubtless important in connection with its occupancy of a wide range of edaphic habitats, also seems to lead to a cessation of physiological activity in winter, serving perhaps as a protective mechanism in sub-continental conditions. Conversely, a tendency in the *Erica* species to remain active in winter may exclude them from this part of the geographical range of *Calluna*, but confer an advantage in the milder winter of the more oceanic districts.

Grasses and other herbs

The competitive balance between *Calluna* and certain grass species is of considerable importance in relation to the use and management of heathlands. Where *Calluna* is associated with grasses, their relative vigour is to a considerable extent related to soil conditions. However, the balance between *Calluna* and grasses can very readily be shifted by grazing. Although under a mild or medium grazing intensity the cover of a *Calluna* stand may be increased, plants are readily damaged or even killed by heavy grazing. In grasses, on the other hand, tiller-production is normally stimulated by grazing. Hence, especially on soils of relatively high nutrient status, heavy grazing rapidly shifts the balance in favour of the grasses. Under these conditions the biotic factor alone may cause a change from heath to a grassland community over a relatively short period of years (Chapter 9). Equally, on protection from grazing changes may proceed in the opposite direction (Fig. 51). An example illustrating these shifts of balance is provided by cover data from a heath stand, 30 years old when burnt, regenerating under the influence of heavy sheep grazing (D. Smart). At the end of the fourth season after burning grasses contributed 88% cover and dwarf-shrubs 11% (*Calluna* 5%). However, in an exclosure which protected a sample area from grazing during the latter two years of this period, the cover of dwarf-shrubs had increased to 23% (*Calluna* 14%). The numbers of rooted individuals of *Calluna* and other dwarf-shrubs per unit area were approximately equal in the grazed and protected samples. This indicated that regeneration had been uniform throughout and that within the four-year period the dwarf-shrubs had not yet been killed by the combined effects of grazing and competition with the grasses, but their above-ground parts had been reduced to almost insignificant proportions. However, rapid recovery from this position was possible.

In lowland heaths, grazing by rabbits also plays a part in determining the balance between *Calluna* and grasses (Chapter 9). Recent work on a chalk heath in south-east Sussex by Grubb, Green and Merrifield (1969) drew attention to increases in the height and spread of *Calluna* bushes following the decimation of rabbits by myxomatosis in 1954–5. Grasses such as *Festuca rubra, Agrostis tenuis* and *Anthoxanthum odoratum* were adversely affected owing to their susceptibility to the competition of vigorous *Calluna*. Apparently the influence of *Calluna* on other species is not constant throughout its life history but varies in relation to the growth-phase (Chapter 7). The three grass species mentioned above are most affected by *Calluna* in the building phase, though there are other species to which this does not apply (e.g. *Filipendula ulmaria, Galium verum*, which persist even in the middle of large building-phase bushes). Watt (1955) showed that the distribution of bracken fronds in a mixed *Calluna-Pteridium* community in the Breckland, East Anglia, was closely correlated with the growth-phase and hence the competitive vigour of *Calluna* individuals. Invasion of *Pteridium* into a *Calluna*-dominated area on the one hand, and of *Calluna* into a *Pteridium*-dominated area on the other, was determined according to which species was represented by the more aggressive growth-phase.

These accounts all suggest that a prime factor controlling the balance between *Calluna* and other species is the dense shade cast by the canopy of this woody dwarf-shrub in the earlier phases of its life history, if height and spread are unaffected by grazing. However, various observations indicate that additional factors are at work. For example, on dry soils in the south of England, even if *Calluna* bushes are rather widely separated, few if any vascular plants or *Calluna* seedlings colonize the intervening spaces, which may remain bare or partly lichen-covered for many years. In the Breckland, Roff (1964) observed partial suppression of grasses and other angiosperms in a ring around the larger *Calluna* bushes, when these were scattered in a *Festuca*-dominated sward. In these 'interference zones', which extended about 50 cm beyond the periphery of the *Calluna* canopy, the cover of *Festuca ovina*, *Agrostis* spp. and several other angiosperms was markedly reduced. Experiments confirmed that this effect was associated with the presence in the soil of living *Calluna* roots. Detailed observations and experiments led to the conclusion that it was not possible to account fully for the interference on the basis of competition for water or for nutrients, though the latter was not without some effect. Although at the outset of the work the possibility of an inhibitory product of *Calluna* roots had, for various reasons, been discounted, it was finally concluded that an explanation along these lines might indeed fit the facts. This outcome is of interest in regard to the problem of interactions between *Calluna* and young trees, which is the next to be considered.

Young trees

A further well-known example of interaction between *Calluna* and other species concerns its effect on young trees. Because of economic implications, this has been the subject of numerous observations and experiments, which are reviewed by Handley (1963). Certain trees, including *Betula* spp. *Pinus sylvestris, Sorbus aucuparia, Quercus* spp., may colonize heathlands relatively freely but the growth of others, notably spruce (*Picea sitchensis*) when planted amongst actively growing *Calluna*, is often seriously checked. The young trees appear unhealthy and show little or no increase in height. This condition can be avoided or removed by destruction of the *Calluna* stand and it therefore seems likely that it arises from interaction between the young trees and living *Calluna*. The symptoms re-appear on regeneration of *Calluna*, but can be prevented by any action taken to suppress its regrowth, such as mulching with cut *Calluna*, the use of the selective herbicides or artificial shading (Leyton, 1954, 1955).

Several investigators attributed the effects to 'competition' and sought to identify a limiting factor for which the trees and *Calluna* might be competing. Weatherell (1953) observed that check in spruce was less severe in the presence of *Sarothamnus scoparius*. His experiments, as well as others by Leyton (1954), showed that trees could normally be released from check by supplying nitrogenous fertilizer even without the removal of *Calluna*, though the beneficial effect was generally short-lived and disappeared within a few years of treatment.

It has, however, been pointed out by Handley (1963) that, unlike most other competitive interactions, the effect of *Calluna* on spruce does not involve the overtopping of the young trees by the *Calluna* canopy. In his view, the explanation must lie in effects on the functioning of the root system of the tree. Evidence was presented that extracts of the raw humus from vigorous *Calluna* stands were capable of suppressing mycorrhizal Hymenomycetes, and it was argued that the inhibitor might be derived from living roots of *Calluna*, perhaps from the endophyte within them. Furthermore, Robinson (1971) has recently reported that water extracts of litter derived from *Calluna* show phytotoxic activity, having the effect for example of inhibiting the root growth of certain crop plants used for test purposes. As yet there is no direct evidence that these substances are responsible either for the check of spruce and other 'sensitive' trees, or for the effects of *Calluna* on species such as *Festuca ovina* which may grow alongside or amongst it in a community. If, however, compounds capable on the one hand of restricting root growth and on the other of suppressing the formation of ectotrophic mycorrhizae are released by *Calluna*, they might play some part at least in these interactions. This is the most that can be said, until such time as it is known for certain whether or not the results so far obtained have ecological significance.

As far as the establishment of spruce plantations on heathland is concerned, all the evidence indicates that elimination of *Calluna* would be desirable, if it were practicable. At the time of planting, it is standard practice to reduce the *Calluna* as far as possible by cultivation, often preceded by burning (Zehetmayr, 1960). Subsequent regeneration, however, is unavoidable, but whatever the possible role of toxins derived from *Calluna* may be, observations in plantations on heaths indicate that where a correctly balanced fertilizer treatment has been achieved the effect of *Calluna* on young spruce is small. Growth in young plantations in these areas is normally successfully maintained by the addition of nutrients, particularly nitrogenous fertilizer, as necessary.

Calluna also affects other species by modifying the soil environment.

There have been numerous references to the acidifying effects of its litter (including Gimingham, 1960; Wilson 1960) and this has been considered in detail by Grubb, Green and Merrifield (1969). These authors found strong correlations between the size of *Calluna* bushes and the soil pH beneath their centres, also between distance from the centre and pH both at the soil surface and below (Fig. 28). Some of the vegetational changes associated

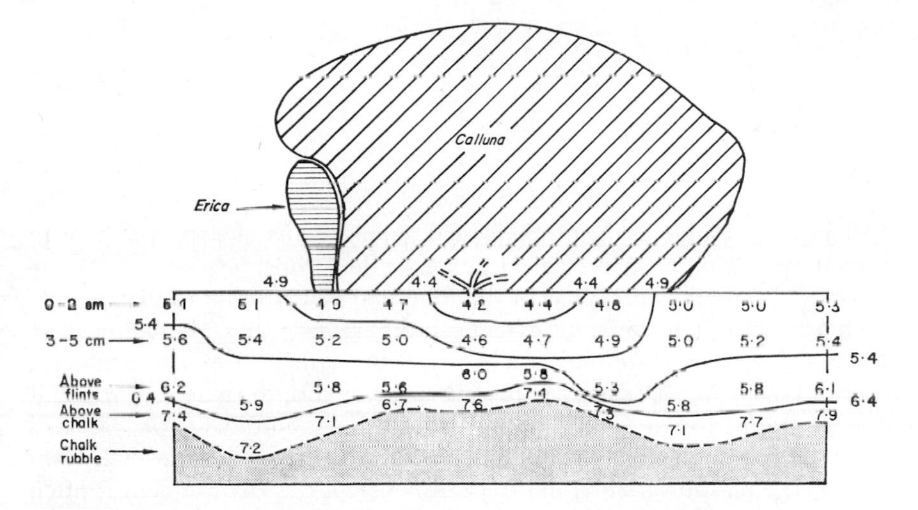

Fig. 28. Variation in pH at various depths along a transect through the middle of a *Calluna* bush (with partly suppressed plant of *Erica cinerea*) on chalk heath at Lullington, south-east Sussex. (From Grubb, Green and Merrifield, 1969.) Height of bush 55 cm; pH determinations 15 cm apart. Horizontal scale = vertical scale above surface; vertical scale below soil surface exaggerated, 2·5 × horizontal scale.

with the spread of *Calluna* may be attributed to increasing acidity and the soil changes consequent upon it. Webley, Eastwood and Gimingham (1952) have shown marked reductions in the populations of soil bacteria and increases in fungi on passing from a fixed *Ammophila* sand dune community, to a dune heath dominated by *Calluna*.

The properties of *Calluna vulgaris* which have been discussed, although representing only a selection of its physiological and ecological attributes, are sufficient to offer some explanation of the width of the ecological amplitude of the species. They also bear closely upon its relationships with associated plants. Although these are complex, recent and current investigations are elucidating some aspects of the observed interactions and have contributed towards a fuller understanding of the effects of heathland management upon community dynamics. The widespread dominance of *Calluna* depends to a considerable extent upon its physiological constitution, but also upon its morphology which is to be considered in the next chapter.

6 Growth-form in relation to community structure and composition

Just as a knowledge of the physiological requirements of a species helps in the interpretation of its ecological range, so does consideration of its morphology. Both are expressions of the inherited constitution of the species; both are variable to different extents in different species. As one facet of morphology, growth-form plays a part in determining the extent to which a species is tolerant, for example, to exposure to wind, the extent to which it exerts dominance in a community or plays a subordinate role, and its ability to co-exist with certain other species. The capacity for growth-form to undergo plastic variation in direct response to the environment is often correlated with the width of ecological amplitude in a species and the range of community-types, in which it is represented.

In certain prominent species of heaths growth-form has been investigated in considerable detail. This has shown very clearly that performance in a variety of conditions, and interactions between species, are closely bound up with growth-form. There is no better example than *Calluna vulgaris* which is treated first, after which other species are considered, particularly with regard to their interactions with *Calluna*. The responses of these species to treatments such as grazing and burning are also closely linked with their morphology, but this aspect of the subject is deferred to Chapters 9 and 10 in which the effects of management are discussed in detail.

The characteristic growth-form of *Calluna vulgaris*

The growth-form most characteristic of mature undamaged *Calluna* is that of a hemispherical dwarf-shrub, generally less than 0·8 m in height though sometimes reaching 1·25 m or more. The stems are branched from near the base (Plate 17); most of them are erect or ascending but the outermost are often procumbent (Plate 20) (Gimingham, 1949, 1960). This is the growth-form described by Du Rietz (1931) as that of a 'semi-sedentary dwarf-shrub'.

The development of this form and of the characteristic branch system is best outlined by following the first few years growth of a seedling, and by describing the sequence of stages in a single year's growth of the peripheral shoots of a mature plant (Nordhagen, 1937; Beijernick, 1940). In greenhouse conditions the young seedling (Plate 10) grows vertically as a single unbranched shoot for 10–16 weeks after germination, producing leaves in opposite and decussate pairs (Gimingham, 1949). The first laterals then begin to appear, usually from the axils of the sixth to tenth pair of leaves and not from those of the lowermost leaves. This fact is significant when responses to grazing and burning are considered, because as a result the young plant carries a reserve of a few dormant axillary buds, near the base of the stem.

Further laterals are then produced in pairs from axillary buds, in a neat decussate sequence (Plate 10). These grow obliquely upwards at an angle of 20–30° to the horizontal, maintaining the gradation in length acquired from the regular periodicity of initiation. The young plant displays a pyramidal form suggestive of a miniature spruce-tree, and usually passes its first winter in this condition. The size of the overwintering seedlings, however, varies considerably according to the date of germination. By the end of the first season, the only departure from the regularity of this branching system is seen in the older and larger seedlings, in which the first-formed and lowest pairs of branches may grow faster than the rest, exceeding the length expected from their position in the sequence of laterals (Plates 10, 11). These branches become procumbent and spread out over the soil surface.

Growth continues from the leading apex in the subsequent season when, increasingly, differentiation between 'short-shoots' and 'long-shoots' becomes evident. Short-shoots are laterals of limited growth, usually shed after two to three seasons, bearing small leaves (1–2 mm in length) which are imbricated on closely-spaced nodes. The main extension growth each year, however, is in the long-shoots on which the internodes are longer than the leaves, and the leaves themselves larger (3–4 mm long). It is in the axils of these leaves, which are active for one season only, that the short-shoots or flowering shoots are produced. A long-shoot by the end of the season usually shows three zones: (*i*) a basal zone of vegetative short-shoots produced early in the season, sometimes themselves carrying short-shoots of the second or even third order; (*ii*) a middle flowering zone often between 3 and 8 cm long bearing flowers on very short leafy stalks, or short-shoots with flowers; and (*iii*) a distal zone of end-of-season short-shoots, usually only a few millimetres long and with very densely packed leaves (Fig. 29 and Plates 12–15.)

During the second season, usually only the 'leader' behaves as a long-shoot and produces flowers, so the pyramidal growth-form is maintained, apart from further spreading growth in the lowermost branches. This monopodial development may persist for a further year or more, but more frequently the leading apex dies or is damaged during the winter. When this occurs, several of the overwintering end-of-season short-shoots, clustered just below the apex, grow out into new long-shoots. Two or three, usually the uppermost, adopt an erect habit and grow on together as leading long-shoots, later becoming lignified from the base upwards. The rest become lateral long-shoots, which adopt a more spreading habit, are shorter, have fewer flowers and are less lignified. Sooner or later these are shed, while the leading long-shoots become permanent branches. Thus, at yearly intervals most of the leading long-shoots are replaced by two or three new

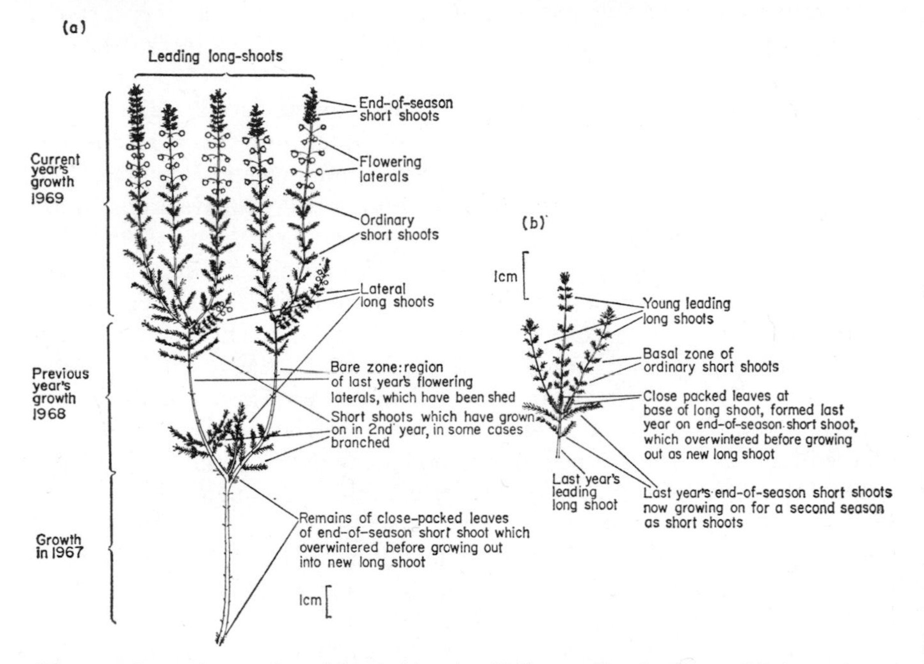

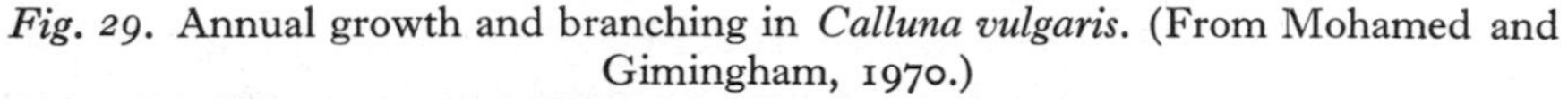

Fig. 29. Annual growth and branching in *Calluna vulgaris*. (From Mohamed and Gimingham, 1970.)
(*a*) Condition at the end of the growing season (early October);
(*b*) shoot tip early in the growing season (June) to show new long-shoots developing from overwintered end-of-season short-shoots.

ones, so giving rise to the regular bi- or tri-furcation normally visible in a mature branch system (Plates 16, 17). This type of branching maintains and increases the shoot density at the periphery, producing the shoot-clustering effect regarded by Ellenberg and Mueller-Dombois (1966 *b*) as typical of chamaephyte dwarf-shrubs (Plate 17).

Early in the life history of *Calluna*, therefore, the monopodial system gives place to a sympodial one. This occurs not only in the main axis, but also in the lower laterals which by this time have grown out as long-shoots and have adopted an ascending habit. Quite rapidly, if the plant is undamaged and not growing in close proximity to others, its form becomes densely hemispherical. Thereafter, this form is maintained by the pattern of growth and branching just described. Every winter many of the long-shoot apices become defunct, and are replaced by two or three new long-shoots derived from some of the end-of-season short-shoots (Plates 12, 13). The close-packed leaves of the overwintering short-shoot remain at the extreme base of the long-shoot as it develops, and their axillary buds remain dormant (Plate 14). A few long-shoot apices may continue growth in a second year, in which event no branching occurs, but the junction between the sectors of the twig formed in successive seasons is marked by the presence of a group of closely-imbricated leaves formed as internode length decreased at the end of the season before overwintering. Again, their axillary buds remain dormant.

Beijerick (1940) describes this as a 'storeyed' system of branching. The pattern is emphasized by the division of each year's increment into the three zones by the flowering portion of the long-shoot. During the subsequent winter the flowers are shed, leaving a bare zone which separates the green zones carrying short-shoots (Plates 12, 13). Usually the latter are retained in an active condition for up to three years. Thus branches of *Calluna*, viewed from the side, during the summer show an upper zone of bright green comprising the current year's growth of new long-shoots, as yet unlignified with their developing short-shoots. These are separated by a bare zone, now lignified, from the rather darker green band of the previous set of short-shoots, each of which is now adding an increment at its tip. Below a second bare zone, another green band is often present if its short-shoots are surviving into their third year. Herbivorous animals feeding on *Calluna* use the current season's unlignified long-shoots and the short-shoots: the growth-habit in which the foliage-bearing short-shoots are retained for two or three years is therefore of significance in regard to availability of edible material.

Owing to this pattern of growth the age of *Calluna* branches can be determined with some certainty up to about six or eight years, first from the sequence of annual long-shoots, each with the series of closely spaced leaves or scars where the apex or end-of-season short-shoot overwintered, and later on by the branching points which represent the junctions between one year's growth and the next. Lower down in old bushes, the sequence may become obscure owing to shedding of the weaker branches. Ring-counts, not always easy to obtain with accuracy, are then the only way of ageing a stem.

Apart from changes in the later stages of the life-history of an individual *Calluna* plant which are detailed in the next chapter, the growth-form described is typical of *Calluna* growing under favourable conditions. However, populations display considerable genetic diversity: some of this is evident in variations in flower structure, flower colour, foliage colour, hairiness, etc., and some in growth habit. Most populations include some individuals which, although producing the characteristic hemispherical bush, have a much less dense branch system and allow more light to penetrate the canopy. At the other extreme, a well-known variant is one in which the majority of branches become decumbent, producing a compact, low hummock of inter-twined shoots (numerous 'dwarf' varieties have been taken into cultivation). The extent to which these variations may be of ecotypic significance is discussed on p. 117, but in regard to the habitat range of the species it is important first to establish the extent to which the basic pattern of growth-form may show plastic modification in response to more extreme environments.

Modifications of growth-form in various habitats

In the foregoing section, the normal pattern of growth has been described as seen in isolated individuals, unaffected by neighbouring bushes. This growth-form, however, is subject to considerable modification both as a result of interference between neighbouring plants, and of certain environmental influences. The capacity of *Calluna* to remain a vigorous competitor despite considerable modification of its normal growth-form may be regarded as contributing to its wide ecological amplitude in comparison with related ericaceous species.

With elimination of about 50% of the daylight illumination of plants growing in the field, by artificial shading, Björkman (1945) found clear indication of etiolation effects, which were still more obvious with elimination of up to 75%. Long-shoots became weak and attenuated and their internodes extended; short-shoots were fewer, shorter and more distant; flowering was reduced. Plants grown for a number of years under the shade of a tree canopy lose their compact form, fewer frame branches survive and these are thin and often straggling. None the less, *Calluna* can persist in this form even where the average light intensity is less than 50% of full daylight, probably down to about 30%. Below this, as Björkman's analyses of plots kept at 25% full daylight show, the quantity of *Calluna* gradually declines.

These results help to account for the striking departures from the typical hemispherical form of an isolated individual, observed among plants growing together in dense stands. The density of the canopy in the building phase is such that the light intensity reaching side branches may be reduced to less than 30% of full daylight, while as little as 1% may reach ground level (Chapter 7). On hill slopes, descending branches of one bush tend to overlie and shade out potentially ascending branches of the next below it, and all plants appear to consist of parallel, more or less decumbent stems straggling in the direction of the downslope and producing a canopy with a wave-like profile (Plate 20). On flat ground, elimination of most of the lateral frame branches may produce a form with rather tall, trunk-like main stems, branching densely near the top to produce a crown of leafy shoots which contributes to a relatively smooth and unvarying canopy

(Plates 18, 19). This form is particularly evident where dense even-aged stands have developed after burning (Chapter 10) and the capacity of the plant to retain dominance in this form no doubt contributes to its universal importance in heaths managed by burning.

Although the maximum height of the canopy may be considerably restricted in unfavourable environments, the general growth-form of the individuals may not differ greatly from the pattern described earlier, and the capacity to become dominant is retained. A continuous decrease in height is often evident on passing up a hillside and many dense stands of low stature (less than 10 cm) have been described (e.g. at 378 m in north-east Scotland; Nicholson and Robertson, 1958). Localized gradients in shoot height occur on small prominences, while on terraced slopes there is a regular transition from tall *Calluna* in the shelter of the 'step' to very short on the 'brow' (Metcalfe, 1950).

Under other environmental regimes, such as severe exposure or repeated grazing, vertically directed long-shoots may be repeatedly destroyed. Growth is then concentrated in branches from the lowermost nodes of the original axis, which, as described on p. 111, are generally plagiotropic. As a result, a spreading procumbent form develops, with branches from adjacent plants inter-twining, often producing an extensive system of contorted stems trailing over the ground (Plate 26). Since effective cover can be established in this form, *Calluna* extends its spheres of dominance to high mountain plateaux (usually above 670 m) and northern coastal cliffs where it gives rise to dwarf turf-like stands seldom exceeding about 8 cm in height (Crampton, 1911, Crampton and Macgregor, 1913, Armstrong, *et al.*, 1930, Metcalfe, 1950). Where the causal agent is wind blowing predominantly from one direction, as often in mountain regions, the creeping shoots are orientated parallel to one another, extension taking place into the wind shadow created by the fringe of densely packed leafy shoots (Fig. 30). Behind this, erosion leaves only the bare stems, rooting at intervals and linking the growing fringe to the original point of anchorage up to 1 m or more behind. An appearance is presented of more or less parallel stripes of vegetation separated by bare ground, each moving in a direction determined by the prevailing wind – a direction which involves an uphill movement as often as downhill or along level ground. In a somewhat similar manner a low, dense and rather turf-like stand may sometimes be formed under moderate to fairly heavy grazing by sheep or rabbits (Gimingham, 1949).

The growth-form also varies in relation to the nature of the soil. Nicholson and Robertson (1958) found in east Scotland that performance, as measured by height, canopy density, mean number of flowers per shoot and weight of the standing crop, on podsol soils was superior to that on wet hill peat (Table 13). Probably optimal development is related primarily to relatively free drainage, since it can be obtained on drained, oxidizing peat

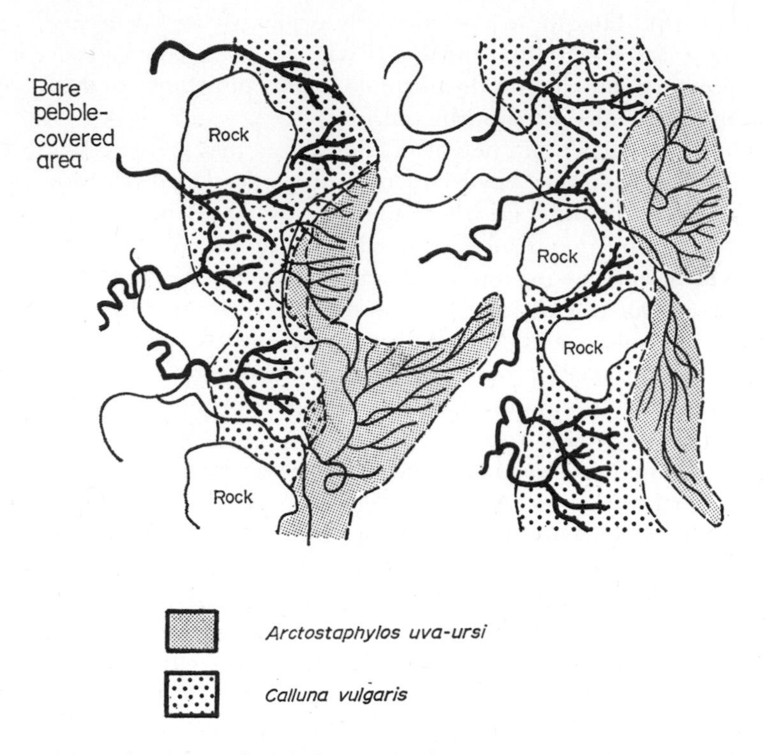

Fig. 30. Map of a mountain heath community (after Metcalfe, 1950: Cairngorm Mountains, east central Scotland). The vegetation forms into successive, parallel strips growing slowly in one direction, determined by the prevailing wind. In this instance, *Arctostaphylos uva-ursi* leads, growing into the wind-shadow created by the strip of vegetation. *Calluna* follows closely: behind, all green shoots and smaller branches are eroded away, leaving the old stems exposed on the bare stony ground. Both species have spread a considerable distance from their original rooting points, but the distal branches may root adventiously in humus which is held down by the compact shoot systems

as well as on podsols. On such soils, accumulating humus or a moss stratum may bury the stem bases and promote adventitious rooting in a layer which retains moisture. However, on substrata liable to waterlogging, as on wet hill peat or in bogs, a weak, sprawling form results, flowering poorly and no longer dominant. When grown in pots with an artificially maintained water-table, active roots are confined to the aerated soil layer above the water-table even if this should be as shallow as 2 cm, and in the field this restriction of the root region to an extremely superficial layer is probably sufficient to reduce performance. The very much better performance of *Calluna* on the tops of old *Sphagnum, Rhacomitrium lanuginosum* or *Molinia* hummocks in bogs, where it presumably has access to a more extensive aerated root-region (Rutter, 1955), is striking. Strangely enough neither root nor shoot production in water-culture seems to be impaired by lack of aeration, or even by bubbling nitrogen through the culture solution (Poel, 1948).

TABLE 13

Performance of *Calluna* on podsols and peat, Glensaugh, Kincardineshire, north-east Scotland; altitude *c* 335 m. (Data from Nicholson and Robertson, 1948).

Soil	Height of canopy cm	% cover	Mean no. flowers/ shoot	Dry weight of standing crop g m^{-2}
Strongly podsolized	37·5	96.8	41·6	1004·0
Weakly podsolized	33·0	81·6	76·0	1056·0
Peat	15·0	49·2	10·4	796·8

Ecotypic variation in growth-form

In communities of severe environments such as those of high mountain and northern habitats, where the typical growth-form of *Calluna* is no longer evident and instead a dwarf, more or less prostrate form is universal, the question arises as to how far this habit arises from the formative effect of the environment and how far it is genetically determined. Grant and Hunter (1962) collected seed from stands of *Calluna* both at low and high altitudes from several localities in Scotland, including Shetland, and analysed the growth-habit of numbers of plants raised from each population.

While in all cases a majority of plants displayed the typical erect habit of growth, completely prostrate or intermediate types were present in significant numbers in the progeny of populations belonging to high altitudes. They were very scarce or lacking among plants grown from low-altitude seed samples, except in the case of the Shetland site where the young plants consisted 'almost entirely of prostrate variates'. The seed from which the latter were obtained was taken from a stand in a very exposed habitat. There is some evidence, therefore, of ecotypic selection of inherited variations tending to produce a low, prostrate growth-form in high altitude or otherwise exposed environments. The capacity of *Calluna* to extend its sphere of dominance to these habitats is no doubt an expression of its ability to continue vigorous growth even when subject to considerable phenotypic modification of form, but it appears that genotypic variation contributes as well.

The growth-forms of associated species

Growth-form, and the extent of possible modification of habit within a species, have a bearing not only upon ecological amplitude, but also upon interactions with other species. Several common associates of *Calluna* are next examined from this standpoint.

Erica cinerea

Details of growth-form in *Erica cinerea* have been less intensively investigated than in *Calluna*. While the very young seedlings are alike in their vertical unbranched habit (Plate 10), subsequent development introduces several important differences. At equivalent age, *Erica cinerea* seedlings are generally slightly taller than those of *Calluna* but, whereas the latter soon produce a regular series of laterals, in *Erica cinerea* branching is delayed until 15–16 weeks after germination (Gimingham, 1949) when the seedlings are often about 35 mm high (Plate 21). Branches are fewer and more widely spaced, growing on indefinitely and assuming an erect habit; the pattern of branching is therefore more diffuse and straggling. Some of the lower laterals develop into main branches, as in *Calluna*, and growth becomes sympodial in the second and subsequent seasons. Bannister (1965) describes the annual sequence of shoot growth which broadly resembles that of *Calluna*. Axillary buds in the lower (older) part of the leading shoots become active and produce short-shoots, but as their stem development is extremely limited they give the appearance of little more than clusters of leaves. Above these is a raceme-like flowering region and beyond, in the youngest part, is the apical bud with a few lateral buds. The former often fails to survive the winter, and the three uppermost lateral buds produce new leading shoots. These may either grow on together or else one may develop dominance.

Although the pattern of growth resembles that of *Calluna*, branching is always more open and diffuse (Plate 21) and the canopy is never so dense. The result is that when growing alongside *Calluna* of equal age *Erica cinerea* almost always shows an erect, straggling growth-form and fails to develop into a hemispherical dwarf-shrub as it can when growing in isolation. Further, the branches very readily become semi-prostrate, and as the species is more shade-tolerant it is well adapted to the formation of a second stratum, subordinate to the *Calluna* canopy (Chapter 3). The lowermost, prostrate branches root adventitiously, and produce more numerous short, erect laterals. All these features obscure any regularity of annual growth zones, which is so evident in the 'storeyed' growth of *Calluna*.

Erica tetralix

This species, which scarcely overlaps in habitat range with *E. cinerea* but is the chief associate of *Calluna* on many of the wetter sites, has a growth-form basically similar to that of *E. cinerea* (Bannister, 1966). It differs, however, in the absence of short-shoots, and in the more regular tendency for the lower branches to become buried in moss or litter layers, producing adventious roots and behaving essentially as rhizomes. As with *Erica cinerea* when *E. tetralix* occurs along with vigorous *Calluna* its form is erect and straggling, and it fails to produce a compact bush. However, where the latter is absent or reduced in importance *Erica tetralix* produces a much branched dwarf-shrub up to 60 cm in height. The buried creeping stems are usually out of range of fire damage and *E. tetralix*, when present, is usually the quickest and most vigorous to regenerate vegetatively after moor burning, sometimes becoming dominant in the post-fire period.

Empetrum nigrum

Empetrum nigrum is another ericoid species regarding which some information exists, contributing to an interpretation of its interactions with *Calluna*. In this case, there are greater differences in growth-form since *E. nigrum* is a slenderer, more straggling plant than *Calluna*, and although sometimes forming dense prostrate mats does not reach the height nor adopt the form of a bush.

Empetrum nigrum belongs essentially to northern heath communities, particularly those of rather oceanic districts such as the Faroes, Norway, north Britain, west Sweden and Denmark and parts of the Netherlands. Although widely associated with *Calluna* in these areas, it is in the heath communities on freely drained sandy soils on the one hand, and wet peats on the other, that it is best represented. The moisture regimes of these habitats are rather widely distinct, and although heaths on podsols and related soils may be ranked as intermediate in this respect, *E. nigrum* is in general less prominent in these.

Although the ecological amplitude of *E. nigrum* and its relationships with *Calluna* have not been fully worked out, and as yet few data are available on its 'water relations', part of the explanation seems to lie in the conditions required for germination, and part in the subsequent morphological development. There is little doubt that an open habitat is required for establishment, since seedlings or young plants may be found in the open patches in a low dune heath or on exposed peat, whereas they are extremely infrequent in a dense stand of *Calluna*. Under experimental conditions germination is very slow (Grevillius and Kirchner, 1925; Hagerup, 1946), but appears to be favoured by light.

There is some evidence, which requires verification, that growth is better on a moist mineral soil (e.g. sand) than on purely organic substrata. During the first season following germination only the two cotyledons and a bud are present, but usually during the second year a vertical stem about 1 cm long is produced. At the end of this season's growth, as in subsequent years, a group of two or more branches is produced and these grow on in the subsequent year, giving rise to an irregularly whorled arrangement of branches. In an open situation, all branches normally become prostrate, perhaps merely because of the weight of the distal crown of branches produced each year. Under these conditions, the plant may spread outwards

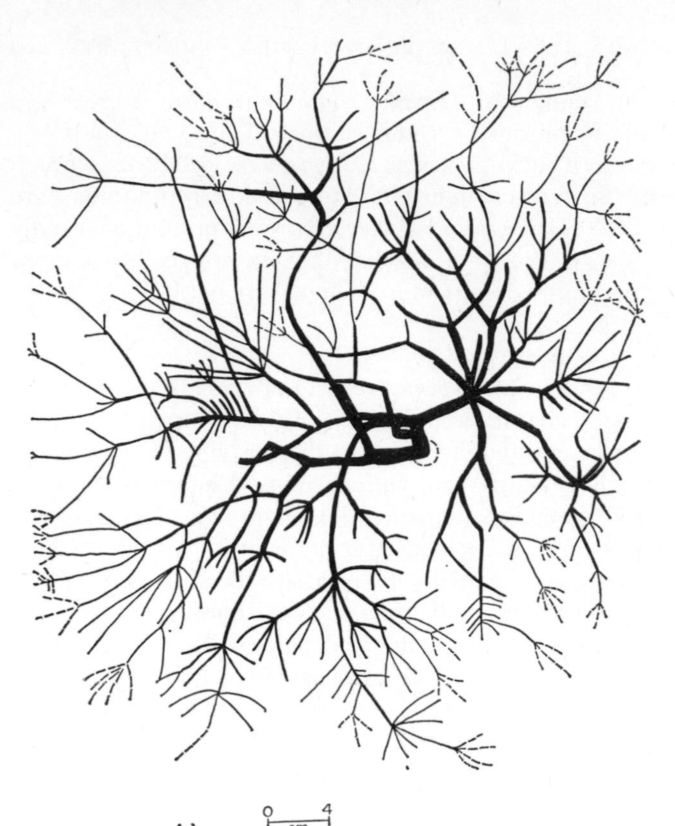

(a)

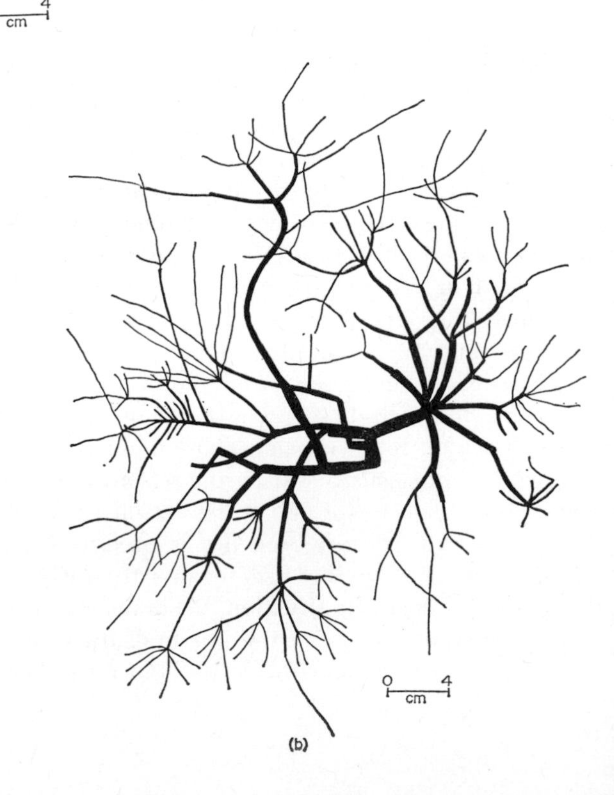

(b)

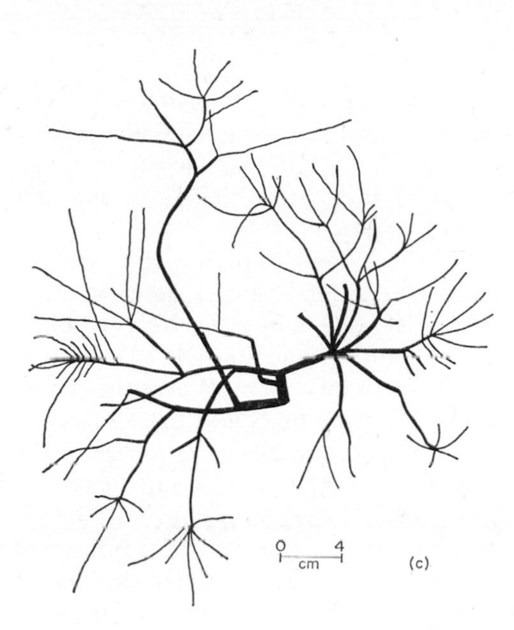

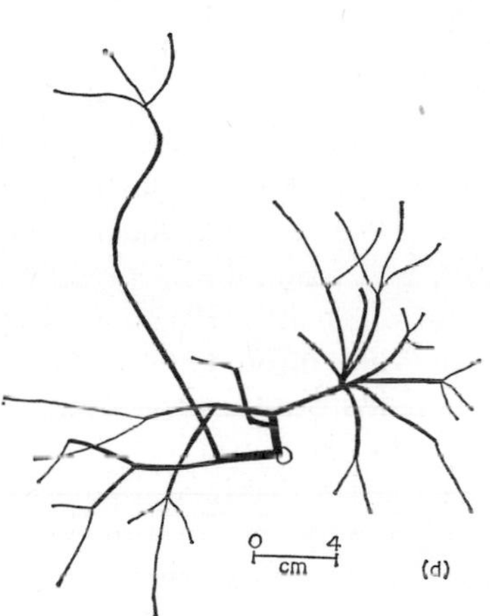

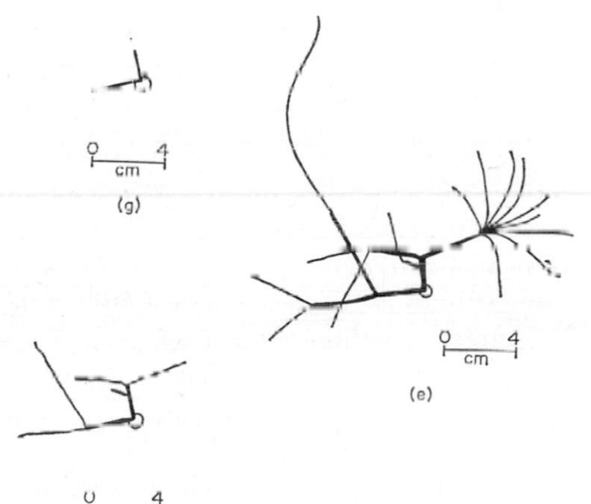

Fig. 31. Growth-form of *Empetrum nigrum* on dune heath in north-east Scotland. (Figure by Mrs. S. Y. Landsberg.)

(a). Diagram of branch system of complete circular patch (= single individual), at the end of eight years growth.

By repetition of the diagram with progressive elimination of one year's growth increments at a time, the patterns at the end of seven years (b), six years (c), five years (d), four years (e), three years (f) and two years (g) are illustrated.

in all directions forming a circular patch or a dense cushion (Fig. 31). In this form it may be found growing vigorously amongst the low vegetation of old, acid, fixed dunes, amongst lichens in the *Empetrum*-lichen heaths on mountains, and with a rather more straggling form amongst *Sphagnum* and other species of bog communities.

However, when growing in partial shade cast by the canopy of other species, e.g. *Calluna*, the number of surviving branches at each distal cluster is reduced normally to one or two. Hence, a more extensively creeping, less compact habit of growth results, and at the same time partial etiolation may increase the length of each annual shoot increment. Where the shoots are supported by a dense framework of *Calluna* branches, *Empetrum* may keep pace with the growth of a *Calluna* bush, producing its foliage at the same level as the *Calluna* canopy. Its growth-form thus permits it to survive and maintain its presence in a *Calluna* community, but only in the form of greatly elongated straggling shoots, occasionally bifurcating and ramifying through the supporting structure of *Calluna*. Hence, except where it spreads into a gap in the *Calluna* canopy, its contribution in terms of cover and density is inevitably restricted.

In all probability, therefore, the limited representation of *Empetrum nigrum* in habitats occupied by *Calluna* is to be explained at least in part by its morphological reaction to growth in mixture with *Calluna*. *Empetrum* spreads rapidly in these habitats on release from the dominance of *Calluna*, as can be observed in the first few years after burning a community in which both are present.

Vaccinium myrtillus and *V. vitis-idaea*

Some investigations bearing on the ability of these species to contribute to heath communities have been carried out. Both are more shade-tolerant than *Calluna* and hence are frequently woodland species, but both occur along with *Calluna* in certain types of heathland (Chapter 4). Both are also more prominent in the dwarf heaths at high altitudes on mountains, particularly where snow-cover provides effective winter protection.

However, neither of these species play more than a subsidiary part where *Calluna* is vigorous. Presumably owing to their shade-tolerance, they are capable of contributing to a second vegetational stratum below the *Calluna* canopy: this role is particularly marked in *V. vitis-idaea*. Their rhizomatous habit enables them to occupy effectively any gaps in the *Calluna* canopy and, as with *Empetrum*, release of the *Calluna* dominance by burning may permit a temporary increase in their representation. Both are plants of soil which has a well-developed raw humus layer, but also may grow vigorously among boulders or at the margins of screes, and in very exposed situations. In such habitats *Vaccinium myrtillus* may replace *Calluna* as dominant, as for example in the '*Vaccinium*-edge' communities of the Pennines (Pearsall, 1950).

Their areas of distribution are northern, *V. myrtillus* being listed by Matthews (1937) in his Continental Northern Element and *V. vitis-idaea* among Arctic-Alpine species. (Both show increased seed-germination after cold treatment.) However, *V. myrtillus* has in general a rather more oceanic distribution than *V. vitis-idaea*, and in agreement with this it has been shown to be somewhat more easily killed by drought (J. Lockie).

The morphology and growth-form of *Vaccinium myrtillus* have been studied in detail in relation to the age-structure and dynamics of pure stands of this species, by Flower-Ellis (1971). It was shown that stands may consist of expanding clones based on a sympodially branching rhizome system. Primary rhizomes decay, releasing their branches and thus giving rise to numerous rhizome units which produce aerial shoots distally. Each rhizome unit may be 2 m or more in length. Its branches fan out, overlapping with adjacent fans. Thus there may be an advancing front extending at a mean rate of about 4 cm per year, while the branches from the primary rhizome repeat the developmental sequence after a time lag and resemble a series of 'waves' of young rhizomes, spaced at intervals corresponding to periods of about three years. Behind is a hinterland apparently containing a mosaic of units of different stages, recalling the pattern described by Watt (1947) for *Pteridium aquilinum*. Calculations suggested that the age of clones might range from 40 to 100 years, and their diameters from 6 to 10 m. Such extensive clones are probably more characteristic of forest than heathland communities, but the analysis is valuable in demonstrating the system of vegetative spread of this species from which its behaviour in association with *Calluna* and other heathland species can be interpreted.

Arctostaphylos uva-ursi

This species is an important associate of *Calluna vulgaris* in the species-rich heaths of west Norway and at medium altitudes in the eastern central Highlands of Scotland, as well as in more southern heath-types in south Sweden, Denmark, etc. (pp. 73, 79, 80). It plays a part which is complementary to that of *Calluna*, in that it is a creeping species with little shade-tolerance, occupying the ground where gaps occur in the *Calluna* cover. Even in favourable habitats it is therefore sparse in dense even-aged stands of *Calluna*, as developed after burning, but much more abundant where the cover is discontinuous. Branches in *Arctostaphylos uva-ursi* adopt a prostrate habit, usually with rather regular laterals. Plants growing in isolation form intricate mats, but when mixed with other species they can extend monopodially for considerable distances until conditions of improved illumination are encountered. Thus in *Calluna* stands wherever gaps occur, as in the centres of mature and degenerate plants (pp. 127–9), branches of *Arctostaphylos* if present soon extend into the gap and there develop a patch of interlaced laterals. In these communities excavations reveal numerous partially buried woody stems up to 1·5 cm in diameter, now bare of foliage, following a sinuous course between the patch and its rooting point some distance away (Metcalfe, 1950), (Fig. 30). Consequently, where *Arctostaphylos* figures in communities undergoing cyclical processes as described in the next chapter, it plays a prominent part in these (Watt, 1947*b*). The low stature and tendency for stems to become partly buried serve as some protection from fire. *Arctostaphylos* may temporarily extend its cover after heath burning until restricted as described above by the regenerating *Calluna* canopy.

Several dwarf-shrub species are able to co-exist with dominant *Calluna*, including for example *Ulex* spp. and *Genista* spp. as well as those discussed above. Their persistence in this niche is determined to a considerable degree by their respective patterns of growth-form and the extent to which these are adaptable in response to variation in the environment. Furthermore, the effects of heath burning upon these associated species may derive not only from the direct influence of the fire, but from their interactions within the resulting even-aged stand and uniform canopy of *Calluna*.

7 Cyclical processes in heath communities

The two preceding chapters have demonstrated that the widespread dominance of *Calluna* owes much to the morphological as well as to the physiological characteristics of the species. It has also been indicated that during the life-span of an individual plant there are changes in morphology which have a significant bearing on its interactions with other species. It is, in fact, impossible to interpret the structure and dynamics of *Calluna*-dominated communities, and the effects of treatments such as grazing and burning, without reference to the morphological differences between plants at various phases in their life-history. Changes in the pattern of growth take place gradually with increasing age, but the characteristics of each phase are fairly distinctive and recognizable. Watt (1955) has shown that they may effectively be described by his terms 'pioneer', 'building', 'mature' and 'degenerate', which he first applied to the phases of vegetative development of rhizomatous species such as *Pteridium aquilinum* (1945, 1947*a*). The exact age at which an individual plant passes from one phase to the next may depend to some extent upon environmental conditions. Since the influence of the plant upon other members of the community varies considerably with its growth-phase it may therefore often be more instructive to describe a stand in terms of the growth-phases represented than of the actual age of the plants. For this reason an account is now given of the sequence of phases (see also Gimingham, 1960; Barclay-Estrup and Gimingham, 1969), before turning to more detailed considerations of the cyclical processes which result where the full sequence is uninterrupted by burning and grazing.

The four phases of the life-history

(*a*) The pioneer phase

This is the period of establishment and early development, lasting generally for about three to six years (even sometimes ten), before the plant has developed a fully bushy habit (Plate 23). The early stages of growth after seed germination, producing the trim pyramidal shape, have been described on p. 111. Flowering commences on the upper part of the leading shoot usually in the year following that in which germination took place, but is seldom vigorous during the pioneer phase. The change from monopodial to sympodial growth has also been described, with the accompanying gradual replacement of the pyramidal shape by that of a hemisphere, in which several more or less equal frame-branches radiate from a central stem-base. By this time the plant is passing out of the pioneer phase.

Following upon vegetative regeneration from the stem-base after burning, severe grazing or cutting, the sequence of events may be similar, except that young shoots are usually formed in clusters (pp. 177, 193–4). Hence although these begin to branch in the normal regular way, they are already crowded and the pyramidal form more quickly gives place to a radiating system of branches. The pioneer phase in these cases is characterized more by the development of a series of separate cushions, than of numerous independent small pyramids. Usually the former reach the appearance of the next phase rather more rapidly than the latter, and where regeneration is most vigorous a typical pioneer phase is virtually by-passed.

(*b*) The building phase

At this stage, there is usually a clear differentiation between short-shoots and long-shoots. The bulk of the foliage, and hence most of the photosynthetic activity of the plant, is concentrated in the short-shoots. Each year peripheral extension-growth takes place in the long-shoots, and this pattern of growth (described in detail on pp.111–13) is maintained throughout the building phase which lasts normally until the plant is about 15 or more years old. The short-shoots show a characteristically bright green colour, and the resulting canopy reaches its maximum density, permitting little light to penetrate to the ground below. Flowering is usually profuse. In the absence of interference from neighbouring individuals an evenly dome-shaped outline is produced, while the radius of the hemisphere may reach $\frac{3}{4}$ m, or more.

(*c*) The mature phase

During the subsequent period, usually until the plant is between 20 and 25 years old, growth in the long-shoots may show some reduction in vigour, their flowering zones become shorter and the short-shoots take on duller colours, becoming more clustered and obscuring the regular 'storeyed' appearance. These are indications that the plant is in the mature phase, in which it reaches its maximum height (though not necessarily its maximum diameter). This phase also frequently shows the beginning of a tendency for the central, oldest frame branches to spread apart, permitting more light to reach the ground.

1. Upland heath in north-east Scotland. On the slopes burning has produced a patchwork of stands of varying age. The abundance of boulders suggests that surface erosion has taken place. Farmland and a few plantations occupy the valley.

2. Heath vegetation at 3350 m on Mount Albert-Edward, New Guinea. The dwarf shrub is *Styphelia suaveolens* (Epacridaceae) and the grass, *Deschampsia klossii*. Photo: P. S. Ashton.

3. Heathland in the province of Halland, South Sweden. Numerous scattered bushes of *Juniperus communis* form a discontinuous tall shrub stratum. *Calluna vulgaris* is the dominant dwarf-shrub, with *Vaccinium myrtillus* in the foreground.

4. Pattern in a community on blown sand in north Scotland created by the radial expansion of colonizing individuals of *Empetrum nigrum*.

(*a*) On the freely-drained sand of a dune heath.

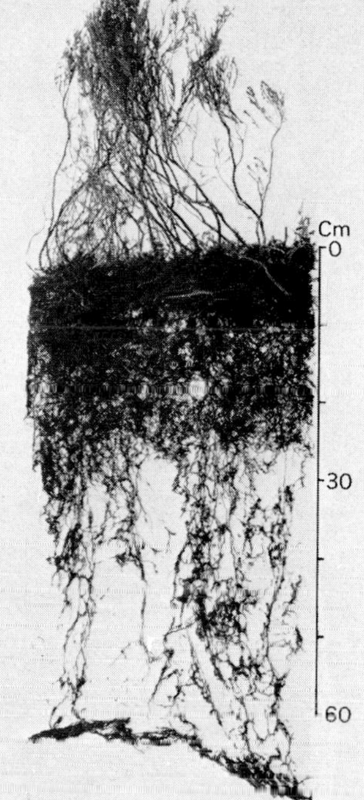

(*b*) On a podsol with iron pan. The deeper roots have formed a mat just above the pan.

(*c*) Old plant showing mass of adventitious roots produced from stem bases buried in litter and humus.

(*d*) Branch of a plant from wet heath, showing procumbent base with adventitious roots.

5. Development of the underground parts of *Calluna* in different habitats. (Scale: × 0·1) Specimens prepared by R. Boggie.

6. *Calluna – Ulex gallii* community, with *Pteridium aquilinum*. Dorset, S.W. England.

7. *Calluna – Ulex minor* community. Hampshire, S. England.

8. *Calluna – Vaccinium vitis-idaea – Empetrum nigrum* community, Denmark.

9. Tip of long-shoot of *Calluna vulgaris*. Axillary buds have formed small end-of-season short shoots. The abaxial (lower) leaf-surface is confined to the groove visible on the leaves nearest to the camera. (Scale: × 10.)

10. Stages in the development of seedlings of *Calluna vulgaris*. (Scale: (*a*)–(*c*) × 0·9
(*d*)–(*f*) × 0·5.)

(*a*) *Calluna* (*left*) shown at 12 weeks after germination alongside a seedling of *Erica cinerea* (*rig*
of the same age, (*b*)–(*f*) *Calluna* seedlings arranged in order of increasing age. (*b*) and (*c*) show
regular pyramidal form (*c.* 15 weeks after germination); in (*d*), (*e*) and (*f*) the first-formed branc
have grown long in proportion to the rest and in (*f*) are adopting an ascending habit (*c.* 30 we
after germination).

11. Young plants of *Calluna* in their second season of growth, showing the pyramidal form with lowest branches extending horizontally. (Scale: $\times$ 1.)

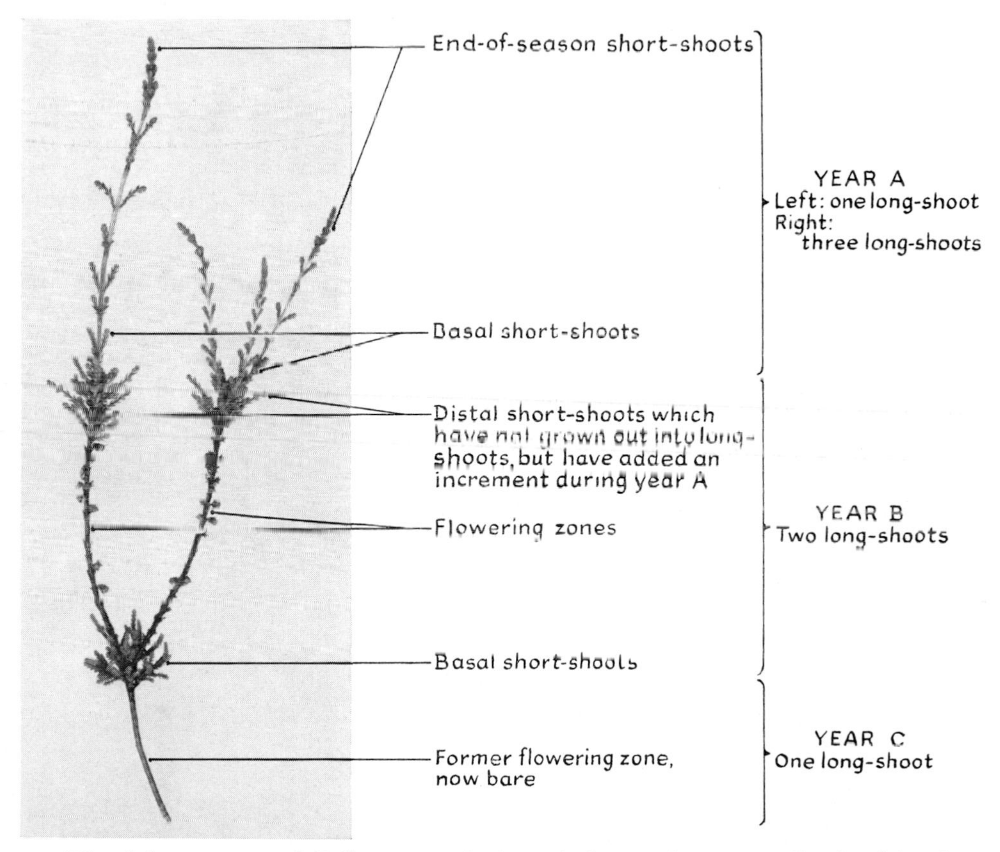

12. Tip of shoot system of *Calluna* towards the end of a growing season. On the right, the long-shoot of Year B has been replaced by three new long-shoots. On the left, the apex of Year B's long-shoot has continued growth in Year A.

There has been no flowering in the current year, but some of the previous year's flowers are still attached in the flowering zones. (Scale: $\times$ 0·7.)

13. As Plate 12, about life-size, showing two years' growth with flowering in both years. Flowers of the previous year have fallen, leaving the flowering zones bare except for widely spaced leaves. Short-shoots produced in the earlier year have added a growth increment in the current year. (Scale: × 1.)

15. Tip of long-shoot at higher magnification (September), showing part of flowering zone and end-of-season short-shoots. After overwintering the uppermost of these will produce next year's long-shoots. (Scale: × 3.)

14. Long-shoots of the current year, slightly magnified (September). Closely imbricated leaves can be seen at the base of the long-shoots. These leaves overwintered from the previous year on end-of-season short-shoots, the apices of which produced the new long-shoots. (Scale: × 1·5.)

16. Typical branching pattern of *Calluna* showing bi- or tri-furcation at the end of each annual growth increment. Some of the branches are weaker than others and will eventually be shed. (Scale: × 0·8.)

7. Complete main branch from vigorous late-building or ma-re *Calluna* bush. (Scale: × 0·15.)

18. Branch from a mature *Calluna* bush which has grown in an even-aged stand. This is undivided for nearly half of its total height. (Scale: × 0·15.)

19. Branch of *Calluna* grown in a stand of high stem-density, undivided up to a point only just below the canopy. (Scale: × 0·1.)

20. Outermost branches of an old, degenerate bush of *Calluna*, growing horizontally and rooting adventitiously in the litter. Plants growing on hill slopes also adopt this form, straggling in the downhill direction. (Scale: × 0·15.)

21. Comparison of young plants of *Erica cinerea* and *Calluna vulgaris*. (Scale: × 0·5.)
Extreme left: 15–16 weeks after germination. *Calluna* (*b*) is already well-branched; in *E. cinerea* (*a*) only one pair of laterals has just begun to grow.
(*c*) *Erica cinerea* and (*d*) *Calluna* towards the end of a second growing season, showing differences in habit.

22. A bush of *Calluna* in the degenerate phase. The central branches are dead, while the outermost ones which have rooted adventitiously in the litter are still active.

23. Colonization of the gap in the centre of a degenerate bush by pioneer plants of *Calluna*, completing the vegetational cycle. The gap is also occupied by *Pleurozium schreberi* and scattered tufts of *Deschampsia flexuosa*. *Parmelia physodes* is seen on the old branches.

24. Imitation of grazing: comparison of the effects of cutting seedlings of *Erica cinerea* (*left*) and *Calluna vulgaris* (*right*) at a height of 1 cm. The lowest branches of *Calluna* continue plagiotropic growth; in *E. cinerea* two branches adopt an erect habit. (Treatment carried out about 10 weeks after germination; plants photographed 16 weeks later.) (Scale: × 1·5.)

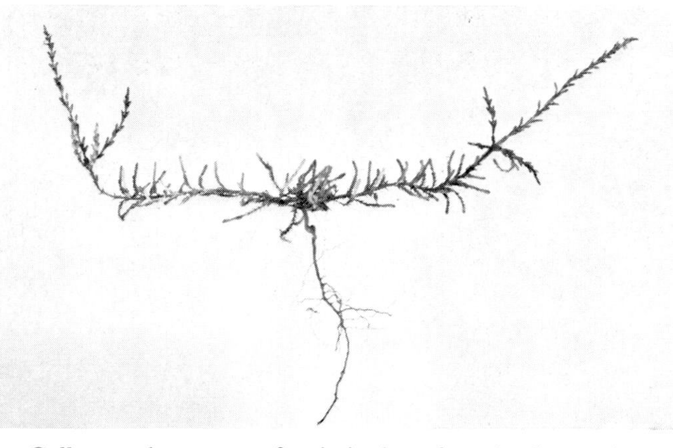

25. *Calluna:* a later stage after imitation of grazing by cutting a seedling at 1 cm. Two branches have spread out horizontally and have become ascending at the tips, producing numerous laterals. (Treatment about 13 weeks after germination; plant photographed 24 weeks later.) (Scale: × 0·5.)

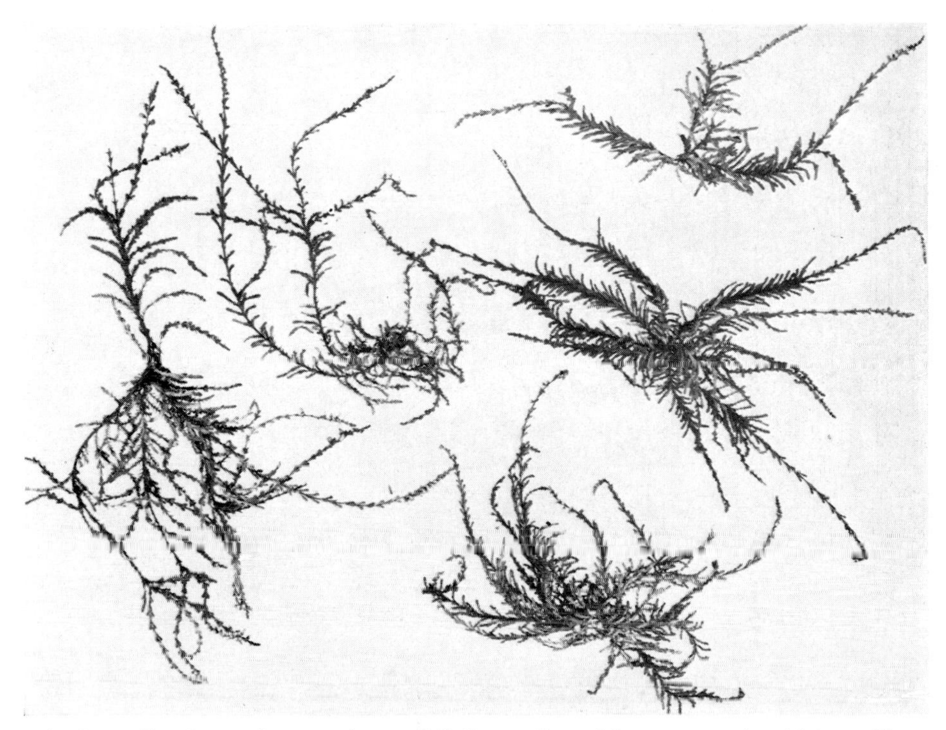

26. Straggling form of young plants of *Calluna* collected from an area in which seedlings were subject to sheep grazing. (Scale: × 0·5.)

27. *Calluna:* form of plant developed after cutting to less than 1 cm above ground surface: a dense cluster of radiating branches. (Treatment at about 20 weeks after germination; plant photographed 20 weeks later.) (Scale: × 1·5.)

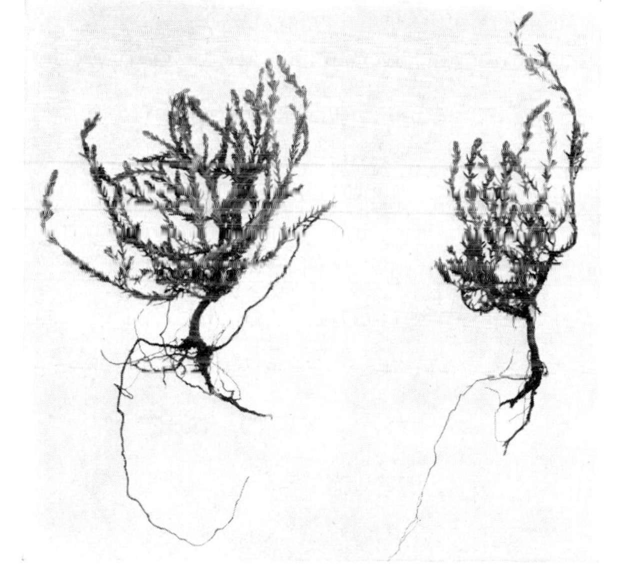

28. *Calluna:* young plants resembling Plate 27 collected from an area subject to heavy grazing. (Scale: × 0·5.)

29. Burning a stand of *Calluna* in the building phase. The fire is started using a special paraffin burner.

30. The fire progressing through the vegetation.

31. Production of clusters of new shoots from old woody stems at branching points, after cutting away or burning the upper part of the *Calluna* plant. (Scale: × 0·5.)

32. Vegetative regeneration in *Calluna*: cluster of new shoots produced near the stem base after cutting or burning above. (Scale: × 3.)

33. A thick deposit of ash on the ground between charred *Calluna* stems after a fire.

34. *Left:* the pattern produced by burning *Calluna* stands in long narrow strips. *Foreground and right:* peat erosion, possibly intensified by injudicious burning.

(*d*) The degenerate phase

This process frequently culminates at an age of between 25 and 30 (–33) years in the death of the central frame-branches, leaving a gap in the centre of the plant which widens as further branches die back. In this degenerate phase, the outer branches of the bush remain alive for a number of years, often lying flat on the ground like the spokes of a wheel, having perhaps been pressed down in winter time by the weight of snow on the shoots at their tips (Plates 22, 23). Frequently, they put out fine adventitious roots into the surface humus, and the effect may be to delay the onset of degeneracy in these branches. Hence the ring of green shoots, formed from these outer branches, may remain for a number of years surrounding the gap produced by the death of the central branches, the remains of which litter the ground. In time, all parts of the plant become moribund, and finally only a few dead branches mark the patch formerly occupied.

Cyclical processes

Changes are bound to occur in the occupancy of any particular patch of ground by plants. A. S. Watt (1947*b*) drew attention to the fact that, where these are not seral, they may in many communities be of a cyclical nature. Among the examples in which he demonstrated cyclical processes was the dwarf *Calluna-Arctostaphylos uva-ursi* heath of mountain habitats (Fig. 30). This was apparent in its simplest form in extremely wind-exposed situations where the semi-prostrate shoots of *Calluna* are able to spread only in one direction, into their own shelter. Often they are preceded by the even flatter shoots of *Arctostaphylos*, and the two form into long double strips of leaf-bearing shoots, growing uni-directionally in the direction of the prevailing wind. Any one spot of ground, at first bare, is successively covered by *Arctostaphylos*, then by *Calluna*. As the latter grows forward, so its branches are eroded on the windward side and the ground below is once more exposed. The trailing stems of both species may remain rooted and anchored some distance behind their slowly advancing front.

Watt also pointed out that in the absence of this environmental 'pressure' from one side, a similar cycle of events could be observed in more sheltered situations in which the components (often including *Cladonia arbuscula*) were arranged in a mosaic rather than in linear series. The interpretation was that the young vigorous shoots of *Calluna* are dense enough to exclude other species, but that as an individual plant matures its central branches tend to spread apart and create a gap in the middle of the bush. Here *Cladonia* can become abundant, and later the creeping shoots of *Arctostaphylos*, where present, will locate and occupy the developing gap in the dominant *Calluna*. (Fig. 32.)

This account recognizes that, except in the most wind-swept habitat, the vegetational changes are intimately associated with progressive changes in the habit of the individual *Calluna* plant at different stages in its development. At the time these were not fully analysed, but the sequence of four main phases as given earlier in this chapter was described by Watt in 1955 and data were presented showing that, in a Breckland heath, the occurrence and quantities of other species, including *Pteridium aquilinum*, varied with growth-phase of the individuals of *Calluna* sharing any particular patch of ground. If, as suggested by these observations, the competitive interactions between *Calluna* and other species change in the course of the sequence of phases, this will in part explain the occurrence of cyclical change in vegetational cover. Other species become associated with *Calluna* as it gets older, and will be available to expand in the gap caused by its eventual degeneration. As a result, re-establishment of *Calluna* on the same site may be delayed for a considerable period. In addition, *Calluna* has its own effects upon the substratum (p. 109), which are generally unfavourable for its own immediate regeneration. However, in the absence of colonization by other species a new generation of *Calluna* may become established in time.

It is probable that any heath community in which the environment is suitable for full expression of the sequence of growth phases, in the absence of disturbance, will exhibit vegetational change. Where this involves entry of species new to the community it is seral (Chapter 3), but otherwise it is cyclical.

A number of different schemes illustrating possible pathways for this type of cyclical change have been proposed on the basis of examination of pattern in the vegetation at a given time (Fig. 32). The species concerned vary according to the floristic composition of the particular community, for example in bryophyte-rich communities the ground beneath a developing gap in a mature *Calluna* bush is usually colonized early on by *Hypnum cupressiforme*. Later, the centre of the circular patch of *Hypnum* may be invaded by *Pleurozium schreberi* or *Hylocomium splendens*, as *Calluna* passes into the degenerate phase. Alternatively, in lichen-rich communities the gap phase may at first be characterized by a collection of *Cladonia* species, with patches of *Parmelia physodes* clinging to the dead or prostrate *Calluna* branches. The role of other vascular plants also depends on the extent to which they participate in the community generally: the gap may, for example, be colonized by any one of the following species – *Arctostaphylos uva-ursi, Erica cinerea, Empetrum nigrum, Deschampsia flexuosa, Vaccinium myrtillus, V. vitis-idaea*. The growth-form of the first three of these (pp. 118–22, 123) is such that new pioneer individuals of *Calluna* may be able

---

Fig. 32. Examples of cyclical change in *Calluna*-dominated heath communities.
(*a*) In a *Calluna–Arctostaphylos* community, east central Scotland (after Watt, 1947*b*).
(*b*) In a *Calluna–Vaccinium* community in north-east Scotland.
(*c*) In a dune heath, north-east Scotland.
(Redrawn from Gimingham, 1964.)

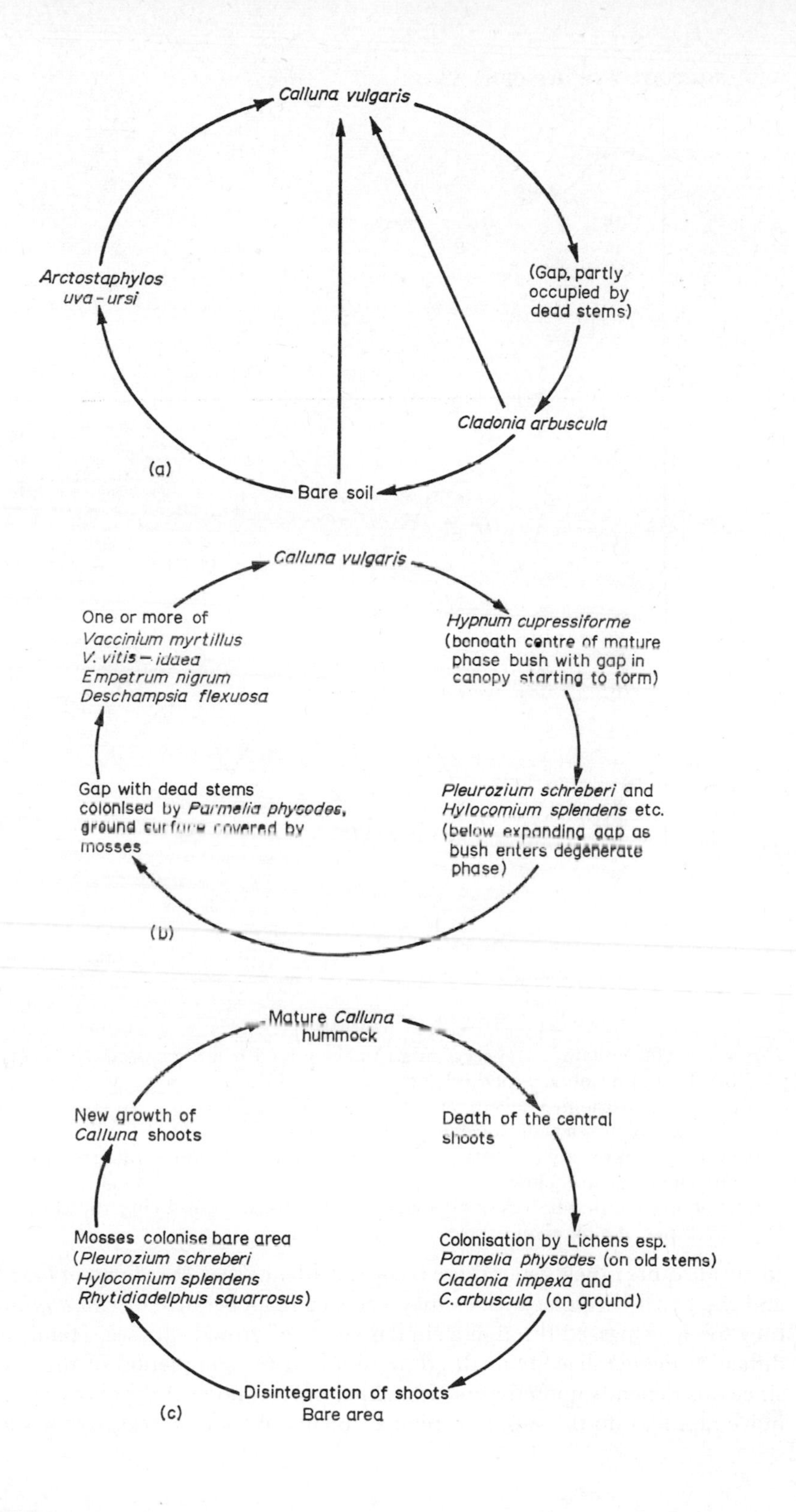
Calluna vulgaris
Arctostaphylos uva-ursi
(Gap, partly occupied by dead stems)
Cladonia arbuscula
(a)
Bare soil
Calluna vulgaris
One or more of
Vaccinium myrtillus
V. vitis-idaea
Empetrum nigrum
Deschampsia flexuosa
Hypnum cupressiforme
(beneath centre of mature phase bush with gap in canopy starting to form)
Gap with dead stems colonised by Parmelia physodes, ground surface covered by mosses
Pleurozium schreberi and Hylocomium splendens etc. (below expanding gap as bush enters degenerate phase)
(b)
Mature Calluna hummock
New growth of Calluna shoots
Death of the central shoots
Mosses colonise bare area (Pleurozium schreberi Hylocomium splendens Rhytidiadelphus squarrosus)
Colonisation by Lichens esp. Parmelia physodes (on old stems) Cladonia impexa and C. arbuscula (on ground)
Disintegration of shoots Bare area
(c)

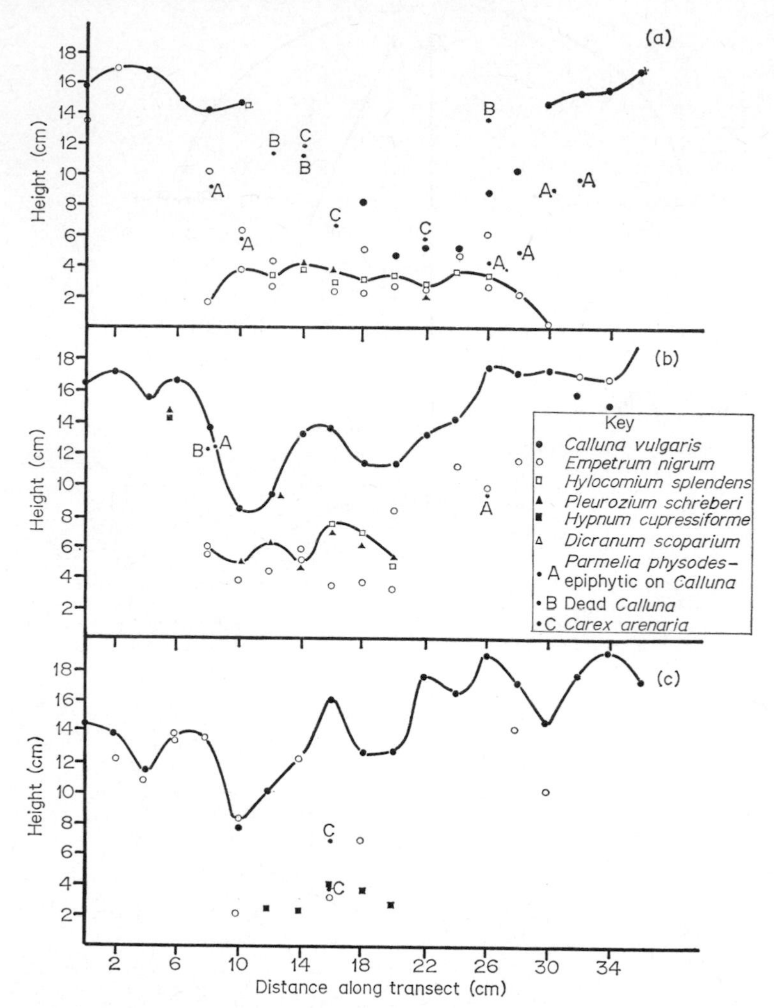

Fig. 33. Profiles of three transects in an uneven-aged heath community, recorded by the point-contact method (cf. Fig. 99.) (Data: Mrs E. M. Birse.)

(*a*) Transect through a degenerate plant of *Calluna*, showing a gap in the canopy, below which is a well-developed patch of mosses.

(*b*) Canopy reforming across a gap as young plants of *Calluna* colonize, leaving a patch of mosses below.

(*c*) Canopy continuous, the only evidence of a former gap being the continued presence of a patch of mosses.

to invade quite quickly, but there is some evidence that *Deschampsia flexuosa* and the two *Vaccinium* species may prevent reappearance of *Calluna* until they too have passed through a similar series of growth-phases. Hence it is difficult to generalize about the time required for completion of the cycle, since this depends upon the species taking part. If none of the above species move rapidly into the gap, new pioneer plants of *Calluna* reappear without the intervention of another vascular plant species, quickly re-establishing dominance of the patch (Plate 23). Their canopy merges with that of adjacent individuals, and the only trace of the former gap may be a surviving circular patch of mosses (often a patch of *Pleurozium schreberi* or *Hylocomium splendens*, surrounded by a peripheral fringe of *Hypnum cupressiforme*). Certain rather pure stands of *Calluna* which have been undisturbed for a number of years show a somewhat undulating canopy (in addition to scattered gaps). The depressions in the canopy level coincide with patches of mosses below, suggesting the positions of recently recolonized gaps (Figs. 9, 33). Such a cycle may well be complete in less than 50 years, whereas the period may be considerably longer when another vascular plant species is involved.

Evidence for vegetational cycles

It has to be admitted that the account of cyclical changes, as presented above, is based largely on the assumption that a repetitive pattern is indicative of dynamic processes. Direct evidence of this ideally requires long-term recording. However, since all elements of the pattern occur side by side in the same habitat, examples of each of these examined over a shorter time period may be expected to show changes which can be pieced together to verify the hypothesis of cyclical processes.

This approach was adopted by Barclay-Estrup and Gimingham (1969), working on a heath in the lowlands of north-east Scotland. The area had been free from burning and grazing certainly for about 40 years, possibly for close on 100 years. Associated with the uneven-aged stand of dominant *Calluna* were *Vaccinium myrtillus*, *V. vitis-idaea* and *Empetrum nigrum*, together with other vascular plant species and a number of bryophytes. The substratum was a shallow peat (about $\frac{1}{3}$ m) overlying glacial till. A marked pattern was evident in the vegetation, clearly associated with the occurrence of *Calluna* in each of its main growth-phases. The size scale of this pattern was generally rather in excess of 1 m^2, hence quadrats of this area could be marked out so that each was occupied largely by *Calluna* at a particular growth-phase, in most cases represented by a single individual bush. Repeated mapping over a short period of years provided direct evidence of dynamic changes (Fig. 34).

Changes in *Calluna* canopy

Cover contributed by *Calluna* increased rapidly in three years in quadrats containing pioneer *Calluna*. In those with building *Calluna* its cover was already in most cases almost 100%, but the canopy was still expanding in the younger building plants. Older building plants showed little change in the density of the canopy during the short period, but quadrats containing mature plants with signs of a gap in the centre all showed rapid enlargement of the gap and consequent reduction of cover in the centre of the bush.

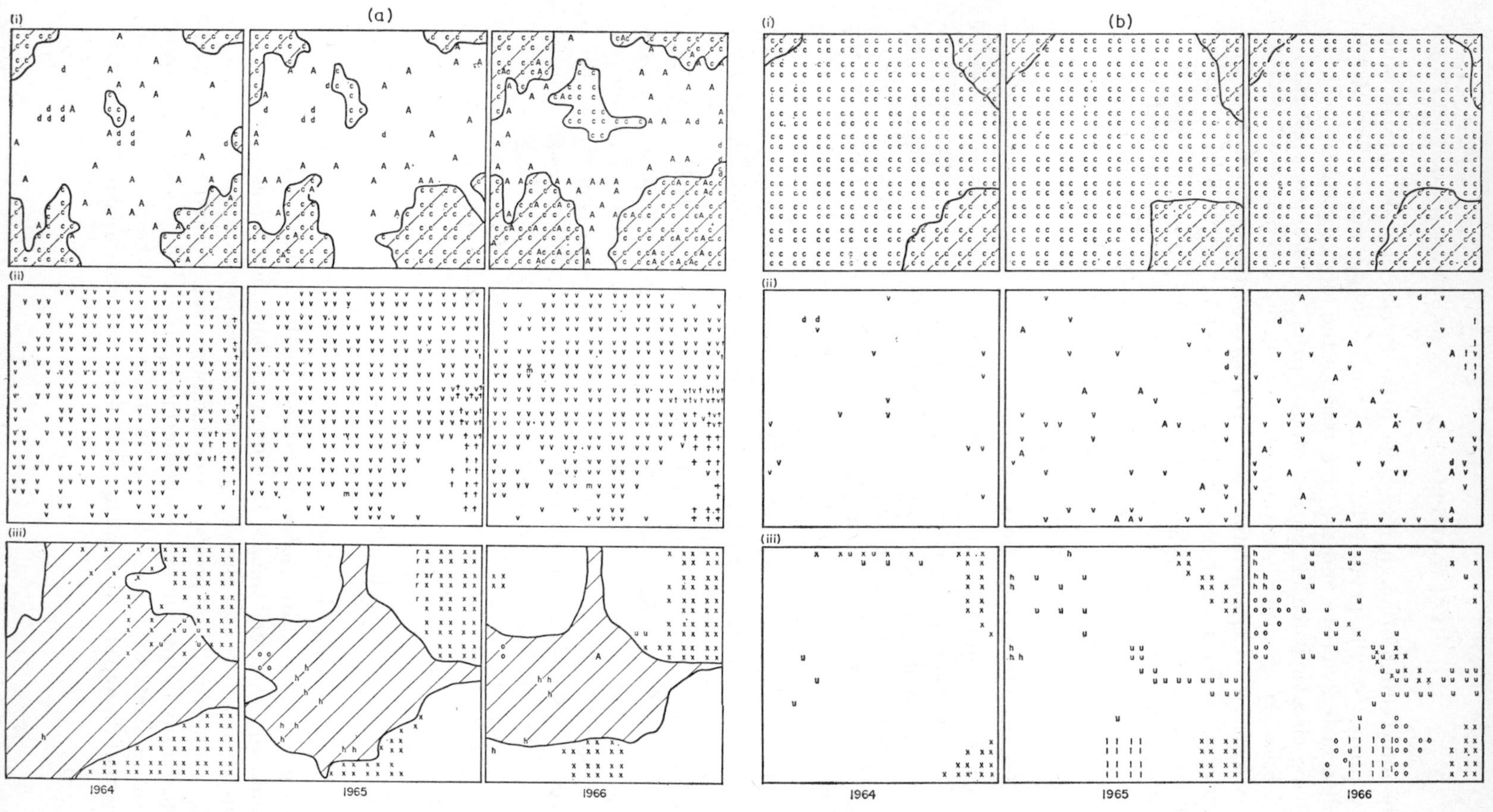

Fig. 34. Caption on page 134.

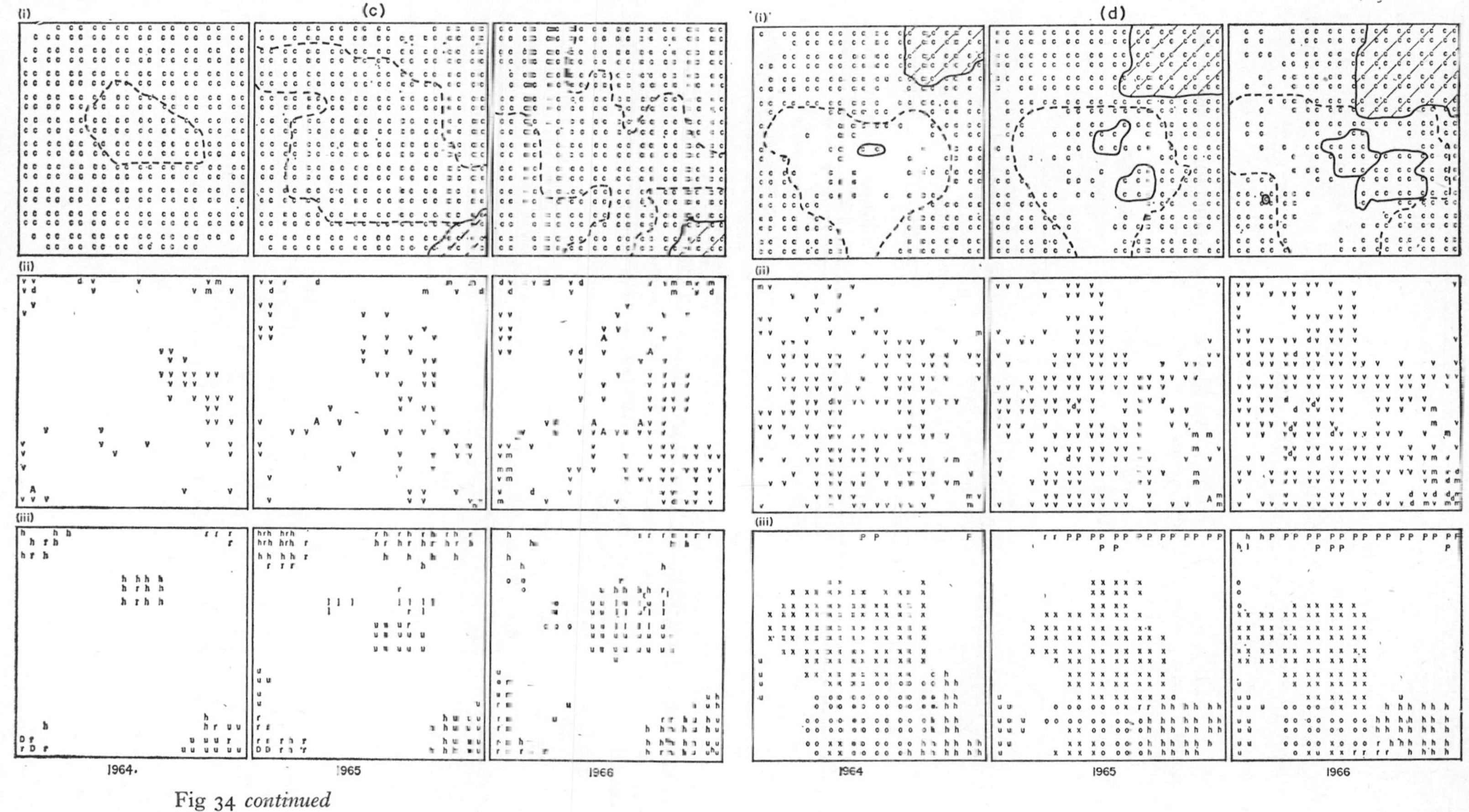

Fig 34 *continued*

This process was accelerated in winter when the central branches were pressed aside by snow. Degenerate plants were all characterized by a large central gap interrupted only by collapsed bare or dead stems. Cover was confined to the periphery and declined slowly.

Fig. 34. Quadrat maps in a *Calluna–Vaccinium vitis-idaea* community to show vegetational change related to the sequence of growth-phases in *Callnua*. (From Barclay-Estrup and Gimingham, 1969.) Each large square = 1 m × 1 m. Plant occurrences were recorded in a grid of small squares 5 cm × 5 cm: each symbol on the maps represents a record in one of these small squares.

Top rows (*i*): *calluna* (also, in (a), Gramineae and Cyperaceae).
Middle rows (*ii*): All other vascular plants.
Bottom rows (*iii*): Bryophytes and lichens.

(a). A quadrat containing pioneer *Calluna*, in three successive years.
Row (*i*): Unshaded outlined areas – pioneer *Calluna*.
Hatched areas – near-by plants in building phase.
Row (*iii*): Hatched area – continuous *Pleurozium schreberi*.

(b). A quadrat containing building *Calluna*, in three successive years.
Row (*i*): Most of the quadrat is occupied by one building phase individual of *Calluna*.
Hatched areas – near-by plants in mature phase.

(c). A quadrat containing mature *Calluna*, in three successive years.
Row (*i*): Broken line marks the area of an expanding gap in the *Calluna* canopy.
Hatched areas: a near-by plant in mature phase.

(d). A quadrat containing degenerate *Calluna*, in three successive years.
Row (*i*): Broken line marks the area of the gap in the centre of the bush, within which a continuous line denotes the area occupied by newly arrived pioneer individuals. Hatched areas: a near-by plant the building phase.

Symbols:
C – *Calluna vulgaris*
V – *Vaccinium vitis-idaea*
m – *V. myrtillus*
t – *Erica tetralix*
A – *Eriophorum angustifolium*
d – grasses and sedges – *Deschampsia flexuosa, Agrostis canina, Eriophorum vaginatum, Carex nigra.*
X – *Sphagnum* spp.
P – *Polytrichum commune*
o – *Hypnum cupressiforme*
h – *Hylocomium splendens*
u – *Plagiothecium undulatum*
r – *Pleurozium schreberi*
L – *Cladonia spp.*
D – *Dicranum scoparium*

Changes in other components of the community

Figure 35 illustrates the marked differences in the cover contributed by

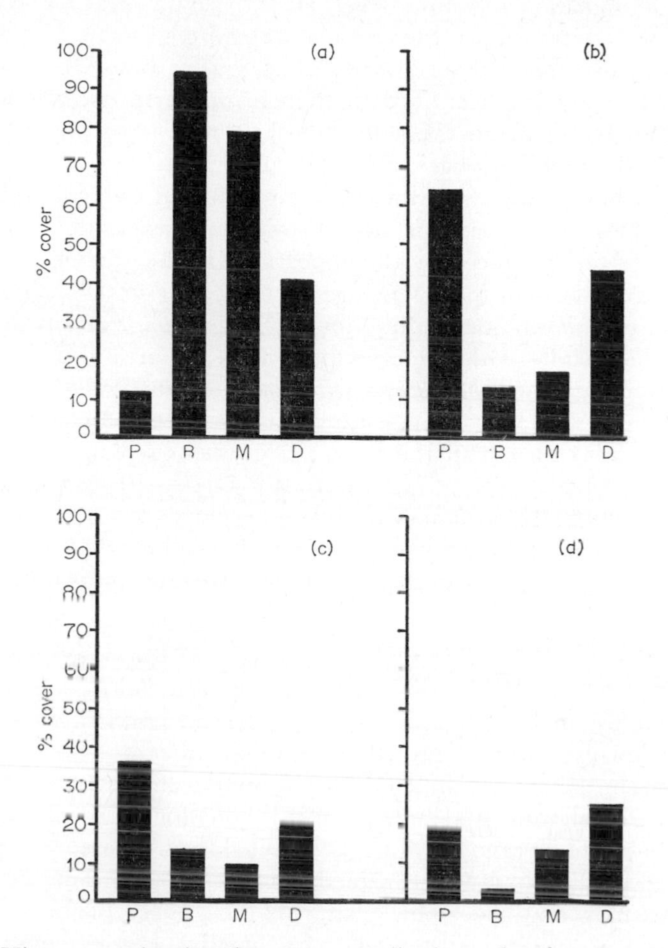

Fig. 35. Histograms showing the cover contributions of various categories of the vegetation in areas occupied by *Calluna* in each of its growth-phases (data from nine 1 m^2 quadrats). (From Barclay-Estrup and Gimingham, 1969.)
(*a*) *Calluna vulgaris*
(*b*) Other dwarf-shrubs
(*c*) Gramineae, Cyperaceae, Juncaceae
(*d*) Bare ground and lichens
P – Pioneer phase, B – Building phase, M – Mature phase, D – Degenerate phase.

other species associated with *Calluna* in its various growth-phases. That these represent points in a series of changes is shown, for example, by the decrease in bryophyte cover during a three-year period in quadrats occupied by pioneer *Calluna*: similarly in one example a thriving colony of

Cladonia spp. had almost disappeared in this time. Bryophyte cover was small under building *Calluna* plants, but was increasing quite rapidly where a gap had started to form in the centre of a mature bush, allowing increased illumination at ground level. Here there was nearly 100% cover in the bryophyte stratum in the late mature and degenerate phases.*

Other vascular plants (other dwarf-shrubs, grasses and sedges) were well represented amongst pioneer heather: there is some evidence of a decline in cover as the *Calluna* plants expanded. By the time *Calluna* was at its most vigorous in the building phase, other vascular plants were much reduced and still declining. However, where the development of a gap indicated a change to the mature phase an increase was recorded from year to year. This increase was produced by the appearance in the gradually enlarging gap of one (or more) of the following: *Empetrum nigrum, Vaccinium myrtillus* or *V. vitis-idaea* – depending upon which species was already established in the vicinity. Other species present in the area also contributed increased cover, e.g. *Eriophorum angustifolium*. Not unnaturally, continuance of these changes was clearly demonstrated in the centre of degenerate plants.

These records substantiate the claim that a series of changes in fact takes place. It is, perhaps, misleading to portray them as separate stages, as in Fig. 35. Undoubtedly, a number of linked developments occur as the gap begins to form in the centre of a *Calluna* bush, including increases in bryophyte (or lichen) cover often concurrent with entry of other dwarf-shrubs.

Changes in biomass, shoot production, cover and litter fall

In the foregoing paragraphs, morphological characteristics have been used to designate the main phases in the life history of an individual plant of *Calluna*. Equally, rates of growth and production of dry matter vary throughout the life of the plant, and may be used to characterize the phases. These in turn determine the biomass of the individual at any point in time and the density of its canopy (which is reflected in any measure of per cent cover). Barclay-Estrup (1970) obtained measurements of biomass and net production in young shoots for plants in each phase (Table 14) using quadrats of $\frac{1}{4}$ m^2, a size small enough to be contained completely within the area of one individual (except in the pioneer phase). For each phase, ten samples were harvested after the end of the growing season, to provide an estimate of total above-ground biomass at the beginning of winter. As an index of net production of young shoot material during the season, the current year's increments in long- and short-shoots were separated in a subsample of about 40% of the total biomass.

In the pioneer phase the biomass of *Calluna* per unit area is low, but in relation to this the rate of production of new shoots is maximal. The actual

* In a series of stands in east Yorkshire representing the complete sequence of growth phases of *Calluna* from pioneer to degenerate, Coppins and Shimwell (1971) found a progressive increase in bryophyte biomass.

TABLE 14
Height, biomass and production of *Calluna* in each of 4 growth-phases.
(Means from ten samples per growth-phase)
(From Barclay-Estrup, 1970)

	Pioneer	Building	Mature	Degenerate
Mean Height (cm)	24·1	52·1	63·2	55·2
Biomass (g m^{-2})*	287·2	1507·6	1923·6	1043·2
Net production of young shoots (g m^{-2} in one year)	148·8	442·4	363·6	140·8

quantity of dry matter laid down in new shoots per unit area is greatest in the building phase, and this leads to a peak in biomass when the plants are mature. With reduced production in the degenerate phase and death of central branches both quantities are reduced.

The effect of these changes upon cover can be seen from measurements made by the point method in areas occupied by *Calluna* at each of its growth-phases (Table 15) (Barclay-Estrup and Gimingham, 1969).

TABLE 15
Mean cover of *Calluna* in each of four growth-phases

	Pioneer	Building	Mature	Degenerate
% Cover	12·0	93·8	78·4	41·3

These clearly demonstrate the rise to maximum cover during the building phase when the greatest production of new shoots per unit area is taking place and the canopy is consequently at its most dense. Through the remainder of the plant's life its cover contribution declines.

From the second or third year of its life, components of the shoot system are shed by the plant and contribute to the accumulation of litter on the soil surface. These include the weaker long-shoots and branches, but the bulk of the litter consists of short-shoots shed after their second or third season's growth and seed capsules with the remains of floral parts. Litter production therefore depends upon the production of short-shoots and upon die-back. Hence, the rate of litter-production is another feature of the behaviour of the plant which varies in relation to growth-phase. This was shown by Cormack and Gimingham (1964), working on dune heath (Forvie, Aberdeenshire) and upland heath (Kerloch, Aberdeenshire). Litter was trapped below a number of plants representing each of the four phases, and was

* In this Chapter, quantities are given as g m^{-2} because the values refer to patches of small extent in a mosaic containing a mixture of growth phases. In later chapters where discussion turns to generalized levels of biomass and production, etc., applicable to larger areas, quantities are given as kg ha^{-1}.

collected monthly, sorted and weighed. It was shown by this method that the amounts of litter deposited annually per unit area increased throughout the life of the plant, to a maximum in the degenerate phase.

A subsequent chapter (Chapter 8) is devoted to a fuller treatment of biomass and production, including further discussion of the amounts of litter shed.

Associated changes in the micro-environment

Although each of the four phases in the life-history of a *Calluna* plant has its own visible and distinctive characteristics, the growth-phases are not to be regarded as sharply delimited either in appearance or in time. But while the divisions between them are arbitrary, the differences between plants at various stages in the sequence are genuine and important. Furthermore, as already indicated there are accompanying changes in the soil and micro-climate. These have been investigated recently in some detail (Barclay-Estrup, 1971), and as they are fundamental to an understanding of cyclic vegetational change in heaths, a brief account follows.

(*i*) Illumination at ground level

In the pioneer phase, individuals of *Calluna* are small and play little part in restricting illumination at ground level. Photocell readings indicate that as much as 75% of the available light reaches the ground among pioneer plants, the exact proportion depending largely upon partial shading by adjacent vegetation. However, under the dense canopy of building plants, illumination is reduced to 2%, or even less, of its value in the open. As the centre of the bush begins to open out when the plant becomes mature, this value rises gradually, remaining for a time between 17 and 20%. In the degenerate phase, illumination in the centre of a large gap reaches a level similar to that of the pioneer phase (Fig. 36).

(*ii*) Temperature

Measurement of temperature at ground level again emphasizes the similarity of the conditions associated with the pioneer and degenerate phases, which lack a continuous *Calluna* canopy. Insolation of the surface of the soil (or low-growing vegetation such as bryophytes and lichens) and heat loss by radiation may each be considerably greater than in the building and mature phases. Hence, areas containing pioneer *Calluna* show the greatest extremes of high and low temperature, being normally the warmest areas during the day (except in mid-winter) and the coldest on clear nights. The highest temperature recorded at the soil surface among pioneer heather in a heath on peat in north-east Scotland in 1964 was 41°C, though considerably higher temperatures have been recorded on open sites in heathland (50–60°C: Vaartaja, 1949; Stoudjesdijk, 1959).

In contrast, the environment below plants of the building phase is least subject to extremes. Summer maxima normally remain below 25°C in the study area in north-east Scotland, and winter minima seldom exceed −4°C. Conditions remain much the same under mature plants, becoming slightly more variable. Hence, in these phases the temperature regime is generally more equable, remaining cooler in warm weather and warmer in

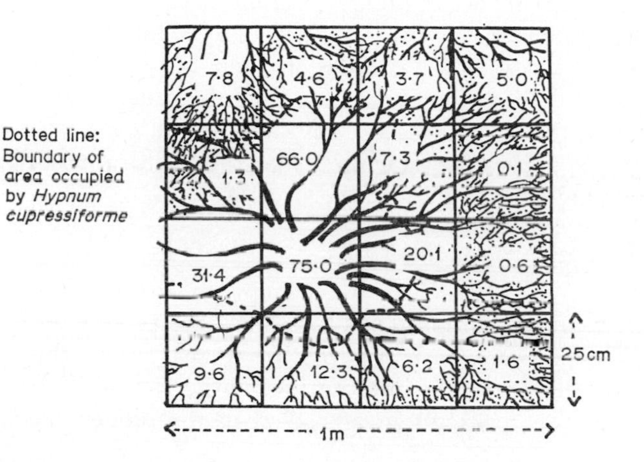

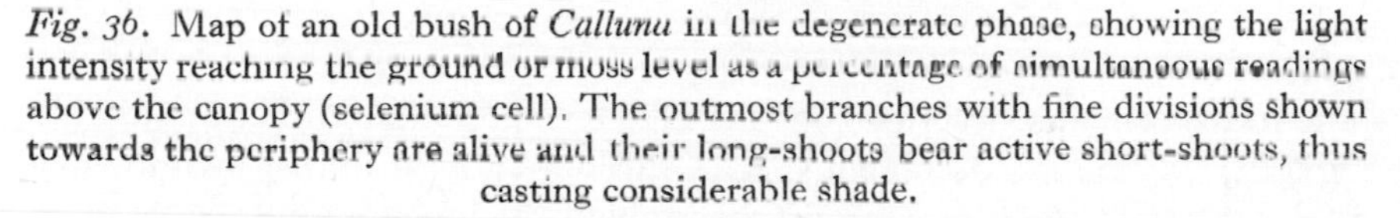

Fig. 36. Map of an old bush of *Calluna* in the degenerate phase, showing the light intensity reaching the ground or moss level as a percentage of simultaneous readings above the canopy (selenium cell). The outmost branches with fine divisions shown towards the periphery are alive and their long-shoots bear active short-shoots, thus casting considerable shade.

cold. In the warmer weather, the air in contact with the soil surface in the pioneer and degenerate phases warms up to temperatures well above those of the ambient air, and the profile shows a relatively steep gradient. Though less marked, the same applies in the centre of the gap in a degenerate plant. In the other two phases, particularly the building, the gradient is much reduced. Sometimes there is a localized warming of the air in the region of the canopy, where it has been in contact with the dense shoots whose temperature has been raised by the radiant energy of the sun (p. 50). All gradients are reduced or absent in cool, dull weather.

(*iii*) Air movement

The irregular canopy of a patchy stand of *Calluna* greatly reduces the rate of air movement at ground level, even in windy weather. It is unusual for conventional anemometers to show any measurable lateral air speed at about 20 cm above the soil surface under a canopy of building *Calluna*.

The shelter provided by mature plants is only slightly less effective. However, where breaks in the canopy occur, corresponding to the incidence of pioneer or degenerate plants, air circulation is detectable.

(*iv*) Saturation deficits

The development of saturation deficits in the air-layer close to the ground is related to the growth-phase of *Calluna* plants in ways which are predictable from the effects already described. In cool oceanic climates such as that of Scotland the incidence of high saturation deficits is relatively infrequent and seldom prolonged. However, deficits of 8–10 mm Hg sometimes occur in the ambient air as on a warm day in June, but under building *Calluna* there is generally little sign of increase from the usual level of 2–3 mm Hg. Exceptionally, there may be a rise to 7 mm, but in contrast in the pioneer and degenerate phases the same layer may show saturation deficits of 10 to over 14 mm, higher than those of the air above.

(*v*) Precipitation throughfall

Differences in the density of the canopy in the different phases are also reflected in the proportion of rainfall intercepted. Only very preliminary observations have as yet been made on this effect, but evidence was obtained that interception by the canopies of building and mature plants was markedly greater than in the pioneer and degenerate phases.

Summary

The measurements described in this chapter show that while, in suitable habitats, *Calluna* is a dominant species this does not mean that its influence upon the environment is constant throughout the life of an individual. The micro-habitat conditions change considerably during the sequence of growth-phases, and this alone affects the occurrence and quantities of associated species. Where an even-aged stand of *Calluna* has developed following burning, the micro-climate will be rather uniform throughout the stand at any one time, and will change gradually as the stand ages. Correspondingly there are gradual changes in the contribution to the community made by associated species, although at any one time they may be distributed rather uniformly throughout. In contrast, an uneven-aged stand is characterized by a mosaic or pattern in the occurrence of associated species corresponding to patterns in the values for microclimatic parameters, since all phases of the *Calluna* cycle are represented side by side.

The dispersion of associated species is doubtless in part a reflection of this pattern in the micro-environment. However, in so far as some of these species depend upon the same resources of nutrients, water, etc., as *Calluna*, their occurrence may be controlled by the competitive vigour of *Calluna*, which changes as it passes from one growth-phase to the next. These influences are difficult to apportion precisely, but their integrated results are evident.

Barclay-Estrup (1971) has summarized the main changes in vegetation and environment during the cycle as follows:

Pioneer phase

Calluna vulgaris individuals up to three to ten years of age; contributing about 10% cover.
Remaining cover from bryophytes (abundant) and other vascular plants, here at their maximum.
Biomass of *Calluna* low, but net production in young shoots, although low per unit area, is high in relation to biomass.
Microclimate characterized by extremes:
illumination at ground level high;
temperatures high on days of strong insolation, night and winter temperatures often low;
saturation deficits sometimes high;
air-movement at a maximum;
through-fall of precipitation maximal.
In the pioneer phase the community dominant has minimal influence upon the rest of the vegetation.

Building phase

Calluna vulgaris individuals 7 to 13 years of age; contributing about 90% cover.
Representation of associated species minimal.
Biomass of *Calluna* ranks second only to that in mature phase; net production in young shoots at maximum.
Microclimate profoundly affected by the closed, dense canopy of *Calluna*:
illumination at ground level reduced to as little as 2% of outside light;
temperatures generally lower than at other stages, but higher during the night and briefly during winter;
saturation deficit always low;
air-movement negligible;
interception of precipitation at a maximum.
In the building phase, *Calluna* exerts its greatest effects upon the environment and excludes most other species.

Mature Phase

Calluna vulgaris individuals 12 to 28 years of age, contributing about 75% cover. Increasing cover of other species, particularly bryophytes.
Biomass of *Calluna* at its maximum, net production in young shoots slightly less than in building phase.
Microclimate still strongly influenced by *Calluna* canopy, but showing some changes:
illumination at surface increases to about 20% of that in the open;
higher and lower temperatures than in building phase, but extremes not yet great;
saturation deficit still low;
air-movement still greatly restricted;
interception of rainfall still considerable.

In the mature phase, the vigour of the dominant *Calluna* is declining and its effects on the micro-climate are consequently lessened. Conditions become somewhat more favourable for other species, which begin to re-invade the area occupied.

Degenerate phase

Calluna vulgaris individuals 16 to 29 or more years of age, contributing now only 40 % cover.

Other species much more prominent, with bryophytes at their maximum.

Biomass of *Calluna* still relatively high compared to the pioneer phase; but net production in young shoots per unit area now at a minimum.

Microclimatic factors approach – sometimes exceed – the extremes of the pioneer phase:

- illumination at the centre of the gap up to 57 % of that in the open;
- temperature amplitude at surface approaching that of pioneer phase, soil temperatures often higher;
- atmosphere close to the ground subject to increased saturation deficits on warm days;
- air-movement greater;
- interception of rainfall reduced to a level approaching that of pioneer phase.

In the degenerate phase a gap develops in the centre of the *Calluna* bush and gradually increases in diameter. The decreasing dominance of *Calluna* results in conditions more suitable to a variety of associated species, and also to the establishment of *Calluna* seedlings which may then initiate a new cycle.

8 Biomass and production

In recent years estimates of biomass and annual production have been made on a number of examples of heath vegetation. These are important quantitative characteristics of the community, necessary for a full analysis of the ecosystem and as a basis for any discussion on the management of heathlands for domestic animals or game, their potential for improvement, or the desirability of conversion to some other form of land-use.

Biomass

Brief consideration was given in Chapter 7 to values for the above-ground biomass of a square metre of *Calluna* in each of its main growth-phases, as calculated by Barclay-Estrup (1970) from samples taken in a patchy,

TABLE 16

Vegetation biomass in relation to phase of *Calluna* life-history (Cammachmore, north-east Scotland, altitude 107 m) (Data from Barclay-Estrup, 1970)

	Phase of *Calluna* life-history represented in areas sampled (in parentheses, mean age of *Calluna* individuals in years)			
	Pioneer (5·7)	Building (9·0)	Mature (17·1)	Degenerate (24·0)
	Biomass (g m^{-2})			
All vascular plants (including *Calluna*) and bryophytes	889·2	1702·0	2305·2	1560·8
Calluna only	287·2	1507·6	1923·6	1043·2
Other dwarf-shrubs and grasses	179·6	41·2	52·0	83·2
Bryophytes	422·4	153·2	329·6	434·4
Height of *Calluna* (cm)	24·1	52·1	63·2	55·2
Depth of moss layer (cm)	6·0	2·4	4·1	6·3

uneven-aged stand on drained peat in north-east Scotland. The rest of the vegetation was also harvested to determine the weights of the various components of the community associated with *Calluna* at each particular phase. These are given in Table 16, where for purposes of comparison the *Calluna* data are repeated (cf. Table 14). Except in the pioneer phase, *Calluna* predominates in all samples, and therefore the total biomass of all plant material contained in the quadrats follows the trend shown by *Calluna* alone, rising to a maximum when *Calluna* is in the mature phase and declining as it becomes degenerate. However, all the associated species are much reduced in the areas occupied by building and mature *Calluna*. Vascular plants make their greatest contribution while *Calluna* is in the pioneer phase whereas the largest crop of bryophytes is found in samples representing the degenerate phase.

From these measurements it is evident that in an uneven-aged stand there is great variation in biomass from patch to patch, according to the growth-phase of the dominant. It follows that the value for biomass characterizing the stand as a whole will be determined by the proportion of the total area occupied by *Calluna* in each of its phases, or in other words by the age-structure of the stand. However, unless the ecosystem is in a 'steady state' and its age-structure is static, this value will change from year to year. An example of a steady-state ecosystem containing *Calluna* is discussed by Forrest (1971). This is a high-altitude *Calluna-Eriophorum vaginatum* community on hill peat at Moor House National Nature Reserve in the Pennines, northern England, and forms a part of the main British moorland site for productivity studies in the context of the International Biological Programme. Although referable to blanket bog and so only of marginal relevance to the present discussion, the biomass figures are quoted in Table 17 because of their interest in representing the steady-state, in contrast to those of Table 16 which vary in relation to the age of the stand. In the community analysed at Moor House, *Calluna* stems of all ages up to a maximum of 35 years were present: their mean age was 11·54 years and the modal age eight years.

The majority of biomass estimations in heath vegetation apply to even-aged stands which are not in a steady-state. This is partly because such studies have often been undertaken in connection with investigations of the development of managed heaths in relation to grazing, and partly because the reduced variance between samples lightens the task of obtaining reasonably accurate estimates (Gimingham and Miller, 1968). The results obtained are meaningful only when related to the age of the stand since burning and compared with estimates from the same stand at several stages of its development, or from other stands of different ages in the same habitat. As a stand ages the individuals of *Calluna* pass through the successive growth-phases, but instead of a patchwork in which all phases are represented side by side (as in Barclay-Estrup's stand) the changes are

TABLE 17

Biomass and production data in a *Calluna-Eriophorum vaginatum* community on hill peat, altitude 550 m, Moor House National Nature Reserve, north Pennines (Data from Forrest, 1971)

	Summer biomass (August 1968) (kg ha^{-1})		Production above ground (kg ha^{-1} year^{-1})	
ABOVE GROUND				
Calluna vulgaris				
Green shoots	3000		1300	
Wood	4400		380	
Standing dead	2290			
Total for *Calluna*		9690		1680
Eriophorum vaginatum				
Inflorescence	7		30	
Leaves	1480		560	
Total for *Eriophorum*		1487		590
Empetrum nigrum				
Green shoots	160		90	
Stems	260		20	
Total for *Empetrum*		420		110
Rubus chamaemorus		10		19
Vaccinium myrtillus		1		1
Listera cordata		3		3
Sphagnum spp.		1000		450
Other Bryophytes		30		20
Lichens		430		30
Total above ground		13 071		2903
BELOW GROUND				
Calluna vulgaris		8070		1830
Eriophorum vaginatum		3340		1620
Total below ground		11 410		3450
Grand total		24 481		6353

contemporaneous and at any given time all the individuals are in the same phase. It may be expected therefore that biomass values for unit areas of even-aged stands will be comparable to those obtained by Barclay-Estrup for the small patches occupied by *Calluna* in the equivalent phases (allowing for variation due to habitat differences, effects of grazing, etc). That this is generally true is shown in Table 18, where estimates of biomass are compared from stands of various age in different parts of Britain, sampled in summer-time. The range of values from 1000 to over 5000 kg/ha^{-1}* for two-year-old stands expresses differences in the rate of regeneration after burning, resulting from numerous factors (Chapter 10). As Chapman (1967) points out, where regeneration is largely vegetative and is rapid and uniform, the stand assumes the characteristics of the

(* see footnote on p. 137.)

TABLE 18

Biomass (kg ha^{-1}) of *Calluna*-dominated heath communities, sampled in summer, in relation to age of stand (years since burning)
The bulk of the biomass is contributed by *Calluna vulgaris* in all cases)

	North-east Scotland		South-east Scotland		North-east
Years since burning	Cairn o' Mount, Kincardineshire Alt. 274 m	Kerloch, Kincardineshire Alt. 152 m (*Calluna* only)	Polworth Moss Alt. 213 m	Listonshiels Alt. 305 m	Blanchland Moor, Hexham Alt. 305 m
1					
2		1010*		3238	5196
3	5070				
4			4011		
5					12 868
6		8360*			
7					
8	10 997		9234	13 366	
9					
10					'Over 10'
11					19 206
12		14 500*			⋮
13					
14					
15	'Over 15'				
16	29 261				
17	⋮	17 400*			
18					
19					
20					
21					
22					
23		18 520*			
24					
25		18 400†			
26					
...					
36					
37		20 960*			
38					
39					
40					
Data from:	Robertson and Davies (1965)	*Miller (personal communication) †Kayll, 1966	Robertson and Davies (1965)	Robertson and Davies (1965)	Robertson and Davies (1965)

England	South England		For comparison: samples from uneven-aged stands	
Teesdale	Dorset	Headland Warren, Dartmoor	(*a*) Cammachmore North-east Scotland. Alt. 107 m. 4 growth-phases sampled, mean ages as given	(*b*) Moor House, North England. Alt. 550 m. All age classes sampled: Modal age 8 years
Alt. 440–550 m (*Calluna* only)	Alt. 60 m	Alt. 425 m		
	1580			
	3370	3260		
	3780			
	4930			
6000	7230	6316	8892	
	8770			13 000
			17 020	
	9820			
		11 152		
	14 280			
20 000	16 690			
	20 080			
	18 153	22 176	23 050	
	23 570			
	21 660		15 608	
	22 220			
········	········	········	········	········
	24 700			
	27 780			
	21 840			
Bellamy and Holland (1966)	Chapman (personal communication) Community included *Ulex minor*, *Erica cinerea*, *E. tetralix*	Chapman (1967)	Barclay-Estrup (1970)	Forrest (1971)

building phase almost immediately and a true pioneer phase is hardly represented. Stands in the building phase show values of above-ground biomass for *Calluna* ranging usually from just over 6000 kg/ha^{-} to about 17 000 (sometimes up to 19 000) kg/ha^{-1}, while figures between 18 000 and 29 000 kg/ha^{-1} generally relate to mature stands.

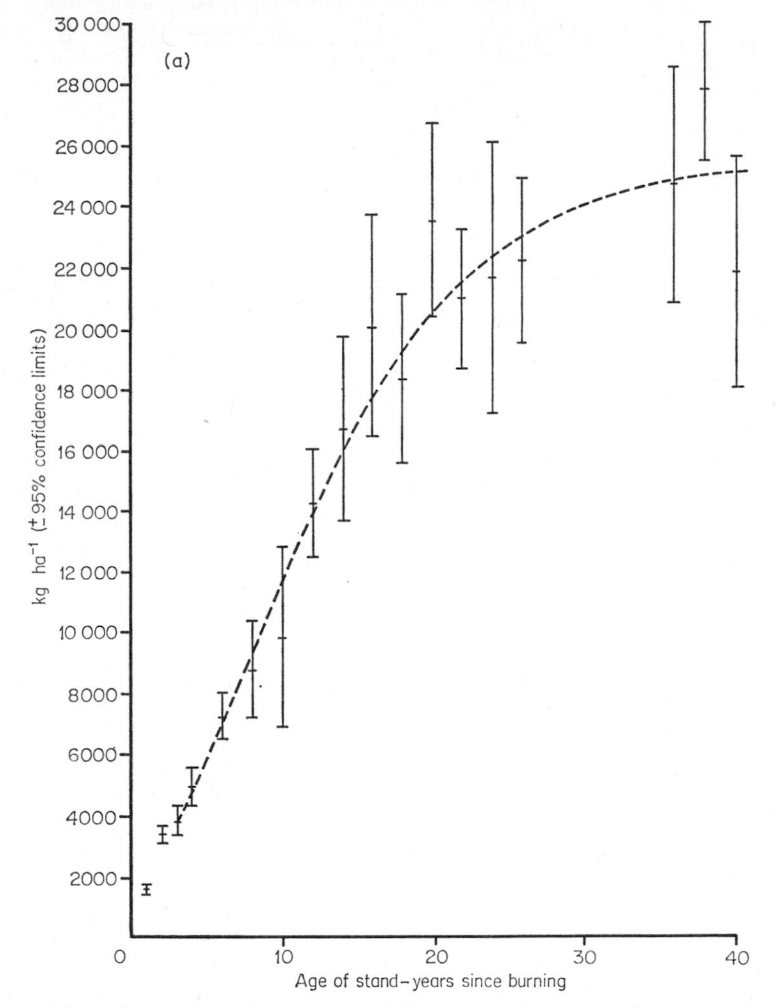

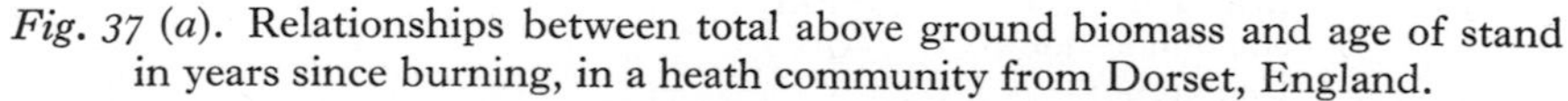

Fig. 37 (a). Relationships between total above ground biomass and age of stand in years since burning, in a heath community from Dorset, England.

Biomass estimations from a series of stands of increasing age in the same habitat provide a 'growth curve' applicable to the community concerned. S. B. Chapman has prepared one of the most comprehensive examples (Fig. 37) from Hartland Heath in Dorset, southern England. The curves reflect the growth-phases through which the individuals of *Calluna* comprising

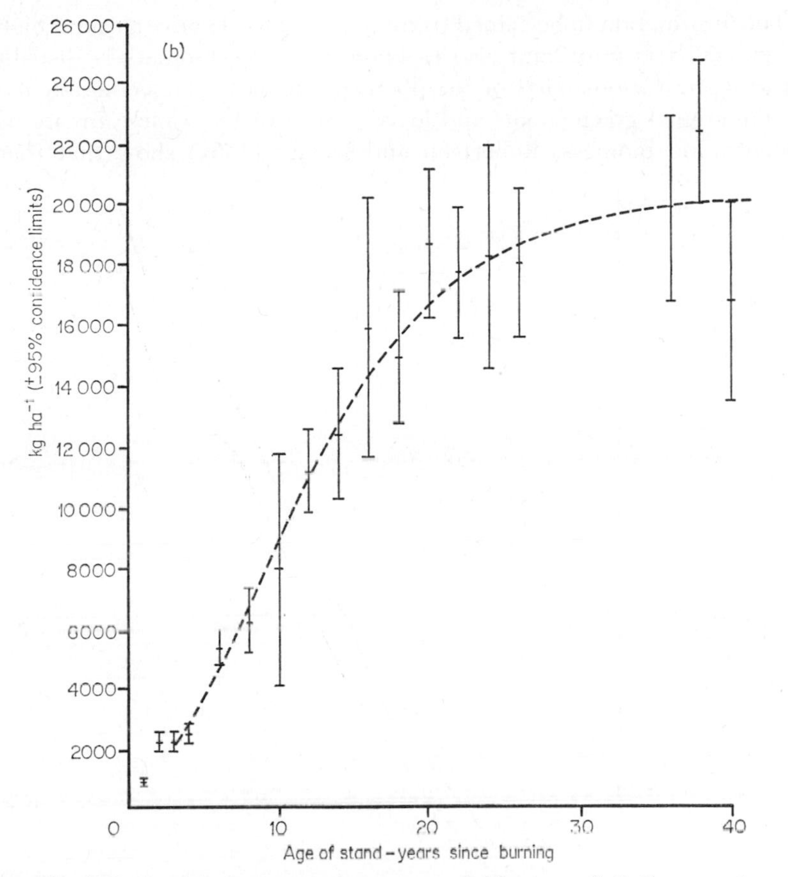

Fig. 37 (b). Relationship between above-ground biomass of *Calluna*, and age of stand, in the same community.

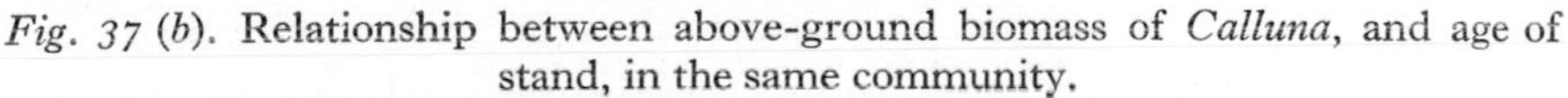
In both (*a*) and (*b*) Gomperts growth curves are fitted to the data from three years onwards, i.e 'after an initial rather rapid rate of increase has declined. (Figure by S. B. Chapman.)

the stands are passing; a true pioneer phase is lacking because of vegetative regeneration after fire but the building phase corresponds with the steep part of the curve, while the mature and degenerate phases are represented by a levelling-out. However, on most heaths, management by burning aims to prevent stands passing beyond the early mature phase. Sequences of biomass estimates have therefore perforce been limited, in many instances, to that part of the curve preceding the levelling-out. Bellamy and Holland (1966) for example, give a graph of biomass against age of stand (Fig. 38) which shows a continuing, approximately linear, increase up to 13 years after burning, corresponding with the part of Chapman's curve which represents the building phase.

The information to be gained from figures of total above-ground biomass is limited. It is important also to know how this biomass is distributed between the various parts of the plant, particularly the woody stems and the unlignified green shoots and leaves, and to have some estimate of the underground biomass. Robertson and Davies (1965) show that there is

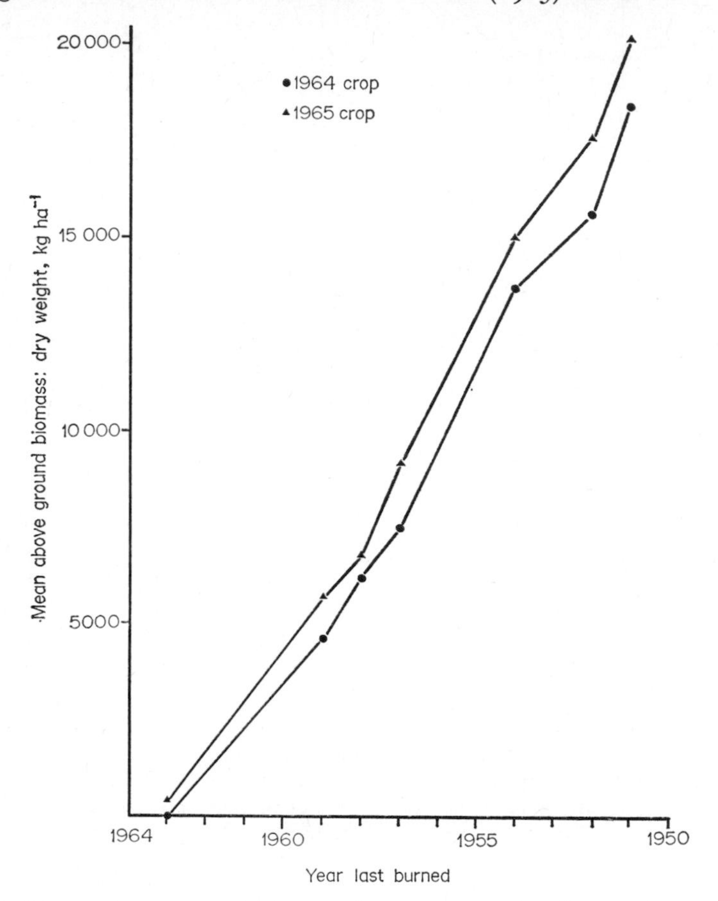

Fig. 38. Mean above-ground biomass from stands of *Calluna* of varying age, harvested in 1964 and 1965, plotted against the year in which the stand was burned. (After Bellamy and Holland, 1966; Teesdale, north England: altitude 440–550 m.)

progressive decrease with age in the ratio of the weight of unlignified green shoots with leaves to that of woody stems, from about 3:1 in young *Calluna* (2 to 4 years old) to about 1:6 in a 15-year-old stand. A fuller breakdown by Miller (personal communication) shows that between the ages of 2 and 12 years the proportion of young shoots (of the current and previous year's growth) drops from 81% to 21% of the total above-ground biomass, while the weight of stem increases from 12% to 62%. By the time the stand has reached an age of 35 years, young shoots account for only 14% of the biomass, while stems reach 75%. The remainder of the *Calluna* biomass is accounted for by dead shoots, rising from 3% to 22% at six years and declining slightly thereafter as the stand ages, and a small proportion contributed by flowers (the samples were taken in October).

Root biomass is notoriously difficult to estimate, but attempts have been made for *Calluna* by Forrest (1971) working at Moor House and Chapman (1970) working in Dorset (southern England). The data for Moor House are incorporated in Table 17 and show that in the wet hill-peat, a habitat certainly not optimal for *Calluna*, 45% of the total biomass of this species was located 'below ground' (this included buried stem bases as well as roots). On the sandy heath soils of the Poole Basin in Dorset, roots (of all species) amounted to about 70 000 kg ha^{-1} or 76% of the total biomass of the community (Chapman, 1970). The actual weights of root extracted per unit area from the soil below heathland stands of varying age showed no significant differences, and therefore as the bulk of above-ground plant material increases with age the ratio of root to above-ground vegetation decreases from nearly 8:1 in the youngest stands to about 3:1 in older ones. Chapman points out that as burning removes only the above-ground parts the root biomass may remain relatively stable throughout the stages of vegetative regeneration (p. 196).

Primary production

It is not as easy to obtain reliable estimates of production in perennial, evergreen, woody vegetation as it is, for example, in annual crops or species where aerial parts die away completely each year. However, during the first few years of regeneration after burning, both litter-fall and wood formation are negligible, so the current year's shoot-growth approximates to annual dry-matter production. At this stage cover is usually incomplete, and great variation may be observed in the figures for annual production by *Calluna*. These depend upon the nature and rate of regeneration (pp. 192–6), which themselves are determined by the age of the stand before burning, the conditions during and after the fire, and the various habitat factors. Miller (1969) notes that where regeneration is slow, production may amount to only 50 kg ha^{-1}, increasing gradually during the first ten years. On the other hand, when regeneration is more rapid the annual yield of shoots is of the order of 200 kg ha^{-1} in the first year, increasing about ten-fold in the next four or five years as cover becomes nearly continuous. However, regeneration is sometimes so rapid that an almost closed community is produced in the first year, as in certain stands investigated by Chapman in the south of England. Here, the dry weight of the shoots of *Calluna* produced in the first growing season was about 2000 Kg ha^{-1}, a figure almost equivalent to the annual production of new shoots and flowers in Miller's four- to six-year-old stands (i.e., late pioneer or early building phase) in north-east Scotland.

A recent experiment shows the effects of soil type (peat or mineral soil) on the regeneration and subsequent production by *Calluna* and other species in the few years immediately following burning on a heathland in north-east Scotland (Table 19). The influence of grazing is also shown, which at this stage reduces both the standing crop and production in *Calluna,* though not in the grasses and sedges. In older stands new shoots formed in the current year account for only part of the net production of dry matter, the rest of which is incorporated, principally as addition to the wood, throughout the stem and branch system established in previous years. However, in *Calluna*, the former fraction is easy to separate and hence to sample and measure. Furthermore, young shoots of the current season's growth comprise most of the food of herbivores grazing on *Calluna*. The dry-weight increment in this category therefore provides an index of the production of edible material, on which several investigations have concentrated. The quantity may be measured directly, and where plants are protected from grazing one set of samples harvested at the end of the growing season is adequate. Under grazing conditions, however, movable cages harvested at not more than monthly intervals may be required. Since variability in the proportion of current growth in whole shoots cut at ground level is low in even-aged stands (10–20% of the mean, Gimingham and Miller, 1968), this proportion may be determined in small sub-samples of the total harvest of clipped quadrats, and the weight of the current year's shoot-growth per unit area calculated from the total biomass of the samples. If a single season's shoot production only is being measured, no separate estimate of litter-fall is required since this is negligible from the new shoots if harvesting is carried out before flowers are shed.

On this basis, Miller and Miles (1969) found that the level of annual production of new shoots (and flowers) reached by the end of the pioneer phase (p. 151) was sustained thereafter. No appreciable decline was observed in stands of up to 30 years of age. His mean figures, shown in Table 20, vary between 2200 and 2800 kg ha^{-1} $year^{-1}$, with differences from year to year due in part to weather conditions during the growing seasons.* For comparison a mean of 2740 kg ha^{-1} $year^{-1}$ calculated from Barclay-Estrup's data for a stand located only a few miles from Miller's study area, shows good agreement. However, in this instance there was evidence of changes in rate associated with growth-phase, suggesting a peak in the building phase followed by progressive decline.

Very similar rates of annual shoot production have been found in heaths from markedly different habitats. Bellamy *et al.* (1969) recorded 2720 kg ha^{-1} $year^{-1}$ in a blanket mire ecosystem in Teesdale, at an altitude of

* A high correlation was found when shoot production was compared with an index of fine summer weather in the period from May to August.

TABLE 19

Effects of soil type and grazing on production and biomass in a heath stand regenerating after fire
Kerloch, north-east Scotland, Alt. 152 m. Burnt Spring 1966
(Data of J. B. Kenworthy, G. R. Miller, Miss A. M. Slater and C. H. Gimingham)
kg ha^{-1}

	Mean net production of *Calluna*					Biomass of *Calluna*								Biomass of *Erica cinerea*						
						Autumn	Spring	Autumn	Spring	Autumn	Spring	Autumn	Autumn	Autumn	Spring	Autumn	Spring	Autumn	Autumn	Autumn
	1966	1967	1968	1969	1970	1966	1967	1967	1968	1968	1969	1969	1970	1966	1967	1967	1968	1968	1969	1970
Shallow podsolic soil*	50	100	670	1200	1470	50	20	120	190	740	900	2[illegible]50	3800	10	10	80	120	410	1380	1450
Peat*	240	1120	1950	3060	2210	240	110	1420	1200	3390	2830	5[illegible]80	7800	10	—	70	20	50	160	210
Ungrazed plots†	110	540	1570	2310	1820	110	70	610	840	2500	2160	4[illegible]50	6500	10	—	80	100	370	1290	1190
Grazed plots†	180	680	1050	1960	1860	180	60	900	550	1780	1570	3[illegible]90	5100	10	10	70	40	90	260	470

	Biomass of grasses, sedges and rushes						
	Autumn 1966	Spring 1967	Autumn 1967	Spring 1968	Autumn 1968	Autumn 1969	Autumn 1970
Shallow podsolic soil*	50	40	400	460	930	860	740
Peat*	+	+	+	20	+	+	+
Ungrazed plots†	30	30	290	320	440	450	510
Grazed plots†	20	10	100	150	490	420	230

*Data for grazed and ungrazed plots combined. †Data from both soil types combined. +Present, but with insignificant biomass.

TABLE 20

Annual dry weight increment in new shoots (current year's growth) in *Calluna* in north-east Scotland (altitude 150 m)
(Data from Miller and Miles, 1969)

A. Mean production (kg ha^{-1} year^{-1}) in stands of different age, measured in 1964-8

	Age of *Calluna* (years)			
	13–17	17–21	23–27	36–40
Shoots	1880	2080	2210	1900
Flowers	290	450	570	460
Totals	2170	2530	2780	2360

B. Mean production in 1964–9 from stands of all ages.

	1964	1965	1966	1967	1968	1969
Shoots	2430	2070	1680	2070	1850	2130
Flowers	380	270	870	310	370	510
Totals	2810	2340	2550	2380	2220	2640

between 440–550 m. This figure represents all species of the community which, although on hill peat, was dominated by *Calluna*. In a neighbouring area where *Calluna* contributed about 70% cover, Forrest (1971) calculated 1300 kg ha^{-1} year^{-1} for *Calluna* alone. Bliss (1956) reported an annual rate of shoot production in tundra heath, dominated by woody perennials, of 2830 kg ha^{-1}.

Where a measure of total aerial production in heath vegetation is required, it is necessary either to obtain a separate estimate of the rate of yearly input to the woody parts of the dwarf-shrubs, to add to the figure for shoot-growth, or else to find the total production by the 'difference method'. Increments in the wood have been calculated from regression of wood dry weight on age for a large number of plants representing a wide age range (Forrest, 1971). The 'difference method' involves harvesting sets of samples at known time intervals, for example at the beginning and end of the growing season in the absence of grazing, or from movable cages at monthly intervals where grazing has an effect. Net above-ground production for a year, or a mean for net production over a period of years, can be calculated from the differences between successive samples. If litter production is also measured, the figures can be corrected to provide an estimate of total aerial production. In the more uniformly even-aged stands the number and size of samples needed for a statistically precise estimate is less than in uneven-aged stands, where the differences may be small in comparison with variance and as a result the number of samples required may be so large as to make the method impracticable. A further method for making an estimate of net annual production depends on the fact that burning management produces a number of even-aged stands of different age in close proximity to one another. Comparisons of the biomass of each stand, estimated at one time, allow calculation of a mean annual rate of dry matter accumulation. This type of estimate, however, is subject to error introduced by the inevitable differences in habitat and past history even between near-by stands.

Some examples follow of the application of these methods to the measurement of production in heath communities. Figures for net above-ground production are given for lowland heaths of the south of England by Chapman (1967) and Bellamy and Holland (1966). The former, working in Dorset, found a mean rate of about 1200 kg ha^{-1} $year^{-1}$ for the first ten years of development after burning, with a decrease to about 250 kg ha^{-1} $year^{-1}$ in stands aged between 20 and 30 years (i.e., mature to degenerate). Bellamy and Holland, referring to *Calluna* only (which however formed the bulk of the vegetation), obtained the rather higher values of 2080 kg/ha/year in Surrey and 2330 kg ha^{-1} $year^{-1}$ in Hampshire. At a higher altitude (440–550 m) in Teesdale, north England, these authors calculated a rate of 1530 kg ha^{-1} $year^{-1}$ for the net aerial production of *Calluna*, growing as the dominant in heath stands aged between 5 and 14 years on hill peat, using the differences in weight between sets of samples taken in 1964 and 1965 (October; Fig. 38). Comparative estimates by an alternative method were possible because the stands were of known age since burning, providing on each occasion sets of samples from a series of stands of increasing age. Calculation on this basis gave figures of 1780 kg ha^{-1} $year^{-1}$ (using 1964 data) and 1770 kg ha^{-1} $year^{-1}$ (using 1965 data). The results from the two methods are in reasonable agreement, bearing in mind that while the latter integrates the rate of accumulation over a period of years (using the straightest portions of the graphs, Fig. 38), the former gives an estimate for a specific time period. There was some evidence that 1965 was a bad year for the growth of *Calluna* in the area, accounting for a lower than average figure.

In the absence of measurements of litter fall, total annual production cannot be calculated from Bellamy and Holland's figures. However, Chapman found that in his Dorset heath the formation of litter increased steadily until the plants were about 20 years old (mature phase), when it stabilized at about 3000 kg ha^{-1} $year^{-1}$. From these data it was shown that the rate of total dry matter production above-ground rose from about 1600 kg ha^{-1} $year^{-1}$ during the first 6 years to about 3200 kg ha^{-1} $year^{-1}$ between 18 and 22 years after burning and about 3400 kg ha^{-1} $year^{-1}$ in stands aged 22–36 years.

In a very different type of community, the *Calluna-Eriophorum vaginatum* community of hill peat at an altitude of 550 m at Moor House, Forrest (1971) found similar rates of production despite a generally lower biomass.

The estimated rate of total dry matter production above ground was about 2900 kg ha^{-1} year^{-1} for *Calluna* alone, the population of which (as mentioned on p. 144) had a modal age of 8 years and contributed 70% cover. The figure for *Calluna* was made up of an estimate of 1300 kg ha^{-1} year^{-1} for 'green shoot' production and 380 kg ha^{-1} year^{-1} input to wood. In addition, below-ground production was estimated at 1830 kg ha^{-1} year^{-1}, giving a total for *Calluna* of 3510 kg ha^{-1} year^{-1}. Production estimates for the whole community are included in Table 17.

Tentative generalizations may be made from these investigations. In the first place, in the British climate, stands of *Calluna* between six and 20 years of age generally produce in a year from 2000 to 3000 kg ha^{-1} or perhaps more of new shoots (young long-shoots, short-shoots and flowering shoots). Figures for total aerial production are somewhat higher because of the additional wood increments. Secondly, although there is some variation in biomass among heaths of different environments, the rates of production are broadly similar. Where, as on the Moor House hill peat,

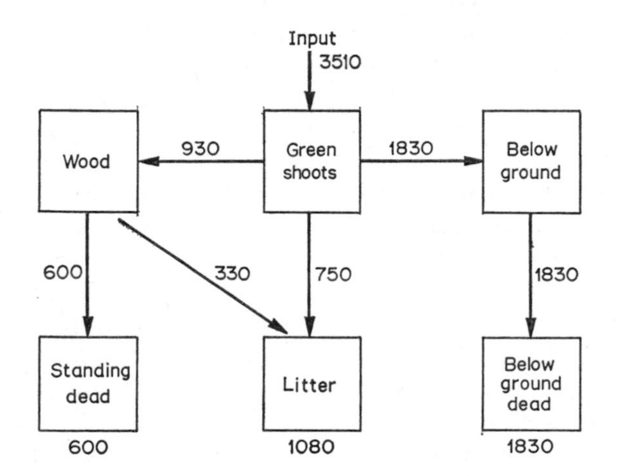

Fig. 39. Flow-diagram showing quantities (kg ha^{-1} year^{-1}) of dry matter transferred among various compartments of a simple model of *Calluna* at Moor House (*Calluna–Eriophorum vaginatum*) community; altitude 500 m. (From Forrest, 1971.)

Calluna itself contributes less than in the majority of heaths, the balance is made up by other species, in this instance *Eriophorum vaginatum* which has a relatively high turnover rate. Further reference is made later to production in heaths at high altitudes.

There is also a substantial annual production below ground, and this quantity is of great importance in regard to the transfer and accumulation of dry matter in various parts of the plant, as illustrated in the flow diagram for *Calluna* from Forrest (1971) – Fig. 39. Further data quantifying this component in dry heath ecosystem will be awaited with interest.

Litter production

Calluna litter comprises four main components: portions of woody stem, long-shoots, short-shoots, flower-buds, and flowers, the last consisting of capsules covered by the persistent calyx. Leaves are rarely shed separately, but remain attached to the long- and short-shoots. Litter fall from young plants is negligible until after the first flowering season (i.e., the second or even the third season of growth). Thereafter, as shown by Cormack and Gimingham (1964), the amount deposited annually per unit area increases as the plants get older and pass through the several phases of their life-history. It was found also in two habitats in north-eastern Scotland (an upland heath at 170 m above sea-level and a maritime dune heath) that, although litter is shed at all times of the year, there are two periods of maximum fall, namely, October–November and February. The peak in autumn is accounted for mainly by an increase in the quantity of short-shoots, and that in late winter by capsules. However, a rather different seasonal pattern of litter fall was reported from the Moor House site by Forrest (1971). Here, in 1968–9, the winter-peak of capsule-fall was somewhat earlier, in January, while the highest rates of short-shoot fall were sustained throughout the summer from June to October instead of showing a distinct peak in October and low summer levels, as in the Scottish results. The contribution of stem material to litter was greatest in January and July. At Moor House no litter fell during February and March owing to snow cover.

Doubtless some of the variation between these two sets of results may be the consequence of habitat differences, although in the earlier investigation it was only in the degenerate phase that there was any significant difference in litter fall between heaths in differing habitats. In this instance litter production was greater in a dune heath than in one on an upland peaty podsol.

Forrest collected 1082 kg ha^{-1} of *Calluna* litter in a year at Moor House in the stand of mixed age-structure, while Chapman records 1020 kg ha^{-1} (largely *Calluna*) produced by a nine-year-old stand, rising to 2881 kg ha^{-1} at 19 and 3180 kg ha^{-1} at 33 in Dorset, southern England (Fig. 40). Of the same order is an estimate by Mork (1946) of 2600 kg ha^{-1} for annual litter production by the *Calluna* dominated stratum of a forest community in Norway.*

Litter is therefore contributed to the soil surface at a rate which increases as stands age, but tends to become steady in the degenerate phase (Fig. 40). At the same time, decomposition is taking place as a result of the activities principally of mites, collemobola, and fungi. The older stems

* Mork assumed, for purposes of estimation, that annual litter production is equal to the dry weight of a year's production of leaves and short-shoots. No sampling of litter fall was undertaken.

may be infected, before death and incorporation into the litter, by parasitic fungi which continue saprophytically in the dead wood. An example is *Marasmius androsaceus* which spreads by means of fine rhizomorphs and

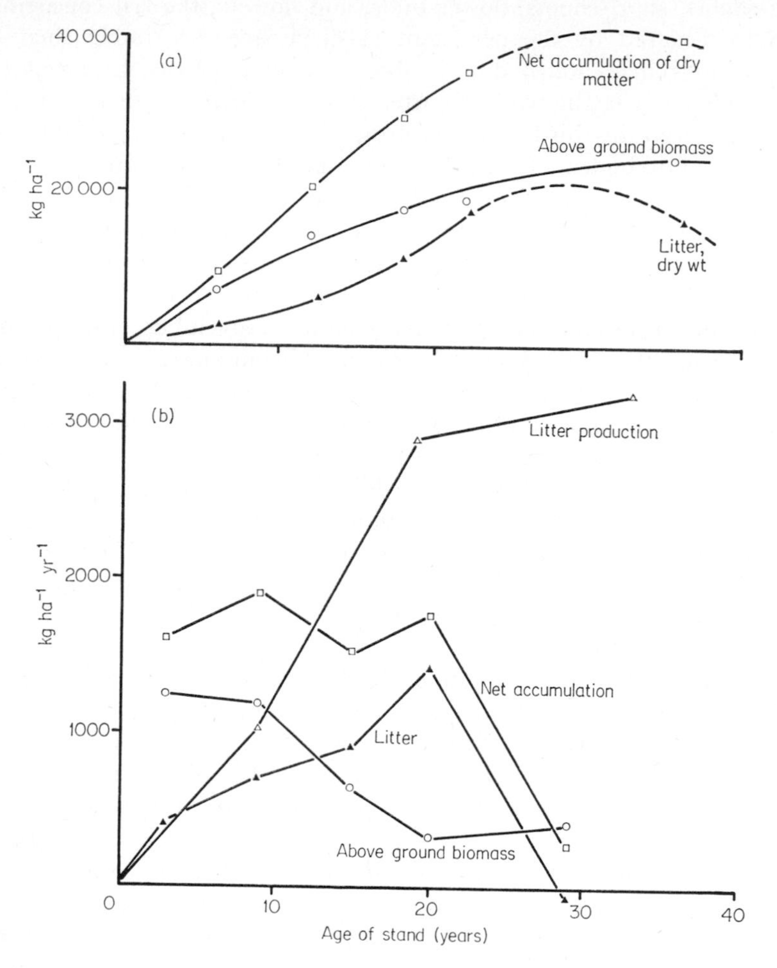

Fig. 40.
(*a*) Above-ground biomass, weight of litter and net accumulation of dry matter in stands of various age in Dorset, south-east England.
(*b*) Rates of accumulation of the above categories, and of rate of litter production. (From Chapman, 1967.)

produces fructifications from old *Calluna* stems near or on the ground (Macdonald, 1949). A measure of the decomposition rate of litter can be obtained by periodic weighing of litter samples contained in nylon mesh bags placed at a number of test sites. In the wet peat of the Moor House blanket bog site, O. W. Heal and P. Latter (personal communication) record the following percentage losses of dry weight in four successive years:

	1st year	2nd year	3rd year	4th year
Calluna shoots	15	29	35	40
Calluna stems	8	16	15	23

Losses varying from 5·5% to 14·5% over one year in a dune heath were found by Cormack and Gimingham (1964), and Chapman (1967) made similar measurements over a similar period under *Calluna* stands of various age in a Dorset heath, as follows:

	In 9-year-old stand	In 19-year-old stand	In 33-year-old stand
% loss of initial dry weight of litter sample in one year	5·4	9·8	12·8

These results show that decomposition rate increases as stands increase in age. Taken in conjunction with the tendency for rates of litter production to level off in the older stands, this indicates that its accumulation in the soil surface slows down, possibly even to the point at which the rate of decomposition equals or exceeds that of deposition (as suggested in Fig. 40). Production is also declining as the degenerate phase is approached, so the increase in biomass proceeds only at a reduced and more steady rate (p. 149). Hence, the net accumulation of dry matter in the above-ground vegetation and litter combined generally declines as stands exceed an age of about 20 years, perhaps ceasing altogether at about 30 (Fig. 40).

All these rates vary according to habitat conditions and no detailed research has yet been carried out on the precise effects of differences in the edaphic environment on litter accumulation, and hence on soil processes. The actual quantities of accumulated litter have been measured in a number of instances, and Table 21 gives an indication of the range of figures obtained. The determinations of weights of accumulated litter quoted here include components derived from all species of the community, although in each example *Calluna* was strongly dominant and contributed by far the larger part. This applies in many heath stands and consequently there is little information on the amounts of litter produced by species associated with *Calluna*. However, appreciable quantities are sometimes contributed by other species, particularly mosses or lichens.

Production at high altitudes

There is some evidence in the literature of a decline in production with increasing altitude. However, the differences are small compared to the

TABLE 21

Dry weight of accumulated litter (kg ha^{-1}) under *Calluna*-dominated communities in relation to age of stand (years since burning)
(In all these communities, *Calluna* contributed the bulk of the biomass and cover)

	North-east Scotland	South-east Scotland		North-east England	South England	
	Cairn o' Mount	Polworth Moss	Listonshiels	Blanchland Moor Hexham	Dorset	Headland Warren, Dartmoor
Age of Stand	Altitude 274 m	Altitude 213 m	Altitude 305 m	Altitude 305 m	Altitude 60 m	Altitude 425 m
2			Nil	Nil		Nil
3	1785					
4		Nil				
5				5367		
6						788
7						
8	4376	4201	3445			
9					4080	
10				'Over 10'		
11				6065		3748
12				⋮	6520	
13						
14						
15						
16	'Over 15'					
17	4848					
18	⋮				11 920	10 828
19					12 940	
33					16 900	
Data from:	Robertson and Davies (1965)	Robertson and Davies (1965)	Robertson and Davies (1965)	Robertson and Davies (1965)	Chapman (1967)	Chapman (1967)

TABLE 22

Production in dwarf-shrub heaths at high altitudes: Cairngorm mountains, Scotland.
(Dry weight of 'current year's growth': kg ha^{-1}).

	Calluna–dominated communities		*Empetrum*–dominated community containing *Calluna*			*Empetrum*–*Vaccinium* community		
	Cairngorm, 855 m	Cairngorm, 915 m	Beinn a' Bhuird, 915 m			Beinn a' Bhuird, 1005 m		
	1970	1970	1968	1969	1970	1968	1969	1970
Calluna vulgaris	1784	1632	228	38	105			
Empetrum hermaphroditum	23		922	624	470	605	437	547
Vaccinium myrtillus	77	43	91	108	165	283	199	230
Vaccinium uliginosum					7	11		1
Herbs*	374	460	166	247	128	610	374	893
Bryophytes†	294	170	136	114	99	97	30	52
Lichens	33	128	414	331	246	489	534	439
Total	2585	2433	1957	1462	1220	2095	1574	2162

*Representing the contributions *Trichophorum cespitosum, Carex bigelowii, Juncus trifidus, Lycopodium selago, L. alpinum.*
†Production figures are an approximation, derived by taking $\frac{1}{3}$ of total biomass at end of growing season (Traczyk, 1967).

variability of the several estimates from different habitats, and it is perhaps more remarkable that on the whole the rates of production of heath communities, and of *Calluna* itself, seem to be rather little affected by the conditions experienced at the higher altitudes. In this context, considerable interest attaches to recent estimates of production in the mountain heath communities of the Grampian highlands of Scotland by C. F. Summers (Table 22), at altitudes between 850 m and 1100 m. Here, in several different community types, leaf and shoot production in a year by all species was comparable with that at Forrest's Moor House site (Table 17) some 300 m lower. Taking *Calluna* alone, its rate of shoot production in communities of which it was the dominant (at 855 m and 915 m) was very close to that found by Bellamy and Holland in Teesdale at 440 m – 550 m, and only slightly less than rates commonly recorded at low altitudes (cf., Miller's figures from north-east Scotland and Chapman's estimates from Dorset, pp. 152–5). In communities where *Calluna* is reduced or absent, other dwarf-shrubs – especially *Empetrum hermaphroditum* and *Vaccinium myrtillus* – make increased contributions, but not to the extent of equalling *Calluna*. Other species, particularly lichens, help to maintain the general level of production in these examples.

That production rates are apparently rather little reduced with increasing altitude suggests a considerable degree of morphological and physiological adaptation in species belonging to these communities, and particularly in *Calluna*. These are currently the subject of further investigation.

Ecosystem modelling

Studies of production and the transfer of materials from one part of the ecosystem to another depend to a considerable extent upon the formulation of appropriate models to illustrate the main pathways of movement. One devised by S. B. Chapman (Fig. 41) for *Calluna* provides an informative example, and could serve as a basis for a variety of investigations.

The rapid advances in techniques of computer modelling are also relevant to the studies discussed in this chapter. As part of a contribution to the International Biological Programme, the possibility of constructing models to incorporate the results of the analysis of the *Calluna–Eriophorum* community at Moor House is being explored (Jones, Forrest and Gore, 1971). The object is partly to facilitate comparisons with other systems, and partly to investigate and predict the controlling effects on the ecosystem of factors such as nutrient supply, or burning.

As far as heathland ecology is concerned this development, of great theoretical and possibly of practical interest, is so far largely confined to the behaviour of *Calluna* under climatic and edaphic conditions on the fringe of the environmental range occupied by dwarf-shrub heaths. As regards climate, these conditions (at an altitude of 560 m) are likened

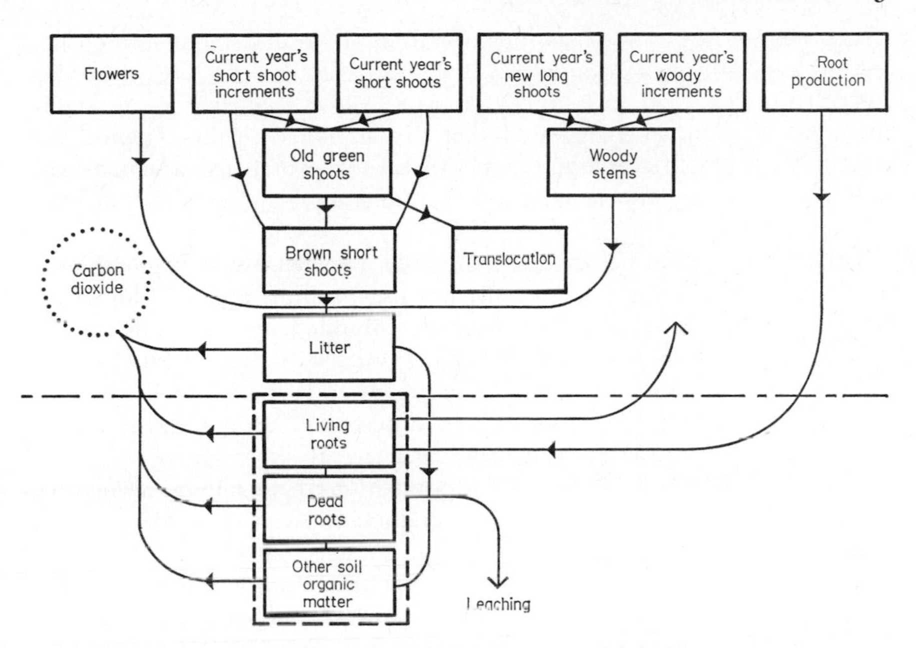

Fig. 41. A possible comprehensive compartment model of a *Calluna* stand, showing transfer pathways. Figure by S. B. Chapman.

by Manley (1942) to those of southern Iceland. The ecosystem analysis here is therefore relevant particularly to similar work in tundra and high altitude formations, with which it is associated in the I.B.P. framework. The results of investigations now in progress on the ecosystem dynamics of heathland vegetation nearer to the centre of its ecological range will therefore make an important further contribution in this field.

Herbivores

Brief mention must be made of the very numerous invertebrate herbivores associated with *Calluna*, although extended treatment is beyond the scope of this book. Species of many different taxonomic groups are represented, some by very large numbers of individuals. However, their biomass is generally small and only in isolated instances is consumption great enough to cause appreciable damage. One such pest of *Calluna* is the 'heather beetle', *Lochmaea suturalis*, which although widespread is localized and periodic in its outbreaks. Leaves, stem apices and the bark of young shoots are eaten by the larvae and also by adult beetles, resulting in patches or more extensive areas of severe damage to the plants (Cameron, McHardy and Bennet, 1944). Individual bushes in the mature and degenerate phases may be killed and attacks seem generally to be worst where

Calluna is not at its most vigorous, for example in old stands and on flat wet ground. However, *Calluna* on dry and freely drained sites may also suffer (Morison, 1963). A moist environment (*Sphagnum* or litter) is required but Morison suggested that dry summer weather favours the completion of the life-cycle and that outbreaks are worst after a sequence of two to three warm, dry summers. Exceptions to this generalization have also been noted.

Other invertebrate herbivores commonly associated with *Calluna* as a food plant include mites, weevils, thrips, psyllids and lepidopterous larvae. A provisional list of the insects, compiled by O. W. Richards, appears in the account of *Calluna vulgaris* for the Biological Flora of the British Isles (Gimingham, 1960). There is much variation in their representation in different localities and in different community-types. This is the subject of several extensive surveys currently in progress.

In contrast to the invertebrates, the populations of vertebrate herbivores on heaths are small, and secondary production is correspondingly low. The chief consumers are domestic grazing animals, particularly sheep and cattle, occasionally (as in the New Forest) ponies. Wild herbivores include rabbits (*Oryctolagus cuniculus*), brown hares (*Lepus europaeus*), and in upland districts mountain hares (*Lepus timidus*), red deer (*Cervus elaphus*) and the game birds, red grouse (*Lapopus lagopus scoticus*) and ptarmigan (*Lagopus mutus*), the latter at the higher altitudes.

In many areas the use of heathland for domestic grazing animals has been abandoned, and even where it continues there is great variation in stocking rates. Furthermore, as Miller (1964) points out, even where heaths provide most of the grazing for sheep or cattle, considerable use is made of other near-by vegetation and of supplementary feeding in winter. The populations of wild herbivores also vary greatly. For these reasons it is extremely difficult to arrive at any estimate of secondary production applicable to heath ecosystems, and this remains a subject which has been little studied. A few estimates have been made of annual cropping of livestock and game from heathland areas, and these may be taken as a very approximate guide to production levels. Miller (1964) calculated that at a stocking rate of one ewe to slightly more than 2 ha (a medium stocking rate for *Calluna* heaths) and with a lambing percentage of 95, the average yield of animal produce would not exceed 17 kg ha^{-1}. For the reasons stated above, only a proportion of this production can be attributed to the grazing provided by heath communities.

Some of the wild herbivores depend rather more exclusively on heath plants, particularly *Calluna*. However, Miller calculates that maximum figures for the 'harvest' of grouse would be unlikely to reach 2 kg ha^{-1} and for mountain hares would seldom exceed 3 kg ha^{-1}. Where heathland contributes to deer 'forest', the annual gross yield of red deer carcass might amount to less than 1 kg ha^{-1}.

These figures, although approximations only, serve to demonstrate that even where *Calluna* heaths are managed for herbivore production the output is not great. Correspondingly, the proportion of the annual primary production utilized by grazing animals is generally small. These and other aspects of the use and management of heaths for grazing purposes are further considered in the next chapter.

9 Heaths as grazing land

The origins of lowland heaths are closely bound up with the provision of forage for domestic herbivores, at first cattle and later sheep (Chapter 2). Their survival at the present day is in large measure related to continuing use for the same purposes, to which may be added the rearing of game birds, particularly the red grouse. Heath plants also provide food for a number of other animals, both invertebrates and vertebrates. An important aspect of the ecology of heathlands therefore concerns the feeding value of the main species and the impact of grazing upon the ecosystems.

Calluna and other heath species as components of the diet of herbivores

Cattle and sheep

Little has been published on the role of *Calluna* and other heath species in the diet of cattle grazing on heathlands. Certainly *Calluna* is heavily grazed, particularly in winter when the biomass of graminoid plants is at a minimum. Where, for example, *Agrostis-Festuca* sward or reseeded pastures are available the cattle concentrate on these in spring and summer, but hill cattle can be maintained throughout the year on areas which are largely heath-covered.

A survey of the rumen contents of hill sheep (MacLeod, 1955) showed that 86% of samples taken fortnightly throughout the year contained *Calluna*. Considering only those samples taken during winter and early spring, the proportion was as high as 93%, compared with 77% in the summer (May–September). In winter and early spring *Erica tetralix* also figured quite commonly in the samples, but *E. cinerea* was seldom taken at any time. These results give a general impression of the importance of *Calluna* in the diet of hill sheep.

Where, in addition to the species of mor soil (*Calluna*, *Eriophorum*, *Nardus, Molinia*), communities of the more palatable and nutritious grasses (e.g. *Agrostis* and *Festuca* spp.) are well represented, stands of

Calluna are less heavily grazed (Hunter, 1954, 1960, 1962). Under these conditions, the use of *Calluna* stands by sheep is markedly seasonal, being concentrated mainly in the winter months. There is sometimes a subsidiary peak in the comparative grazing index (a measure of the relative number of sheep seen grazing a unit area) in late summer (July and August). In winter, *Calluna* not only provides a supply of reasonably palatable and nutritious forage when this is lacking on grass pastures, but holds some of this clear of snow cover except after very heavy falls. The return to *Calluna* in late summer may indicate that the unlignified shoots of the current year are then sufficiently grown to provide reasonable grazing at a time when grasses and Cyperaceae are becoming fibrous (Hunter, 1962).

Grouse and ptarmigan

The feeding habits of grouse were established in considerable detail by Edward Wilson and A. S. Leslie (1911), as a result of analysis of the crop contents of 399 birds. Their work, together with that of Jenkins, Watson and Miller (1963), clearly demonstrates the dependence of grouse on *Calluna* as its major food supply. According to the season, from 50–100% of the crop contents consists of *Calluna*. Other plants, including stems and leaves of *Vaccinium myrtillus*, *Empetrum nigrum*, *Oxycoccus* spp., *Erica tetralix*, *E. cinerea*, etc., contribute only small percentages, rising somewhat in the summer and autumn (July to October). In addition to the green shoots of *Calluna*, flowers and fruits contribute a significant proportion of the diet from September to January.

At altitudes above about 600 m grouse begin to give place to ptarmigan. Although not so exclusively a *Calluna* feeder, ptarmigan take large quantities of this plant where it is available (61·4% October–April; 24·4% May–September; Watson 1964). In Britain, *Empetrum* spp. and *Vaccinium myrtillus* make up the greater part of the remainder of the food, but many other species are also taken (for example, in Iceland *Salix herbacea* is an important food plant).

Rabbits and hares

As with sheep, certain grasses and other herbs are taken by rabbits in preference to *Calluna*, but where the latter is abundant it is eaten freely. Intensive grazing by rabbits can eliminate *Calluna*, which is then replaced by a close-cropped grass turf. This was described by Farrow (1917, 1925) with particular reference to the zones close to burrows. However, his attempt to use this factor to explain the distribution of heath in relation to acidophilus grassland on a wider scale was shown by Watt (1936) to be over-simplified and to ignore the importance of soil differences.

Brown hares will also take *Calluna*, while for the mountain hare of upland areas it provides a major component of the diet.

Red deer

Calluna also provides part of the diet of red deer, especially in winter when the animals occupy lower ground than in summer and when other forage is scarce. Miller (1971) reports that in north-east Scotland (Cairngorm Mountains) heaths, grasslands and *Juniperus* scrub are all used by deer in winter, and that *Calluna* is taken particularly in the earlier part of the winter while it is largely free of snow, and again to some extent in April and May. Grazing pressure is greatest in sheltered localities.

Composition and food value of *Calluna* shoots

Since *Calluna* is an important food plant for several herbivores, some indication is necessary of the concentrations of nutrients in the edible parts of the plant. These are the short-shoots, the terminal and unlignified parts of the long-shoots (usually the growth of the current year only), and to a lesser extent the flowers and seed capsules. Many of the more recent chemical analysis of *Calluna* have been carried out in connection with investigations on the nutrient budget of the ecosystem (Chapter 11) and have therefore been concerned mainly with total quantities of the various elements in the vegetation, rather than with those in the edible parts only. However, Moss (1969*a*) analysed the current year's growth and expressed his results on a dry weight basis, giving figures in substantial agreement with the earlier (and more restricted) analyses of Thomas (1934, 1937, 1956), Thomas, Eskritt and Trinder, (1945), Thomas and Dougall, (1947), Lauder and Comrie, (1936). Lauder and Comrie commented that as regards the content of protein and fat the feeding value, weight for weight, is similar to that of a medium quality hay, and the fibre content slightly less. Moran and Pace (1962) show that the quantity of crude protein (on a dry weight basis) in green shoots of *Calluna* is of the same order (*c.* 7–9%) as that of a barley grain or an average low protein wheat, and the amino-acid composition is comparable with that of hay.

Calluna is also a relatively good source of calcium and magnesium, the concentrations of which range from 0·34 to 0·48% calcium and 0·15 to 0·22% magnesium in the edible shoots. These levels are lower than in hay, for example, but *Calluna* represents a valuable supply of these elements and is considerably superior in this respect to other heathland species with the exception of *Vaccinium myrtillus* (Thomas and Trinder, 1947; Moss 1968). The content of potassium, about 0·56–0·76%, is also appreciable, but that of phosphorus is generally low (*c.* 0·07–0·15%) although variable in relation to the availability of this element in the soil (Moss, 1969*a*).

Analyses of *Calluna* of different ages have shown highest concentrations of most nutrients in the youngest plants and a decline with increasing age* (Thomas, 1934, 1937; Lauder and Comrie, 1936; Thomas and Dougall, 1947; Grant and Hunter, 1968). This is true of N, P, Ca, Mg, Na, and K, but is particularly marked in the first two. Miller and Miles (1969) showed that the concentrations of nitrogen and phosphorus, high in the youngest plants sampled, declined rapidly during the first six years, thereafter remaining fairly steady at a low level (Fig. 42).† Hence, the younger the

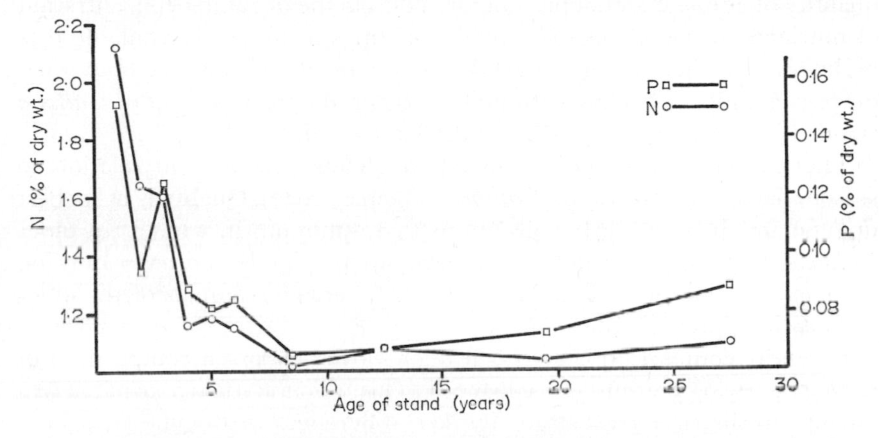

Fig. 42. The nitrogen and phosphorus content in the current year's shoot-tips from stands of different ages. Harvested September 1967; north-east Scotland. (Figure by G. R. Miller.)

Calluna the higher its feeding value – a relationship which is emphasized by the fact that nitrogen and phosphorus are here the elements which are likely be in short supply in the herbivores' diet. However, quality is not the only consideration, and account must be taken of the quantity available in an

* The rapidity of the decline in this instance may in part be due to the fact that the stands sampled were open to grazing. Although new shoots produced immediately after grazing are relatively rich in nutrients (Grant and Hunter, 1966), this effect is short lived as shown by Kenworthy (1964). He found that the concentration of nutrients in samples from grazed areas are generally lower than in ungrazed stands of the same age. However, the conclusion that nutrient content declines with age remains valid.

† It has been pointed out by Kenworthy that to establish this relationship with certainty the plants or stands to be sampled should be protected from grazing, which otherwise may affect the results directly or indirectly. On pp. 178–9 it is mentioned that plants subjected to periodic clipping or grazing produce new shoots containing higher concentrations of nutrients than those of untreated plants. On the other hand, in stands open to heavy grazing, the continual removal of palatable young growth may have the effect that samples of the remaining herbage consist predominantly of older portions of shoots. In this case, chemical analysis gives lower values for nutrient concentration than in samples from ungrazed stands of the same age (Kenworthy, 1964).

area. In relation to unit area, biomass is small in very young stands (pioneer), but increases up to the mature phase when plants are aged about 20 years (Chapter 7). However, as they get older, so a greater proportion of the biomass consists of woody stems and branches. Thomas and Dougall (1947) showed that while the weight of edible material per unit area increases to a peak at about seven years, it is already declining by the time plants are nine years old. Up to the age of about seven, the increased quantity of edible material per unit area offsets the declining concentration of nutrients in the shoots. The total quantities of nutrients available to a herbivore are therefore maximal in stands of about this age, but then decrease rapidly. This has a strong bearing on the management of *Calluna* stands for grazing (pp. 179–181, and Chapter 10).

The nutrient concentrations in edible shoots vary also in relation to season, particularly in young *Calluna* (Thomas, 1937). Quality is at its best in June and July, thereafter declining to a minimum in winter. In older *Calluna*, however, this difference becomes negligible, perhaps offering some explanation of the greater use made by grazing animals of the older *Calluna* in winter (Hunter, 1962).

Recently, comparisons have been made of the chemical composition of *Calluna* growing on different soil types. This was undertaken as part of an attempt to discover reasons for marked differences in the size of grouse populations on certain moors, which could not fully be accounted for by differences in management and their effects on the proportion of young *Calluna*. In two such areas in north-east Scotland (Corndavon and Glen Muick) high densities of grouse are consistently maintained despite a predominance of rather old *Calluna*. Miller, Jenkins and Watson (1966) and Moss (1969*a*) drew attention to the fact that these grouse moors are situated over relatively basic rocks (Corndavon: 70% of the area lies over diorite; Glen Muick: 40% over epidorite, 5% over oligoclase biotite gneiss), whereas other comparable moors in the district with fewer grouse were situated over granite. No consistent differences could be found in the amount of *Calluna* produced or in the amounts available at the end of winter, but the 'rich' moors had two features in common; firstly the phosphorus content of the edible shoots of *Calluna* was significantly higher (differences of the order of 36% being recorded), and secondly there was a greater variety of other plant species associated with *Calluna*. There was some indication that other nutrients followed the same trend: for example in July 1962 *Calluna* at Corndavon contained 19% more potassium, 36% more phosphorus, 140% more cobalt and 21% more copper than at a neighbouring area over granite on the slopes of Lochnagar. The amounts of phosphorus and nitrogen in the early spring diet of grouse may be a limiting factor in regard to breeding success (Moss, 1967). It seems possible therefore, that the higher grouse densities on these 'rich' moors may, at least in part, be related to the consistently higher levels of phosphorus in the diet, perhaps of other nutrients as well. The greater variety of species other than *Calluna* may be a contributory factor, for some of these produce new growth earlier in the year than *Calluna*. Among these is *Vaccinium myrtillus*, which was more abundant at Corndavon and Glen Muick than elsewhere, and has concentrations of both phosphorus and nitrogen in its leaves and young stems higher than in *Calluna* (Moss, 1968).

Although these findings were not related to soil analyses, it is reasonable to suppose that the observed differences were related to differences in the amounts of available plant nutrients in the soil.

Selective feeding

It is generally stated that cattle, when grazing on heathland, are not strongly selective but graze most of the components of the community. The effects upon botanical composition are then determined more by the general grazing intensity, rather than by any differential impact upon the various species. Sheep on the other hand are strongly selective, particularly when stocking rates are such that, at least in summer, the amount of herbage consumed is small compared to its production (e.g. less than 30%, Eadie, 1967). Nicholson, Paterson and Currie (1970) describe how, in experimental plots on a pasture dominated by *Nardus stricta*, sheep seek out the preferred species of grasses, e.g. *Anthoxanthum odoratum*, *Molinia caerulea*, from all over the plot. As the leaf-tips of these become depleted the plants are grazed progressively closer to the ground, while at the same time an increasing proportion is taken of other species which come next in the sheep's order of preference, e.g. *Festuca ovina*, *Agrostis canina*, *Deschampsia flexuosa* and *Nardus stricta*. Grant and Hunter (1968) found that where sheep were free to graze in plots of *Calluna* of various age, they chose the younger stands in preference to the older. This selectivity was most in evidence where grazing pressures were low, and declined as they increased.

Another example of a selective grazer is the ptarmigan. In Scotland, Watson (1964) has shown that *Calluna*, *Empetrum hermaphroditum* and *Vaccinium myrtillus*, which are the most abundant species in the zones frequented by the birds, account for 60% of their food in summer and 90% in winter. However, in Iceland where the same species are abundant, the major components of the food are *Salix herbacea*, *Polygonum viviparum* and *Dryas octopetala* which contribute only 2%, 1% and 20% respectively of the total vegetational cover. Berries of *Empetrum hermaphroditum* are also taken. Moss (1968) concludes that the birds are selecting a diet of high nutrient content. C. F. Summers has obtained experimental evidence that birds become conditioned to choose species which have for a time been predominant in their diet.

There is some suggestion that the herbivores, in general, show preference for the more nutritious foods from the range of palatable species. This is borne out by the observation, referred to above, that grouse introduce *Vaccinium myrtillus* into the diet early in the year before new growth of *Calluna* is available. Similarly, *Erica tetralix* is taken at this time of year probably because, although no richer in nutrients than *Calluna*, new growth starts earlier in the year (Moss, 1969*b*). For the most part, however, grouse subsist upon *Calluna* and it is therefore of interest to determine whether the birds feed indiscriminately on all the edible shoots available, or make some selection.

To test a suggestion that there might be selection on the basis of the age of the *Calluna* plant, birds confined in pens on the hillside were offered turfs taken from stands of varying age (Miller and Miles, 1969). A clear preference for three- to four-year-old *Calluna* was shown. This is evidently not a straightforward choice of food of the highest nutritive quality, which

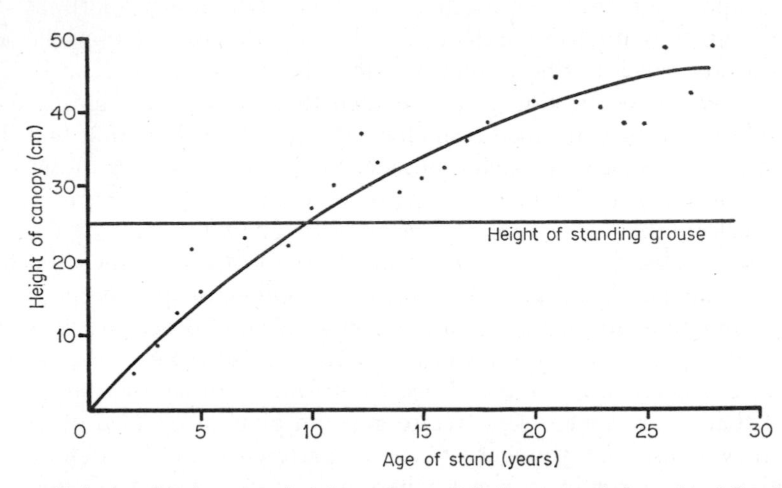

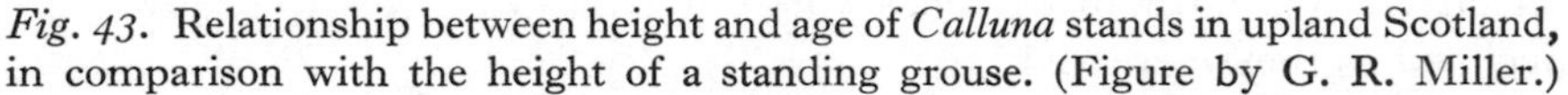

Fig. 43. Relationship between height and age of *Calluna* stands in upland Scotland, in comparison with the height of a standing grouse. (Figure by G. R. Miller.)

would have resulted in the greatest attention being paid to the youngest plants and progressively less to samples of increasing age, as obtains in the case of mountain hares tested in the same way. Possibly the structure of the stand affects the feeding behaviour of grouse, such that the short stature and lower cover value of very young stands make them less attractive (Fig. 43). This hypothesis was tested by offering shoots cut from plants of different age, at a uniform level in a wooden holder. Under these conditions there was some suggestion of preference for shoots from two- to three-year-old plants. Conversely, offering shoots of uniform age at different levels above the ground produced the result that, whereas those at 10 or 15 cm were freely used, those at 5 cm were untouched.

Apart, however, from the effect of the accessibility of the edible shoots, there would seem to be a positive choice of the more nutritious food, although the mechanism by which it is recognized is unknown. Moss (1969*b*) has obtained additional evidence of this by killing birds which had just been feeding and comparing the chemical composition of the shoots in their crops with a random collection made at the same time from the feeding area. In spring (23 April 1967) when concentrations of nitrogen and phosphorus in *Calluna* shoots were lowest, the mean concentrations in randomly selected edible shoots were 1·44% N, 0·104% P; whereas analysis of shoots from the birds' crops (after correction for salivary nitrogen) gave 1·86–1·90% N and 0·153–0·178% P. Apparently, therefore there is positive selection of shoots relatively rich in nitrogen and phosphorus, for estimates of other nutritive components such as soluble carbohydrates and crude fat revealed no differences, nor were any detected in the possibly unpalatable soluble tannins.

Fertilizer treatment

The use of fertilizer to increase the concentrations of nitrogen and phosphorus in the food available to the herbivores offers a means of testing their responses, and the subsequent effects upon their density.

It has already been shown (p. 170) that a relationship exists between the concentration of these nutrients in *Calluna* shoots and their availability in the soil. Furthermore, detailed growth analysis of young *Calluna* plants by D. K. L. MacKerron revealed considerable differences in performance related to soil nutrient status (Fig. 44). Fertilizer treatment may therefore be expected to produce appropriate differences both in quality and quantity of the edible material. Increases in the phosphorus content of shoots following the application of phosphatic fertilizer have been demonstrated by Thomas, Eskritt and Trinder (1945), Fairbairn (1959) and Kenworthy (1964). An additional effect is the stimulation of flowering (Kenworthy, 1964; Robertson and Nicholson, 1961). Nitrogen dressings sometimes produce a marked stimulation of growth even within a single season. Miller (1968) found that on plots treated in June with ammonium nitrate at rates of 20 or 60 kg ha^{-1} of nitrogen a significant increase in growth was evident by October, particularly at the higher level of application. In the subsequent year there was again increased growth on the treated plots, but thereafter growth response disappeared and the fertilizer was presumably exhausted. The application of calcium di-hydrogen phosphate produced no such growth response, but both nitrogen and phosphorus dressings changed the composition of the edible shoots. Ammonium nitrate increased their nitrogen content, while phosphate increased the content of phosphorus and calcium and possibly also the nitrogen content.

Miller also assessed the response of herbivores to changes in the nutrient content of the shoots. Following the application of fertilizers to small plots, the faeces of grouse, and of rabbits and hares (grouped together),

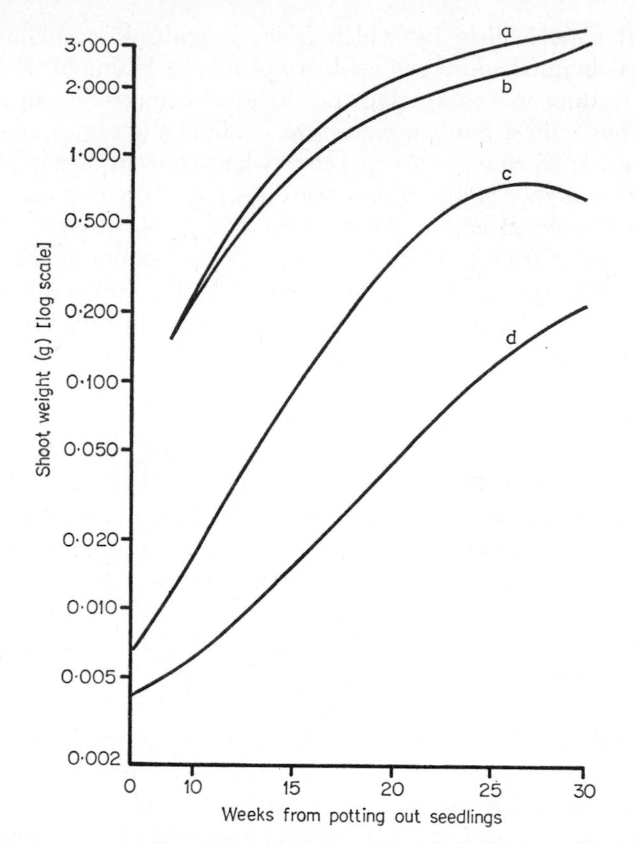

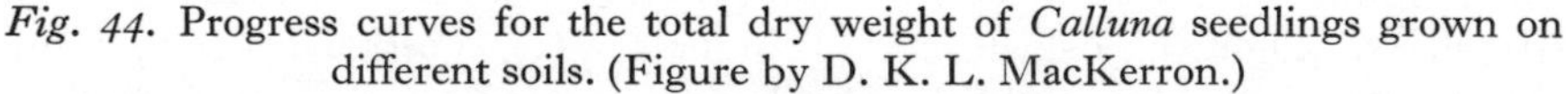
Fig. 44. Progress curves for the total dry weight of *Calluna* seedlings grown on different soils. (Figure by D. K. L. MacKerron.)

(*a*) On soil composed of 2 parts peat (from a heath on drained peat) to 1 part John Innes Compost no. 3.
(*b*) On humus-containing layers of a podsol under a plantation of *Picea sitchensis*.
(*c*) On brown earth soil from *Betula – Corylus avellana* wood.
(*d*) On soil from grouse-moor – thin raw humus mixed with leached gravel.

were collected and weighed. Evidence was obtained that from October to March grouse select *Calluna* which is rich in nitrogen, though there was no indication that this applied in summer. Hares and rabbits select shoots which are richer in nitrogen both in winter and summer and also, in winter, those with a high phosphorus content, but only in the first year after fertilizer application. All the animals were evidently able to detect and return to small 1 m^2 fertilized plots in the middle of an expanse of heath. Calculation of correlations showed that, as regards the nitrogen treatments, herbivores were attracted by differences in the chemical composition of the shoots rather than simply by the increased amount of available food. Subsequent experiments of the same kind have repeated these results, and have indicated that effects are maximal in the second season after application and are still appreciable, though declining, in the third.

In a large-scale experiment (Miller, Watson and Jenkins, 1970) one half of a uniform stand of pure building-phase *Calluna* of about 32 ha in area was treated with calcium ammonium nitrate at a rate of 500 kg ha^{-1}, after shooting all resident birds. Increases both in growth and nitrogen content of the young shoots were recorded, and these lasted into the third season after application. In the year after treatment, incoming grouse settled with the same density of breeding pairs on treated and untreated areas although differences both in the quantity and quality of available food were well established at the time of their arrival. However, the breeding success of birds whose territories were in the fertilized area was markedly higher than in the others and in the subsequent season an increased number of grouse took territories in this area. Breeding was again somewhat better but by the third season after treatment the breeding density on the fertilized area was declining again and no difference was observed in

TABLE 23

Response of red grouse to the fertilizing of *Calluna*

Fertilized area treated in May 1965 with calcium ammonium nitrate at 500 kg/$ha.^{-1}$

All grouse shot in August 1965.

(Data from Watson and Miller, 1970)

	1966		1967		1968	
	Fertilized area	Control	Fertilized area	Control	Fertilized area	Control
Breeding stock of grouse (birds/km^{-2})	44	44	112	44	81	38
Breeding success (ratio of young : 1 old bird)	1·8	0·6	0·0	0·0	0·0	0·0

breeding success (Table 23). A slightly higher density was still observed in the fourth year, but by the fifth differences were no longer detectable.

Effects of grazing on plant morphology and competition

The importance of heaths as grazing land depends not only on the feeding value of shoots of *Calluna* and its associated species, but also on their morphological reactions to partial defoliation and the general effects of grazing on the botanical composition of the community.

The early stages of post-fire regeneration are usually subject to grazing, since there is seldom any means of excluding herbivores, whether wild or domestic. A knowledge of their effects on young plants of *Calluna* and other heath species is therefore important.

Very heavy grazing pressure at this stage, particularly by sheep, can have disastrous results, especially if for other reasons regeneration is slow. In addition to continual removal of new growth, the pulling action of grazing sheep frequently uproots seedlings or, where vegetative regeneration has taken place, drags young shoots away from their attachment to older woody stems. The re-establishment of cover may be long delayed, leading either to sheet erosion of the surface humus or to the spread of unwanted species such as *Pteridium aquilinum* or *Nardus stricta*.

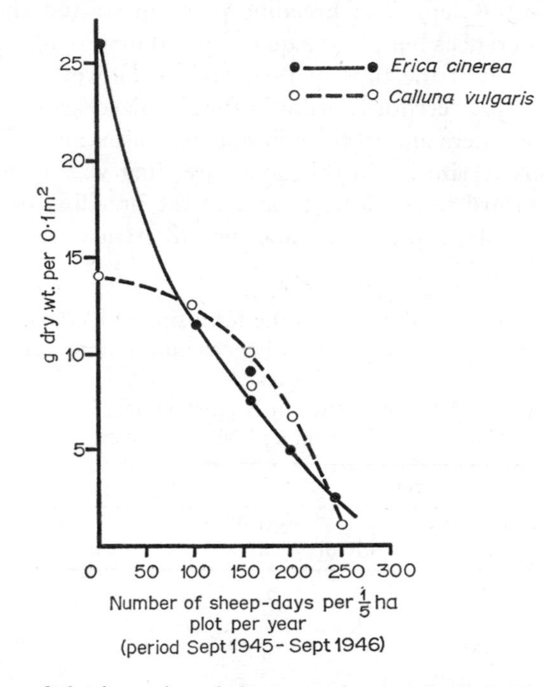

Fig. 45. Effect of the intensity of sheep-grazing on the balance between *Calluna vulgaris* and *Erica cinerea*, regenerating after burning. (From Gimingham, 1949: north-east Scotland.) Plots were burned in spring 1944, ungrazed for one and a half years (except by rabbits); then grazed by sheep at the intensities shown for one year before sampling.

However, at less severe intensities, grazing may have the effect of hastening the development of a closed community. It is certain that it has different effects upon the several species of a community, and determines their subsequent behaviour and quantitative proportions. In a comparative study (Gimingham, 1949) of the effects of grazing on the balance between *Calluna* and *Erica cinerea* regenerating largely from seed, it was shown that a medium grazing intensity can rapidly shift the balance in favour of *Calluna* (Fig. 45). The study area was a hillside in Kincardineshire, north-east Scotland, where a thin, freely-drained podsolic soil overlies a shallow acid glacial till. Here, *Erica* seedlings outnumbered those of *Calluna* between two and five times, and on ungrazed plots four seasons after burning the biomass of *Erica* was considerably in excess of that of *Calluna*. However, by the end of the third summer after only one years' grazing *Calluna* showed the greater weight in all grazed plots, except those receiving the heaviest grazing (i.e. about 200 sheep days per 0·2 ha plot per year). In the latter the damage to both species was so severe that biomass was reduced to an approximately equal, low level.

This result indicated a differential effect upon the two species, such that at medium pressure the performance of *Calluna* became superior to that of *Erica*. Experimental imitation of grazing by clipping young plants at 1 cm above the soil surface after four months growth helped to explain the difference. The morphology of seedlings has been described in Chapter 6: the effect of cutting *Calluna* at 1 cm is to remove most of the leading shoot with its regularly-arranged short laterals, except the first formed and lowest pair which normally emerge from the main stem within 1 cm from its base. These, as mentioned on p. 111, are usually by this stage longer than other laterals and have become more or less procumbent and spreading. After removal of the leader, these branches continue plagiotropic growth, only later becoming somewhat ascending at their tips (Plates 24, 25). Some of the short-shoots along their length may begin to extend vertically, but the general response at first is to produce a plant of prostrate, creeping habit. Should further grazing remove upright branches, this tendency would be perpetuated. If this kind of response occurs under field conditions, and observations suggest that it does where plants in their first few years of growth are subject to grazing (Plate 26), the percentage cover contributed by *Calluna* will be increased.

However, a different response was elicited in about 20% of *Calluna* seedlings which produced no plagiotropic branches, forming instead a dense tuft of up to nearly 20 upgrowing branches (Plate 27). Further experimentation showed that cutting at rather higher levels (e.g. 1·5–2 cm) produced only the spreading response, and that the bushy form resulted from more severe cutting-back. The interpretation was that the latter treatment had, in fact, removed the node from which the plagiotropic branches are formed, leaving only those dormant buds in the axils of the first-formed four or five leaf-pairs, which do not normally give rise to laterals (p. 111). When all nodes above are removed it seems that these buds will grow simultaneously, producing the bushy plant as described. Such individuals may also be found in the field (Plate 28) where grazing on young plants has been severe (sometimes this effect can be attributed to rabbit grazing).

Whichever of these two kinds of response is shown, further grazing serves only to intensify it. In either case the resulting growth-form is strikingly different from that produced in *Erica cinerea* by similar treatment. If seedlings of this species are trimmed to a height of about 1 cm growth usually takes place in two buds just below the cut. Branches from these assume an erect habit of growth and add little to the cover contributed by the plant (Plate 24). In *Calluna* on the other hand, either type of response results in increased cover, which may enable the plants to compete more effectively than the thinner, more erect and less branched individuals produced by similar treatment of *Erica*. Certainly *Calluna* shows the better performance under these conditions. Even so, productivity in the years immediately following a fire is less in plots open to grazing than in fenced plots, although the difference is not so great as in *Erica cinerea* (Table 19, p. 153).

Grant and Hunter (1966, 1968) also showed that, in young plants, clipping (to simulate grazing) caused reduced production, and that clipping in summer (July or August) was much more severe in its effects on the plant than a similar treatment in winter (late December). However, there was evidence to suggest that the effect of continued clipping leads after a time to an increase in total production in the more frequently cut plants (e.g. those cut once every year). This increase was shown more quickly in plants cut in winter (after three and a half years) than in those cut in summer (probably not until after eight years). Clipping increases the compactness of growth, so that as an individual develops under the influence of annual clipping (or grazing) it adopts a cushion-like form with densely packed thin branches. The treatment was designed to remove on each occasion about 80% of the current year's growth. By removing this proportion of new long-shoots, which otherwise would eventually contribute to the woody framework of the plant (p. 112), the build-up of lignified branches is inhibited. Instead a progressively greater number of shoot apices begin growth. Mohamed and Gimingham (1970) examined the effect of decapitating the chief long-shoots of established plants which have passed the pioneer phase. New growth then takes place either in several of the uppermost undamaged lateral short-shoots which grow out as long-shoots, or else in clusters of new shoots which appear from the neighbourhood of branching points lower on the stem (Plate 31). The extent to which the latter occurs depends upon the amount of damage at the periphery and its frequency, but the result is an increase in the total number of green shoots and hence, in the long-term, improved production of edible material. This was indicated in Grant and Hunter's work by a gradual increase, with repeated cutting, of the production of green tissue relative to the woody material.

Furthermore, the green shoots produced under the influence of clipping treatments invariably showed higher concentrations of the major nutrient elements than those of control plants, the greatest difference being in the nitrogen content which was increased on average by 49%. Since, as already shown (p. 169), in the absence of grazing the concentration of these elements normally declines rather rapidly over the first four to six years, clipping can be described as having a 'rejuvenating' effect doubtless connected with the changed distribution of dry matter between green and woody tissue. Consequently, both the productivity and quality of edible material are improved by a grazing regime which removes say 60% of the new growth annually (Grant and Hunter, 1966). In this way individual plants are being maintained more or less indefinitely in a condition similar to that of the early building phase, and plots which have been maintained in this way for quite long periods may be as attractive to sheep as young *Calluna* (Grant and Hunter, 1968). This result reinforces observations made much earlier by agriculturalists (e.g. Linton, 1918; Wallace, 1917) concerning the ability of *Calluna* to withstand moderately heavy grazing and to respond with enhanced production of new growth of good nutrient quality. The practical difficulty is to maintain a sufficient stocking density on hill farms to achieve this level of utilization, in view of the difficulty of

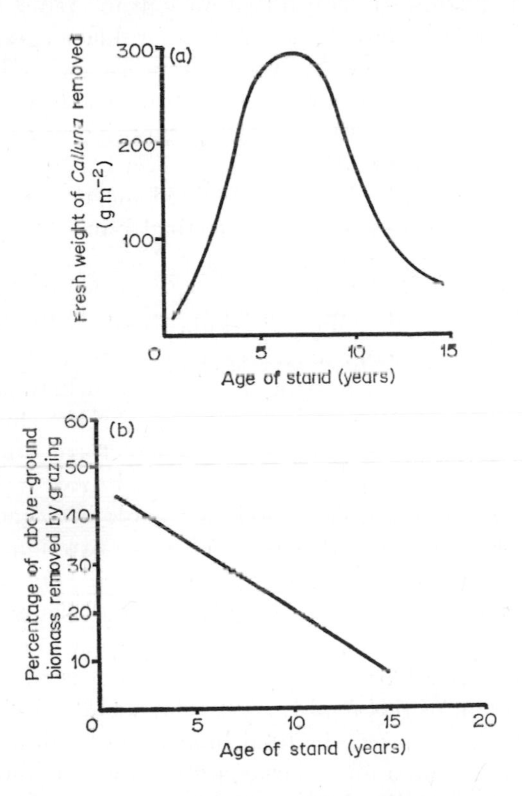

Fig. 46.
(a) Relationship between fresh weight of *Calluna* taken by grazing sheep and the age of the stand.
(b) Percentage of available above-ground biomass removed by grazing sheep in relation to age of stand. (Figures by J. B. Kenworthy; Glen Clova, Central Scotland.)

providing adequate winter 'keep'. Experiments by Kenworthy (1964), in which samples were taken from sheep-grazed *Calluna* stands of varying age for comparison with samples from ungrazed enclosures, confirmed the preference of grazing animals for the younger heather. However, while the greatest percentage removal of forage was from the youngest stands (Fig. 46), the greatest total loss to the system due to grazing occurred between seven and ten years after burning, the loss in dry weight being of the order of 50%. Although some of the nutrients ingested are returned to the ecosystem in urine and faeces, they may at this stage be particularly liable to loss by leaching. Kenworthy concluded that nitrogen, phosphorus and calcium are removed from the system in considerable quantity by grazing, yet only phosphorus is important when compared with the total calculated input (see pp. 213, 218, 220).

Populations and nutrition of herbivores

The foregoing discussion has shown that on soils incapable of maintaining productive grassland, *Calluna* heath provides valuable grazing, and that the *Calluna* plant stands up effectively at least to moderate grazing. However, a community of perennial dwarf-shrubs differs from a grass sward in that a large part of the biomass at any time consists of inedible woody material. Robertson and Davies (1965) separated 'leaf' from 'stem' in samples of *Calluna* stands, and obtained nearly 5000 kg ha^{-1} dry weight of 'leaf' or about a quarter of the total biomass in a 15-year-old stand in north-east Scotland. It is not stated, however, exactly what fraction of the whole plant was regarded as 'leaf'. Probably this included all short-shoots, together with unlignified long-shoots of the current year's growth. Many short-shoots grow on for two or three years, and not all this material is taken by or accessible to herbivores. Estimates of the dry weight of the current year's increment only, in short-shoots and long-shoots, have been made by Moss (1969*a*) and Miller and Miles (1969) both on the open heath and in ungrazed enclosures and these vary around 2000 kg ha^{-1}. The quantity of edible material available in a stand of building heather in this region probably lies between these two figures.

The *in vitro* digestibility of this material has been estimated at 25–26% using sheep rumen liquor. This figure may be somewhat exceeded in very young heather. Data from Moss (1969*b*) indicate that under normal conditions grouse also digest about 25% of the dry matter of *Calluna* shoots. The nutrient content of this food is highest in stands which are less than about six years old. However, the total biomass per unit area is less than in older stands, although a much larger proportion is edible (up to about 75% in two-year-old *Calluna*). Pioneer heather is sought out by sheep, grouse and other herbivores, but combining considerations of quantity with quality, on an area basis, the provision of food is best in stands of seven or eight years of age.

This daily intake of dry matter by sheep has been estimated by Eadie (quoted by Rawes and Welch, 1969) at between 0·5 and 1·5 kg, amounting to 180–450 kg year^{-1}. Accepting 2000 kg ha^{-1} as an approximate estimate of the annual rate of production of young shoots by *Calluna*, it might be imagined that the supply could support between four and ten sheep per hectare. This, however, would be conditional on the food being made available progressively throughout the year, which is unrealistic. Hence, on heathland it is hardly surprising that actual stocking rates usually vary between one sheep to 1·2 ha and one to 2·8 ha, even though species other than *Calluna* contribute to the forage. During the winter *Calluna* may be the only species bearing edible material. Removal both by grazing and winter browning (p. 98) reduces this to a low level in early spring, which is lambing-time when the nutrient requirements of ewes are maximal. Winter feed thus sets a limit on the carrying capacity of heathland, although to some extent this can be raised by the provision of supplementary feeding stuffs or by wintering the animals elsewhere. None the less, the limits on the sheep stock are usually such that there is more than enough edible *Calluna* available at the end of the summer and, from the shepherd's point of view, the herbage is undergrazed at this time of year. Certainly the intensity of grazing is insufficient to maintain stands indefinitely in the building phase by means of grazing alone (see pp. 178–9 and 184). A closer approach to this objective may be effected in some of the rather few areas where substantial numbers of cattle are grazed in addition to sheep. On suitable soils a further consequence may be the partial replacement of *Calluna* by grass species.

It remains to consider the grouse, which in terms of individuals is probably the commonest vertebrate herbivore of British heathlands. Special interest attaches to this bird because of its almost exclusive dependence upon *Calluna* making it a particularly valuable species for the investigation of relationships between population density and food supply in a herbivore. For this reason it has been the subject of very comprehensive study by the Nature Conservancy's Unit of Grouse and Moorland Ecology (now incorporated in the Mountains and Moorlands Habitat Team) working in Kincardineshire, north-east Scotland. Early in their inquiry, it was established (Jenkins, Watson and Miller, 1967) that the population of birds on grouse moors was not determined either by shooting, or by predation, as there were always sufficient surplus birds (which otherwise would die or move away from the area) to take the place of those killed. There was, however, a close relationship between the average breeding density of grouse and the quantity and age of the heather in the study areas (Miller, Jenkins and Watson, 1966), suggesting some effect of food supply on numbers. Fluctuations in breeding densities from one spring to the next, however, were not correlated with the growth of *Calluna* in the intervening season.

On the other hand, breeding success in any year (as measured by the average number of young reared per adult bird) appeared to relate both to growth of *Calluna* during the previous summer and the degree of die-back in winter. Breeding success then influenced the size of the next year's breeding stock, which was therefore related to the performance of *Calluna* between 12 and 21 months previously (Fig. 47).

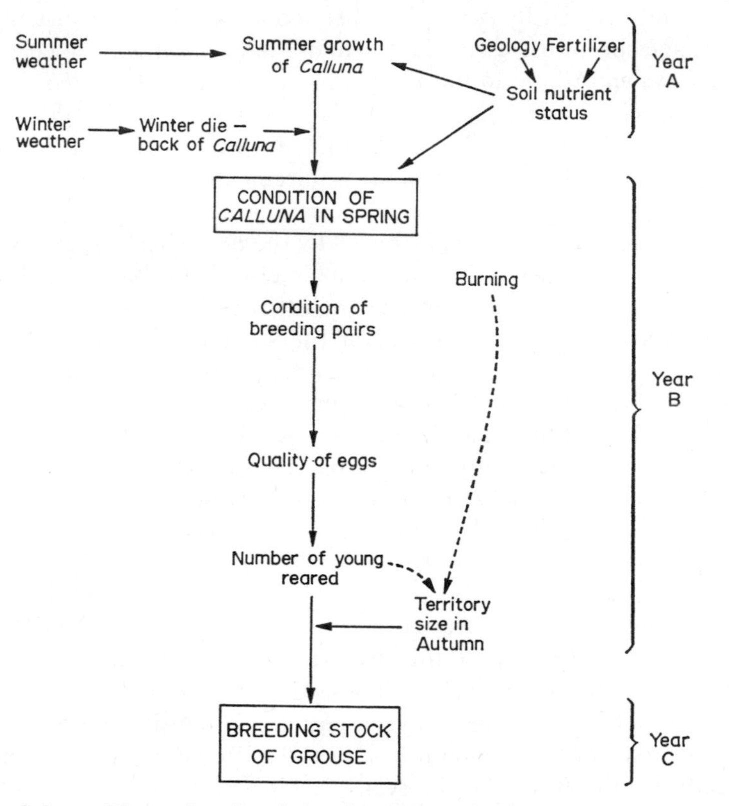

Fig. 47. Scheme illustrating the chain of influences linking grouse numbers to the condition of *Calluna* twelve months earlier. Also shown are some of the factors affecting the condition of *Calluna* in spring. (Modified from Watson and Miller, 1970. The sequence as a whole is hypothetical, but experimental evidence is available for a number of the links.)

Other observations showed that grouse, even at maximum densities of about one pair to two hectares, take only about 5% of the available food. It was difficult therefore to postulate that, even in late winter and early spring, there would be any critical shortage in the amount of food. At this time of year, however, grouse show particularly strong preference for young *Calluna*, or newly grown shoots (Moss, 1969 *a*, *b*). Evidence has accumulated that the quality of the diet in early spring influences breeding success. This is summarized by Jenkins, Watson and Miller (1967) as follows: 'Food may influence grouse breeding success in three ways: (1) heather growth in the previous summer, which affects the amount of green heather in winter, and is significantly correlated with breeding success in the following summer (Miller, Jenkins and Watson, 1966); (2) browning of heather over winter, which reduces the amount of green heather in spring (Watson, Miller and Green, 1966) and is also significantly correlated with subsequent breeding success; and (3) the timing of plant growth in spring.' Breeding success has been shown to depend upon hatching date, clutch size and chick survival. Generally early hatching, large clutches and good survival are correlated and associated with the higher quantities of green *Calluna* shoots available in spring. In some years, however, chick survival is poor despite early breeding and large clutch size. This suggests some independent variation in egg 'quality', possibly introduced by a delay in the onset of new growth of *Calluna* shoots, caused by late frosts. It becomes possible therefore to construct a scheme (Fig. 47) illustrating the postulated chain of effects and suggesting that variations in the production, availability and nutrient content of the edible shoots of *Calluna* play an important part in regulating population density, despite the relatively small proportion taken of the total food material produced during the year.

It is consistent with this finding that the average size of the population of grouse on a given estate is related both to the total area of *Calluna* and to the proportion of this total contributed by young stands. However, in addition to food the birds require cover and protected nest-sites, for which stands of older and taller *Calluna* are required. Breeding density is therefore determined not only by previous breeding success in the area, but by a sufficient admixture of young and old heather for both to be included in all grouse territories. Hence, although a high proportion of young heather is advantageous for the provision of ample good-quality food, it is undesirable for this to occupy extensive areas unbroken by patches of older stands. Therefore management for grouse, while aiming to maintain a high proportion of young *Calluna* seeks to produce a patchwork of stands of varying age. On the other hand, the needs of sheep farming are most readily satisfied by maintaining extensive, uniform stands in a young condition. The management policies most suited for grouse are therefore not quite the same as those which are best for sheep although a compromise can be achieved.

Grazing as a cause of vegetational change

Changes in the botanical composition of heathland vegetation have frequently been ascribed to heavy grazing. However, as most such changes are consequent upon the combined impact of grazing and burning, it is seldom possible to distinguish effects due to one or other factor alone. It is theoretically possible by means of carefully managed, moderately heavy grazing to maintain a *Calluna* stand more or less indefinitely in a productive condition, without need for burning. However, this is a balance almost impossible to strike in practice. At times, *Calluna* is subject to very heavy grazing, which may so reduce its productivity in comparison with other components of the community that they rapidly take its place. Where, as on the relatively fertile brown earth soils, grazing promotes the replacement of *Calluna* by *Agrostis-Festuca* grassland with abundant *A. tenius*, such a change is generally advantageous. There is, however, an ever-present danger of the entry and spread of *Pteridium aquilinum* where *Calluna* is weakened by heavy grazing on freely drained, moderately fertile soils. The abundance of this species in areas formerly occupied by heathland is often attributed to the effects of management for sheep grazing, but again doubtless stems from the combination of burning and grazing.

Some podsolized soils can give rise under grazing to rather similar grassland, though generally *Festuca ovina* is predominant, with *A. canina* exceeding *A. tenius* in importance (Nicholson, 1964). The less fertile peaty podsols, however, are the most widespread, and here heavy grazing generally results in change to a vegetation greatly inferior to heath in grazing value. According to soil conditions, this may be dominated by *Deschampsia flexuosa*, *Juncus squarrosus*, *Eriophorum vaginatum*, or *Nardus stricta*. Much has been written about the spread of the latter species under the influence of heavy sheep grazing in the wet and cool climate of northern Britain (Fenton, 1936, 1937 *a* and *b*; Heddle and Ogg, 1933; Nicholson, 1964), and there seems little doubt that in some areas (e.g. the Southern Uplands of Scotland) the extensive dominance of *Nardus* is a direct consequence of sheep grazing. None the less, in some habitats even heavier grazing pressure may reduce the vigour of this species.

Similar changes have been attributed to rabbits. Farrow (1925) and Fenton (1940) have given evidence of the replacement of *Calluna*-dominated communities by *Agrostis-Festuca* swards, at least on a small scale, where rabbit pressure is high. The latter may sometimes be replaced by *Deschampsia flexuosa*, or where grazing is very intense the grass turf may be largely destroyed and replaced mainly by lichens (*Cladonia* spp.).

Experience of a long history of the grazing of domestic herbivores on heath vegetation shows that it is seldom possible to achieve a balance in which the grazing animal alone is sufficient to maintain *Calluna* indefinitely at maximum productivity. Such a system was probably most nearly approached in medieval times with a moderate intensity of cattle grazing, or mixed cattle and sheep, throughout most of the year. It is probable, however, that even then occasional burning was required to prevent the entry of shrubs and trees, and to rejuvenate the stands of *Calluna*. When, with the advent of hardy breeds of sheep such as the Cheviot and Blackface, it became more profitable for landowners to graze large sheep flocks on the heaths than to let them for small-scale mixed farming, there followed a period of heavy grazing. Both the biomass and productivity seemed high, and there was little warning that deterioration of the botanical composition of the communities and depletion of soil fertility might result from heavy grazing by sheep alone. In time, stocking densities declined and the grazing pressure became inadequate to prevent *Calluna* passing into the mature or degenerate condition. The importance of keeping as much as possible of the area of *Calluna* in the young and most nutritious stage of growth was recognized and burning, formerly spasmodic, became a regular and traditional component of management. This practice was just as necessary when management for grouse was added to the older uses of heathland, and has remained a characteristic element of the system ever since. It is therefore necessary to consider in detail the history and ecological consequences of burning as a factor affecting heathland ecosystems.

10 Heathland management by the use of fire

Historical

The earliest inroads made by man into the forests of western Europe were evidently reversible (Chapter 2). None the less, even these transitory clearings permitted the Gramineae and Ericaceae to spread (pp. 20–26). However, for heathlands to be maintained indefinitely some factor which prevents the return of trees and shrubs must operate. As discussed earlier, in certain cases this may have been a climatic or edaphic factor, but in the great majority it appears at first to have been domestic grazing animals, supported when necessary by fire. The available evidence suggests that it was Bronze Age and Iron Age men who began this process of replacing forest by communities of a different, simpler structure. Apart from the cultivated land, areas from which forest was removed were allowed to be colonized by naturally occurring species, many of them ones which were already frequent in glades or openings in the forest, or above its altitudinal limits on the hills. On the poor soils and in the exposed situations these communities were heaths; elsewhere grasslands developed. Evidence in the form of charcoal deposits indicates in many instances the use of fire in helping to clear patches of forest. The introduction of cattle and goats to the cleared areas may have been largely sufficient to destroy tree seedlings: as their numbers increased they certainly roamed into the surrounding forest and may have prevented regeneration in marginal zones, thus allowing heath or grassland to encroach.

Changes of this kind continued during historical times, but at intervals were accelerated by the felling of trees for other purposes. These included use of the timber for fuel and for constructional requirements, both in buildings and ships. At certain periods wood was in great demand for ship-building, particularly in the maritime countries where heath so readily replaces forest. Further extensive fellings took place in the fifteenth and sixteenth centuries to make charcoal for smelting iron-ore, especially in England. As the English woods were depleted in Elizabethan times, the practice moved to Scotland despite the passage of laws designed to prevent it. Iron ore was imported even to the highland areas for smelting in kilns where fuel was still available, with the result that even the remoter glens suffered this onslaught.

Deforestation was most extensive in Britain, but proceeded apace also in northern France, the Netherlands, Denmark and to a lesser extent in southern Sweden. At first the cleared areas were used mainly for cattle grazing, to supply the needs of the gradually increasing populations. As cattle are not highly selective in their use of forage, this grazing regime was probably largely adequate to prevent the return of trees, though it may have been supplemented by occasional burning as necessary. In upland areas, such as north-west Scotland, a balanced use of herbage was obtained by grazing cattle at high altitudes in the summer and returning them to the glens in winter. Although production would have been best in grassland areas, some breeds of cattle (such as the Highland) thrive well on a diet in which *Calluna* plays a large part. Accessible stands of *Calluna* were invaluable to all stock as a source of food throughout the winter.

At this stage the suitability of these extensive areas of open countryside in western Europe for sheep, and their profitability, became apparent. Spreading northwards, sheep were grazed increasingly on the heaths of the French and north German coastal plain, in England and later in Scotland, also in Denmark and to a lesser extent in Sweden. In the late eighteenth century, landowners in Scotland started to introduce the hardy Cheviot sheep to the highland area. Already the glens were overpopulated with small farmers, while the labour force required to manage large flocks of sheep was small. These factors led to the notorious highland evictions which uprooted large numbers of families from their homes in the glens. Some settled in coastal areas, but many migrated overseas. Sheep now entirely replaced cattle as the domestic herbivore. After some years of high levels of production, the available evidence suggests some decline and it seems probable that sheep, which are highly selective grazers, were not able by themselves to keep the herbage in a productive condition (Chapter 9) and prevent invasion by woody species and other pests such as *Pteridium aquilinum*. Increasingly, it became necessary to manage the vegetation by regular burning as well as grazing. On heaths, burning served the additional purpose of promoting the growth of young, nutritious shoots. Northern British heathlands, therefore, have a history of regular burning from about 1800 and possibly earlier, preceded by more occasional and haphazard use of fire for varying periods. The same probably applies in other parts of the heath region.

The success of sheep farming led to further forest clearance, so the total area of heathland continued to expand. In Denmark and Sweden it reached a maximum about 1860, in Scotland perhaps a little later. Shepherds adopted the principle of a ten-year burning rotation, so that by aiming to burn one-tenth of their total area of *Calluna* each year, every part would be fired once in ten years. This ideal was not always achieved, but there was nothing to prevent burning quite large tracts on each occasion.

Although sheep have now almost completely disappeared from the Scandinavian and other continental European countries of the heath region, in Britain, hill sheep farming with its accompanying system of management by burning has survived, although declining in profitability, throughout the 200-year period since its introduction on a large scale. However, a significant by-product of the increased area of heathland was a rise in the populations of red grouse which feed very largely on *Calluna* (Chapter 9). Land-owners began to appreciate the sporting value of this indigenous game-bird, and obtained an income from renting shooting rights which at least in part offset declining returns from the sheep flocks. Both activities could in fact be combined on the same ground, though increasing attention was paid to grouse-rearing from about 1850.

At first the gamekeepers discouraged the shepherds' practice of burning very large tracts of *Calluna* heathlands (Lovat, 1911). Better shooting was obtained where birds could remain in tall *Calluna* cover until flushed by a row of 'beaters', and consequently burning in small patches was favoured. The result, however, was a rapid fall in the acreage of *Calluna* burnt each year, and an increasing proportion of the total area of many estates was covered by ageing stands. The large numbers of birds available for shooting in the 1850s were nowhere maintained, and populations fell to a disastrously low level in 1872 and 1873 throughout much of northern Britain. The immediate cause of this 'crash' appeared to be disease attributed to parasitic trichostrongyle threadworms. However, a committee of inquiry under the chairmanship of Lord Lovat correctly diagnosed the ultimate cause as inadequate moor management, and urgently recommended a return to the practice of burning on a rotation as nearly as possible approaching ten years. The reason advanced was that this practice acted as a form of control of the parasite, but it now seems certain that high incidence of parasitism is a consequence rather than the cause of a declining population. Recent work has emphasized correlations between population size and the proportion of the total area occupied by pioneer and building *Calluna,* thus confirming the importance of maintaining an adequate burning programme such that all stands are burnt at least once in every 12–15 years. The concentration of nutrients is highest in the young shoots of pioneer and early building *Calluna* (Chapter 9), and this factor may be related in some way to the breeding success of the birds (pp. 182–3) which in turn controls population numbers in the late summer when shooting takes place. At the same time, it has also been shown that to maintain a maximum density of breeding pairs, large fires are inappropriate and a pattern of small ones is preferable (p. 183).

At the present day, grouse moors represent a very valuable source of income to the estate owner. To some extent the interests of sheep farming are served by the same type of management, and the same area of ground may be used both for grazing and sport. Generally, however, the emphasis is on one, while the other plays a subsidiary part. However, although the pattern of burning may then differ in detail from place to place, it remains the generally accepted system of management on British heathlands (Plate 5, 29, 30). Grouse shooting on this scale is unique to Britain, and at the same time Britain is the only country in which sheep farming on heathlands has survived. These are areas of marginal land which can yield a financial return only when used extensively rather than intensively, and only when management costs are minimal. Burning is an inexpensive tool for the management of heathland in Britain, whereas it has largely disappeared elsewhere.

Owing to almost infinite variations of detail in treatment, it is impossible to ascertain exactly the past history of any particular example of *Calluna* heath in Britain. None the less, the outline given above indicates a long history of grazing, supplemented probably from early times by burning. Up to about 1800 this may have been merely sporadic, but since that time very many areas will have been subjected to some 10–15 cycles of burning at more or less regular intervals. This treatment must have had profound ecological effects which it is necessary to examine in more detail.

Temperatures reached in heath fires

The first essential in understanding the impact of fire upon heathland ecosystems is to obtain some measure of the temperatures generated and their duration. This information, however, was not available until 1961, when Whittaker reported the results of using simple 'thermocolour' pyrometers in heath fires. Thermocolours are heat-sensitive paints, each of which undergoes a permanent change of colour at a different temperature. Arranging these in a series, provision can be made for colour changes throughout the range from 40° to 900° in steps which, though variable, are seldom greater than 50°C. Whittaker painted strips of each member of this series on to thin pieces of mica measuring 4 × 9 cm. which were wired together in pairs, face to face, to protect the colours. When exposed to a fire, the maximum temperature reached by the pyrometer lies between the fairly narrow limits given by the highest member of the series to show a colour change and the lowest to remain unchanged. The units are inexpensive and can be made up in large numbers, to be placed in a variety of situations in or on the soil and vegetation in an area about to be burnt.

In a series of spring fires in north-east Scotland, Whittaker found that at a height of 20 cm in the vegetation temperatures of between 500°C and 840°C were reached. At ground level, even in the hotter fires, the temperatures were generally in the range 300–500°C. The gradient in some instances was steep, involving differences of as much as 500°C between these levels. Under these conditions 'clean' burns resulted, removing the greater part of the above-ground vegetation except for charred *Calluna* stems not exceeding 15 cm long. No indication is given by this method of the duration of these temperatures, or the time taken for them to build up and decline. Observations showed, however, that the fire seldom took more than two minutes to pass over a particular spot, its speed of passage being generally about 1·8 m min^{-1} though sometimes much faster on exposed heaths (Whittaker 1960).

Using a different method, Kenworthy (1963) working in Glen Clova, central Scotland, obtained results which confirmed and extended Whittaker's. Before burning, thermocouples were placed in the vegetation at intervals along the expected path of the fire. These were connected by buried leads to a portable recorder, and continuous readings from each thermocouple as the fire passed indicated both the temperatures reached and their duration. The highest temperature recorded was 940°C and the average maximum in 35 fires was 670°C. Depending upon the intensity of the fire, high temperatures (e.g. 400°C or more) were maintained usually only for a half to one minute and had declined again to ambient by two and a half minutes after the initial rise. In both autumn and spring fires rather low maxima were recorded by Kayll (1966), also using sets of thermocouples. Readings of about 500°C (highest 545°C) were obtained in the *Calluna* canopy in an autumn fire, and 250°C (highest 360°C) in several spring fires (thermocouples were situated at levels from 10–45 cm above ground). Temperatures at ground level seldom exceeded 100°C, generally reading about 60°C for half a minute or less. At each recording site temperatures were held above normal for no more than one to one and a half minutes: this agrees closely with Kenworthy's finding. Kayll also measured the rates of advance of these fires through the vegetation, which varied between 2 and 7 m min^{-1} (cf. Whittaker's estimate above).

A number of factors may be concerned in determining the intensity of a heath fire. Whittaker emphasized the effects of environmental factors, especially wind and the moisture content of the vegetation and soil. These were also regarded as of prime importance by Kayll, who showed that the cooler fires consumed only about 30% of the available fuel instead of over 90% in the more satisfactory burns. Management can also control both rate of spread and fire intensity: all the fires investigated by the three authors quoted were burned 'with the wind' in accordance with usual practice in Scotland, but a more slow moving fire is sometimes employed by 'back burning' against the wind.

Clearly, the quantity of available fuel must also have an important effect. This quantity is a function of the age of the stand, for as already pointed out the biomass per unit area increases at least until stands are about 20 years old. Furthermore, throughout the life of the plant the proportion of woody material also increases and this is low in water content and readily combustible. Kenworthy's results demonstrated that, in the absence of major variations in weather, both the maximum temperatures generated (Fig. 48) and the duration of high temperatures increase with advancing age of the vegetation, so reinforcing a conclusion reached by Fritsch in 1927. The most successful fires are those in which the temperature at ground level does not exceed 400°C, for such temperatures will not be lethal to the stem bases of at least the younger plants (p. 192). At the same

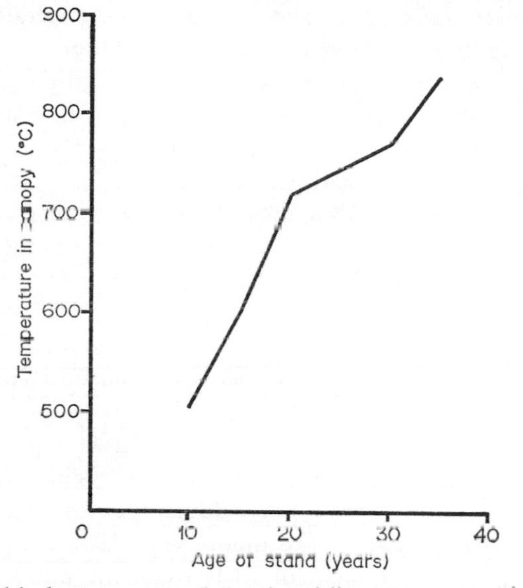

Fig. 48. Relationship between age of stand and fire temperature in a series of fires under similar weather conditions in Glen Clova, central Scotland. (Figure by J. B. Kenworthy.)

time, it is desirable that most of the above-ground parts of the plants should be burned off, while keeping the loss of nutrients in smoke to a minimum. These conditions are satisfied when temperatures not exceeding about 500°C are developed in the canopy. The aim of shepherds and gamekeepers is to achieve fires which have these characteristics.

Stem bases are frequently further protected by partial burial in humus, peat, litter or mosses and lichens. Most of these materials are effective heat insulators: the greatest protection being afforded by humus, peat and the denser types of cushion-moss such as *Leucobryum glaucum* (which also has a large reservoir of water). Rather less effective are the more open moss or lichen strata and *Calluna* litter. When, for example, a temperature of 400°C was registered at the surface of a peat substratum, the accompanying increase at 1 cm below was less than 30°C (Whittaker, 1961). The protective action of shallow accumulations of this kind around the lower parts of old stems is often seen on tracing the origin of new shoots formed after a fire, which emerge from up to 1 cm below the surface, or from the lower side of a horizontal portion of stem. Where stem bases have become surrounded by litter or luxuriant damp mosses, rather hotter fires can be permitted without inhibiting regeneration. However, when surfaces are dry or conditions of wind and fuel give rise to excessively high temperatures, the surface humus or peat may catch alight and smoulder for days, sometimes destroying the whole organic horizon and exposing the uppermost mineral material.

It is instructive to compare the temperatures generated in heath fires with those produced by burning other vegetation types. The highest recorded temperatures in forest wildfires, spreading through the tree crowns, are only slightly in excess of the maxima in heath fires (e.g. just over 1000°C., Ahlgren and Ahlgren, 1960). Temperatures in grassland fires in West Africa fall generally between about 150°C and 650°C at 0·5 m above soil surface, and 90°C–340°C at ground level (Pitot and Masson, 1951; Rains, 1963; Hopkins, 1965), but the duration of the higher temperatures is said to be short (about half a minute). Hopkins found that at ground level temperatures were generally over 540°C and about 84% of the available fuel, in terms of dry weight, was burnt. In the Sudan, temperatures of 715°C and 850°C were recorded near the soil surface in fires on grass savanna (Guilloteau 1956, quoting work by Masson). Similar results have been obtained from Japanese grasslands: e.g. in the *Miscanthus*-type grassland nearly 200°C were recorded at soil surface and nearly 800°C at 17 cm (Iwanami, 1969). In *Zoisia*-type grassland, the temperature did not exceed 500°C (Iwanami and Iizume, 1969). Generally temperatures above 200°C last for little over one minute. Structure and biomass in these communities are comparable with those in heaths, and the temperatures produced on burning, and their duration, are very much of the same order as in heath fires.

Effects on the regeneration of *Calluna*

Vegetative regeneration

Tests carried out on potted plants of *Calluna* by Whittaker (1960) indicated that temperatures above 500°C, applied for periods of one minute to the stem base at ground level, were generally lethal, irrespective of the age of the plant. In most normal fires such temperatures are very seldom reached at ground level and the basal parts of the stems commonly survive. It has, however, frequently been observed that the capacity to regenerate after fire bears a close relationship to age. Many authors have suggested about 15 years as the age limit above which regeneration is either very poor or fails altogether (Lovat, 1911; Wallace, 1917; Fritsch, 1927; Gimingham, 1960).

Under field conditions this may in part be due to the higher temperatures normally generated when burning the older stands (Kenworthy, 1963). Experiments by Kayll and Gimingham (1965) on plots of heather aged about 12, 17 and 24 years eliminated this effect by first cutting away the vegetation at about 2 cm above the ground, then using a propane torch to produce a controlled temperature of 400 ± 25°C at ground level for two minutes. Compared with plots in which the *Calluna* was clipped but not heated regeneration was always poorer, but in both cases there was a marked decline with increasing age (Table 24). These results confirm the

TABLE 24

Effect of age on regeneration of *Calluna* after burning
(% of stem bases regenerating after subjection to a temperature of 400° ± 25 °C at the soil surface, for two minutes)
(From Kayll and Gimingham, 1965)

	Age of *Calluna* stand		
Replicates	12–13 years	16–18 years	23–25 years
1	56	9	6
2	80	13	7
3	56	9	17
4	80	10	0
5	46	29	20
6	28	16	11
Mean:	57·7	14·3	10·2

generally held belief that *Calluna* older than about 15 years progressively loses its ability to regenerate freely. Experiments by Miller and Miles (1970) have confirmed these results, showing that, whether *Calluna* is burnt or clipped, regeneration is best when the plants are aged six to ten years and declines thereafter to a minimum in the 26–30 year-age class.

The causes of this declining capacity for regeneration have also been the subject of investigation. The effect of burning is to destroy the whole of the upper part of the plant, but in younger plants this evidently leaves unharmed sites lower down on the woody branches, available for the production of clusters of shoots. Experiments by Mohamed and Gimingham (1970) showed that these sites are related to the points of origin of existing or former main branches (Plate 31), possibly the location of persisting dormant buds on the close-packed nodes just below the base of each long-shoot (i.e. nodes on the overwintering end-of-season short-shoot, which in spring grows out into a new long-shoot, p. 113). Clipping or grazing, which removes only part of the branch system, may leave a number of these sprouting positions at several levels (p. 178). Even after burning some remain undamaged near the stem bases, particularly at points which have become buried in litter and hence protected from high temperatures. Indeed, on occasion sprouts are produced only from the lower side of a procumbent stem base, curving round the stem before emerging and growing upwards; any buds located on the upper side of this part of the stem presumably having been killed by the higher temperature. It is upon the capacity of *Calluna* to develop these clusters of shoots from the stem base that its suitability for management by burning rests (Plate 32).

Sections through stems regenerating in this way show the origins of these clusters of new shoots lying deep in the wood. Mohamed and Gimingham suggested that the continued addition of secondary xylem may eventually engulf the potential growing-points, so that as the plant ages vegetative regeneration from the lower parts of the stem is prevented.

This implies a gradual deterioration in the capacity of any stem base to produce sprouts after a fire, with increasing age. Miller and Miles (1970) have pointed out that superimposed on this is a general decline in the density of stems in *Calluna* stands as they become older, and this no doubt is more immediately and directly responsible for poorer regeneration in relation to unit area. At Kerloch the mean stem density in six- to eight-year-old stands is 1108 per m^2 (these are stems emerging from the ground, not necessarily all representing separate individual plants). The process of 'self-thinning' as the stand ages causes a progressive reduction in density to about 184 stems per m^2 at an age of 34–37. The number of possible sites for vegetative regeneration is reduced in direct proportion.

In Scotland the period during which heath burning is permitted by law runs from 1 October to 15 April (with the possibility of special permission to extend the period to 30 April in a wet season or even 15 May at altitudes above 457 m), but there has always been a controversy concerning the relative merits of autumn and spring burning. In the north of Scotland burning is carried out largely in spring (mainly March and early April), but elsewhere use is also made of the late autumn after the flowering of *Calluna* is over. However, observations by Miller and Miles (1970) in north-east Scotland have shown that regeneration is more vigorous after autumn burning than after burning in spring. (Regeneration after cutting was, however, best in spring.) The reasons for this difference remain obscure. It has been observed that after grassland fires new shoots may appear several weeks earlier than usual (Daubenmire, 1968). If this response is paralleled in *Calluna*, the new growth after a spring fire might be subjected to adverse weather conditions, whereas after an autumn fire the onset of winter would hold it back until the normal time for resumption in spring. The latter, however, is conjecture, but it seems clear that a tradition of spring burning, although perhaps in some areas dictated by availability of suitable weather conditions or by other factors, does not produce the best vegetative regeneration.

Regeneration from seed

If it is necessary to burn old stands, or in the event of fire which has got out of control and has produced excessive heat, vegetative regeneration will fail and reliance must be placed on regeneration from seed. Under these conditions six or more years may be required before complete cover is re-established, whereas with vigorous vegetative regeneration this may be achieved by the fourth or fifth year after burning (occasionally three years). However, even when vegetative regeneration is successful, it is reinforced by seed germination and the establishment of seedlings, which may effectively fill gaps between the groups of shoots clustered at the old stem bases. Hence it is important to determine the effects of burning on regeneration from seed.

Large numbers of seeds are always present in the soil and litter below *Calluna* stands, and it has been shown (Chapter 5) that they retain viability for several years. Moreover, many seeds must blow on to areas cleared by fire after the next flowering season. It has been found that storage at a relative humidity of 100% is followed by rapid germination when other conditions become favourable, and it has been shown (p. 51) that under dense *Calluna* the atmosphere at ground level (particularly if there is an additional stratum of moss or lichen) remains saturated for the greater part of the year. The rate and amount of germination are improved when temperatures fluctuate rather than remain constant, and when seeds are exposed to light. These conditions obtain in a bare area after a fire, which is therefore generally favourable for effective germination. Furthermore, a well-managed fire normally leaves a more compact seed-bed than is provided by loose, fresh litter.

There is also some indication that the germination of *Calluna* seed may be improved by the passage of fire across an area. This is suggested by the results of experimental heat pre-treatment of seeds, discussed in Chapter 5, p. 91. These showed that while 200°C is normally lethal, short periods of exposure to temperatures between 40° and *c.* 120° may increase the percentage and rate of germination. In properly controlled heath fires the temperatures reached at ground level normally remain below 200° (p. 190), and their duration is usually between a half and one minute. Germination would thus be stimulated, and even when higher temperatures are experienced the insulating properties of the surface soil or litter are such that only seeds lying on the surface would be killed. Many others buried just below the surface would receive a pre-treatment possibly resulting in improved germination. In this way *Calluna* may be regarded as fire-adapted. When weather conditions are suitable in the few weeks after a fire seedlings may appear in large numbers. Generally these spring from seeds which were lying in the soil surface, sometimes from seeds buried at depths of up to 1 cm if those at the surface were killed by unusually high temperatures.

From the foregoing considerations it appears that both the shoot morphology of *Calluna* and its physiology of germination contribute to the success with which it has for generations been managed by burning. There is, however, no intrinsic advantage in burning as a method of management: it serves merely to remove much of the canopy and branch system, thereby promoting production of the clusters of new shoots. Better results can be obtained by cutting which generally leaves a greater number of potential sprouting centres undamaged. Attempts have been made to take advantage of this by cutting *Calluna* stands instead of burning, using equipment designed for the purpose (for example the 'swipe', which employs a swiftly rotating chain). The disadvantages of this method, however, are that, once cut, the branches either lie where they fall and interfere with regeneration by suppressing the young shoots or, if removed, carry away from the system a considerable quantity of mineral nutrients. Further, there are some limits to the type of terrain over which tractors can operate whereas fire can be used almost anywhere. The 'swipe' also tears rather than cuts the stems, so possibly causing rather greater damage to the plants.

Effects on production

Production rates in stands regenerating after fire, and their subsequent changes with increasing age, have been discussed in Chapter 8. In terms of the individual plant, a new curve of production with increasing age is started after burning, but this is unlikely to follow the same path as the curve resulting from the growth of a seedling. A plant regenerating vegetatively shows a faster rate of production, because numerous leading long-shoots are produced in the clusters in contrast to the single leader of a seedling and these draw upon a complete mature root system. In the context of whole stands regenerating vegetatively after fire, this effect can be seen in Chapman's curve of production in relation to age of stand (Fig. 37).

Effects on nutrient loss

The effects of burning on the total fund of nutrients in the ecosystem will be examined in Chapter 11; here reference will be made only to the immediate fate of the nutrients contained in the aerial parts of the vegetation at the time of a fire. In fires which successfully clear most of the vegetation, deposition in the ash and loss in smoke (Plates 29, 30) must account for the total quantity of the contained elements. For some nutrients, losses in smoke are important: in addition to the particles which blow away, nitrogen, for example, is readily volatilized, being present in the plant mainly in organic compounds. Similarly, potassium and to a lesser extent other elements may be lost in smoke, particularly at the higher temperatures reached in heath fires.

Calluna has generally been used for experimental determination of the loss of nutrients on ignition. Although not the sole component of any heath community it generally accounts for over 90% of the dry matter, while the composition of other species seldom departs materially from that of *Calluna* (Thomas and Trinder, 1947). Kenworthy (1964) demonstrated the increasing quantities of nitrogen released from samples of *Calluna* ignited for five minutes at 300°, 400°, 500°, 600°, 700° and 800°C. These rose steeply above about 300°C, reaching 80% at about 800°C. Potassium and iron were lost in significant quantities above 400°C, and smaller amounts of phosphorus above 600°C. In the case of nitrogen, the amount released was shown to be a function of the duration of high temperature, and this may apply to the other elements as well. Although in heath fires the maximum duration of high temperatures is usually less than five minutes, Kenworthy's results were closely paralleled by Allen (1964) and Evans and Allen (1971), who burnt samples of *Calluna* under conditions approximating to those of a heath fire.

Allen compared the effects of burning at temperatures characteristic of a normal fire (550–650°C) with those of a severe burn (800–825°C) (Table 25). Of the elements which form volatile compounds, carbon, sulphur

TABLE 25

Losses of nutrients in smoke from burning *Calluna*
(percentages of the original content, corrected for dry weight)
(From Allen, 1964)

	Total losses	
	'Normal' burn 550°–650°C	'Severe' burn 800°–825°C
K	1·4	4·9
Ca	0·1	2·4
Mg	0·4	2·1
C	60·5	67·5
N	67·8	76·1
P	0·6	3·5
S	50·2	56·3

and nitrogen, over 50% of the quantity contained in the original samples was lost, even at the lower temperatures, while the loss of nitrogen approached 80% in the hotter fire. According to temperature, from 1·4 to 4·9% of the potassium and 0·4 to 2·1% of the magnesium was lost, and another experiment indicated a substantially increased loss of potassium at 900°C. Significant losses of calcium and phosphorus (2·4 and 3·5% respectively) occurred at about 800°C, though little was lost at the lower temperatures.

When *Calluna* was burnt both under simulated field conditions and in a laboratory experiment, Evans and Allen (1971) found losses varying between 10% and 20% for elements other than N, S and C, in the order K > Zn > Na > Cu > Fe > Mg > Ca > P > Mn and Si. Experiments to determine the amounts recoverable from smoke showed that only small quantities of Ca, Mg, Mn and P are lost. On the other hand, smoke losses of Na, K, Fe, Zn and Cu may be relatively high.

Not all nutrients contained in the smoke from a heath fire are necessarily lost from the area. Some particles may be deposited in the neighbourhood and some fractions of the smoke may condense not far from the fire (Allen, 1964); but Evans and Allen (1971) found that only a very small proportion of the quantity of nutrients incorporated into smoke was returned to the ground in particles deposited within 120 m of the fire. They comment that convection currents would carry fine particles into higher air streams, which might remove them beyond the limits of the heathland. Consequently, losses must occur and these certainly increase as the temperature of the fire increases. If the temperature in the canopy can be prevented from rising much above 400°C the losses of important nutrients, apart from nitrogen, are minimized. This provides a further reason, additional to the effects upon regeneration, for carefully controlling the intensity of heath fires and avoiding circumstances which lead to high temperatures.

The remainder of the nutrient elements contained in the vegetation is deposited on the soil surface in the ash (Plate 33). Depending upon the solubility of these minerals, they are liable to extraction (leaching) in rainwater. Experiments by Kenworthy (1964) and Allen (1964) demonstrate that potassium is taken into solution much faster from ash than it is from unburnt heather litter, whereas the reverse seems to hold for phosphorus. Kenworthy suggests that burning has the effect of 'fixing' phosphates in some manner, rendering them less soluble. The rates at which calcium and magnesium are dissolved from ash are intermediate between those for potassium and phosphorus.

Effects on community composition

Regular burning has a profound effect upon the floristic composition of heath communities, partly through the progressive elimination of species directly susceptible to burning, and partly as a result of the increased dominance of others. As already indicated, all tree species are eliminated because of susceptibility, at least when young, to fire. *Juniperus communis*, once widespread on British heaths as indicated by the occurrence of pollen and macroscopic remains in peat and humus, probably owes its decline to this cause; for it has survived in situations (such as islands in lakes, and amongst rock outcrops) which escape regular burning. It is also widespread in the heaths of Scandinavia and north-west Europe where burning has never been regularly practised.

Other species, particularly those which lack persistent underground systems capable of vegetative regeneration or the capacity for rapid establishment of seedling populations after a fire, are also progressively reduced or eliminated. On the other hand, for species possessing one or both of the properties mentioned each burn may result in an opportunity for expansion, depending upon other habitat factors: examples are *Calluna vulgaris*, *Erica cinerea*, *Erica tetralix*, *Vaccinium myrtillus*, *Pteridium aquilinum*, *Deschampsia flexuosa*, *Trichophorum cespitosum*, *Nardus stricta*, *Molinia caerulea* and others. These are all vigorous species in heathland habitats, and with the exception of *Pteridium* have a dense growth form, either that of the dwarf-shrub or the caespitose graminoid. Rapid regeneration of one or more of these species after a fire places them at a competitive advantage over many other species of heathland communities. Where regeneration of *Calluna* is rapid a dense, even-aged stand is formed reconstituting a closed canopy in four to six years. From this stage onwards the whole area of the burn is occupied by building-phase *Calluna*. It has been shown in Chapter 7 that in this phase the influence of *Calluna* upon the environment at ground level is such as to exclude all but a few scattered bryophytes and sometimes limited quantities of *Erica* spp. or *Vaccinium myrtillus* forming a subordinate stratum. Since the objective of good burning management is to keep as large a proportion as possible of the total area of heath under building-phase *Calluna*, the associated flora is inevitably restricted by the competitive vigour of the dominant, in addition to the direct effects of fire. The average number of species including vascular plants, bryophytes and lichens in 4 m^2 of *Calluna*-dominated heath managed by regular burning lies between 13 and 17, as compared to 23–4 in undisturbed *Calluna* communities (Gimingham, 1964*b*).

When, under circumstances to be considered below, burning leads to invasion by other species such as *Nardus stricta*, *Trichophorum cespitosum* or *Pteridium aquilinum* instead of *Calluna*, these have much the same effect in restricting the diversity of the associated flora.

Secondary successions following burning

Burning management aims to promote the re-establishment of dominance by *Calluna* but the time required for this after creation of a bare area on a mature soil naturally allows prior entry of other species. Secondary successions are therefore initiated by the fires, and the course they follow is determined by the nature of the habitat and the rate of recovery of *Calluna*. As long ago as 1915 Fritsch and Salisbury, describing the succession following burning on Hindhead Common, Surrey, showed that the exposed surface humus is first colonized by algae and lichens. They drew attention particularly to *Cystococcus* and other gelatinous algae. Another frequently encountered species occupying bare areas of humus or peat, often after burning, is *Zygogonium ericetorum*, a member of the Chlorophyta. In places this alga may form quite extensive felts covering the surface.

Damp *Calluna* litter overlying peat is often extensively colonized by *Lecidea uliginosa*, sometimes in association with *L. granulosa*. Although frequently overlooked, *L. uliginosa* has been found in north-east Scotland by G. A. M. Scott (personal communication) to cover large areas of burnt ground during the season immediately following the fire. The lichen forms a thin skin loosely binding the particles of litter and humus, and producing groups of apothecia by which it may be recognized. Although no measurements have yet been made, it seems probable that the physical properties of the surface may be considerably affected by the lichen, particularly in regard to the penetration and retention of rain water in the surface. Observations of the development after fire of a kind of 'skin' which discourages downward drainage, may sometimes refer to lichen colonists of this type.

The areas occupied by these species have been shown to reach a maximum in the first or second season following burning, after which they give place to other colonists. In the course of a detailed investigation of Dinnet Moor, Aberdeenshire, Ward (1970) showed that subsequent developments follow a somewhat different course on soils with strong development of the organic horizon (Ao) from that on predominantly mineral soils. On the former, in the period from 5 to 13 years after burning many stands are rich in certain species of *Cladonia*, particularly *C. chlorophaea, C. subulata, C. fimbriata, C. floerkeana, C. glauca, C. pityrea, C. squamosa* and *C. uncialis*. Associated with these are a limited number of bryophytes, particularly *Polytrichum piliferum, P. juniperinum*, and *Cephaloziella* spp. Such stands are associated here with early stages of recovery after fire, though they share many species of *Cladonia* with other lichen-rich heath communities in which the *Calluna* canopy is also low and irregular though not because the stand is young (Fig. 8, p. 44).

As the re-growth of *Calluna* proceeds and a denser canopy is formed, there is a shift from the species listed above towards increased importance of various mosses, at first *Hypnum cupressiforme* and *Dicranum scoparium*, later *Pleurozium schreberi, Hylocomium splendens*, etc. This, too, is the stage at which *Cladonia impexa*, unlike the other *Cladonia* species mentioned, may become abundant. This pattern of change, fully documented by Ward, agrees in most respects with that outlined for the Surrey heath by Fritsch and Salisbury and is probably very general for organic soils.

Where, however, the organic layer is thin or lacking, although there may be a phase of colonization by *Cladonia* and *Polytrichum* species, this is more quickly curtailed owing to the entry of a variety of herbaceous and grass species, for example *Agrostis tenius, A. canina, Festuca ovina, Sieglingia decumbens, Carex pilulifera, Campanula rotundifolia, Antennania dioica*,

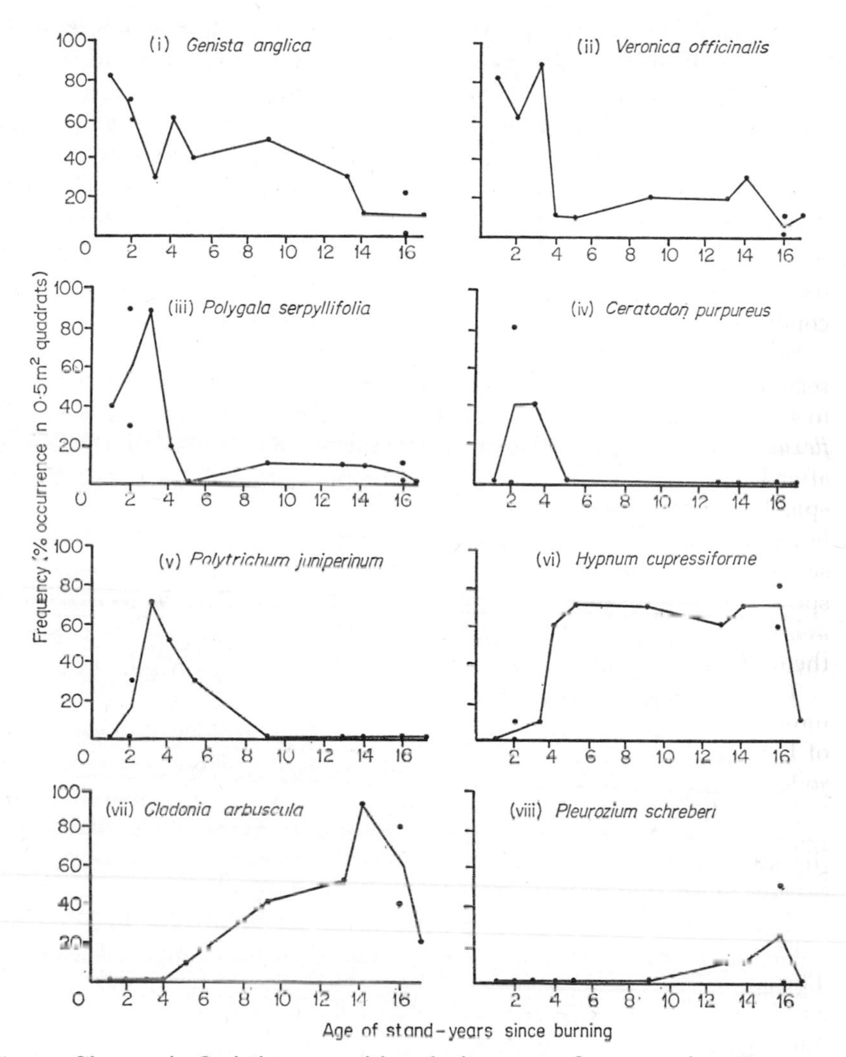

Fig. 49. Changes in floristic composition during a post fire succession. Frequency of selected species in 0·5 m² quadrats in stands of various age. (Figures by Miss E. M. French.)

Galium saxatile, Lotus corniculatus, Lathyrus montanus, Potentilla erecta, Polygala serpyllifolia, Veronica officinalis, Viola riviniana. These flourish in the early years after burning, though later they too may decline or disappear as a dwarf-shrub canopy develops, whereupon the proportion of bryophytes may increase as on the more organic soils. (Fig. 49.)

Whittaker (1960) pointed out that the rate at which these secondary successions take place is determined by the rate of regeneration of *Calluna* and other Ericaceous species, and this in turn is influenced by the age of the stand before burning, the intensity of the fire and the environmental conditions during and after burning. Management aims at rapid regeneration, minimizing the seral stages described above. On the drier ground there are few species, other perhaps than *Erica cinerea*, which will regenerate faster than *Calluna*, but on wet peaty soils *Erica tetralix*, *Juncus squarrosus*, *Trichophorum cespitosum*, *Eriophorum vaginatum*, *Nardus stricta* and *Molinia caerulea* may increase to some degree. All these possess either rhizomatous structures or a dense caespitose habit and are consequently little damaged by fire. While *Calluna* may sometimes re-establish its former cover, increases in the other species are liable under these conditions to be permanent (cf. p. 199).

Similarly, on the more freely drained sites any factor which retards the regeneration of *Calluna* may lead to a temporary or prolonged increase in any other species which is capable of fairly rapid re-growth. *Deschampsia flexuosa* very frequently becomes dominant in the first or second season after burning; on the richer soils *Agrostis* and *Festuca* spp. become conspicuous. However, these grasses give place in time to *Calluna* unless the latter is adversely affected by grazing. In other communities similar sequences involve a rapid spread of one or more of the other Ericaceous species such as *Vaccinium vitis-idaea*, *Empetrum nigrum* or *Arctostaphylos uva-ursi*, with a gradual decline as the *Calluna* canopy begins to over-top them after periods of from three to six years.

The secondary succession may be even further prolonged in cases where most of the original stand is killed out by the fire. Here re-establishment of Ericaceous species is entirely from seed. On the predominantly mineral soils, seedlings of *Erica cinerea* often outnumber those of *Calluna* (p. 102; Gimingham, 1949) and dominance may be exerted by *E. cinerea* for six years or more. Repeated observations on fenced plots have shown, however, that the proportion of *Calluna* in the vegetation gradually increases until eventually it exceeds that of *E. cinerea*, and that this change is hastened by grazing.

Correct management by burning

The results and considerations discussed in this chapter lead to the formulation of a set of principles which must govern heathland management by burning if it is to achieve the desired aim of maintaining the production of edible material by *Calluna*, while at the same time causing minimum deterioration of the environment. These are as follows:

1. Stands should not be allowed to pass into the mature or degenerate phase. Burning should therefore be carried out in the late-building phase. Because the age at which this growth phase is reached varies in different habitats, a better guide to readiness for burning is the average height of the stand. When this approaches 30–38 cm the

stand should be burnt: this may often be at an age of twelve to fifteen years but in some situations a shorter rotation may be necessary, whereas elsewhere (particularly in exposed localities) a longer period should elapse before burning.

2. Burning should be carried out in the conditions which produce a fire hot enough to consume most of the above-ground parts of the plants, but preferably not exceeding a temperature of 500°C in the canopy. Temperatures at ground level should generally remain below 200°C.
3. Numerous small fires are usually preferable to fewer large ones (Chapter 12). Each patch burnt should not normally exceed 2 ha in area, the best size probably lying between 0·5 and 1 ha. It is more satisfactory to burn long narrow strips (e.g. 32 m wide) than to burn square or circular areas (Plate 34).

A comprehensive guide to the practical management of *Calluna* is given by Watson and Miller (1970).

Risks of mis-management

The scope for mis-application of fire, whether in ignorance or by accident, is considerable (Gimingham, 1971*a*). It has been pointed out (p. 202) that on poorly drained ground, burning may lead to the partial or complete replacement of *Calluna* by other species. McVean and Lockie (1969) express the view that in the more highly oceanic conditions of western Britain where peat forming-vegetation is widespread, management by burning is generally inappropriate because it has been responsible for the extensive spread of species of low nutrient content and palatibility to grazing animals, such as *Nardus stricta* and *Trichophorum cespitosum*. Further, regeneration is invariably slow, introducing a danger of peat erosion. In places this is severe, leading to the formation of peat 'hags' or islands of peat separated by gullies down to the underlying mineral material (Plate 31). Despite this danger, burning, often of large blocks, is still frequent in the west of Scotland in the mistaken impression that even there it improves the forage for red deer and sheep. Management by burning is more appropriate where rainfall is rather lower, for example, less than 125 cm year^{-1}, and where the annual potential water deficit (Chapter 2) exceeds *c.* 12·5 mm. However, even under these conditions mistakes are made, and any circumstance which results in delayed regeneration may have damaging consequences. Delays in regeneration may occur if weather conditions have been misjudged and the fire becomes excessively hot, or if the *Calluna* being burnt exceeds about 15 years in age (p. 193). Sometimes these circumstances cannot be avoided: the wind may change or freshen during burning, and it is often necessary to burn old stands because increasing costs and rural depopulation make it impossible to maintain the recommended frequency of fires. The results are two-fold. First, the surface humus may be exposed to erosion by run-off of rainwater (Plate 1) or, on the steeper slopes, to gully erosion (both are common and are the more severe because the free-range system of sheep husbandry is responsible for further delay in the re-establishment of cover). Second, the extended period during which the competitive vigour of *Calluna* is low permits unwanted species to extend. Notorious among these is *Pteridium aquilinum* and particular care has to be taken with burning if the spread of this plant is not to be encouraged.

On occasion, a fire gets completely out of control. Not only is an excessive area burnt but in dry conditions the surface humus or peat may burn and continue to smoulder for days or weeks. Sometimes the humus layer is entirely destroyed, exposing the leached A_2 of the mineral profile. Even if this extreme result is averted, most of the buried seeds are destroyed and for years the colonizing community may be confined mainly to lichens and mosses such as *Polytrichum juniperinum*. Such fires may occur, rarely, if normal burning gets out of control, but more often they are started accidentally, as for example in the past by stray sparks from steam railway engines or by ill-guarded picnic fires. The re-establishment of closed *Calluna*-cover may also be prevented by excessively frequent fires of lesser intensity, and although not so common a fault as insufficient burning this sometimes stems from aimless firing of heathland in ignorance of the erosion and destruction of wild-life which result.

The monoculture of *Calluna*

The outcome of effective management of heathland by fire is to all intents and purposes a monoculture of *Calluna*. Dense, even-aged stands of building *Calluna* show all the well-known properties of monocultures, including on occasion susceptibility to outbreaks of pests such as heather-beetle (*Lochmaea suturalis*). As shown elsewhere, *Calluna* has a marked effect upon the micro-environment (pp. 109, 138–40), particularly on the soil, and these effects are at their maximum in the late building phase. Litter accumulates in quantity on the soil surface (up to about 17 000 kg ha^{-1}, Chapman, 1967). The bulk of this litter consists not of leaves alone, but of leaf-bearing short-shoots, and a substantial proportion comprises also woody long-shoots, small branches, and at certain times of the year, flower parts. This material is rich in lignin: consequently breakdown is slow, mor humus accumulates and podsolization processes are promoted. Insufficient work has, as yet, been done to quantify the effects of monoculture of *Calluna* in hastening podsolization, but observations suggest that these are significant. Thin, hard iron pans in the B horizon are frequent in the heath soils of northern Britain, sometimes impeding soil drainage to the extent that the upper horizons became waterlogged. Peat development follows, and the composition of the community may then change in the direction of acid bog. Such considerations must be taken into account in weighing up the ecological consequences of long-continued heathland management by burning. Recent research seems to suggest that carefully controlled burning may maintain a good *Calluna* crop without significantly depleting the ecosystem of nutrients, with the possible exceptions of available nitrogen and phosphorus. It may, however, reduce the variety of plant and animal life and may slowly, over long periods of time, encourage podsolization. Where control is less effective, positive signs of erosion may be detected quite quickly.

11 The nutrient budget

Fire causes extensive temporary changes in the structure of the ecosystem. One of its effects, among others, is to return to the soil that fraction of the total fund of nutrients which is normally contained in the aerial parts of the vegetation, less any which have been lost to the system in drifting smoke. Considerable controversy has surrounded the problem as to whether or not the *status quo* is resumed as the community regenerates, and whether or not a regime of regular burning causes long-term change in the ecosystem.

Some preliminary lysimeter-type experiments by Elliott (1953) suggested that considerable quantities of the nutrients contained in ash deposited on the soil surface during a fire could be dissolved in rainfall. These might be lost to the system by downward drainage through the soil profile or in run-off from the surface. He calculated that 2·5% of the total fund of nutrients contained in the vegetation and upper horizons of the soil might be lost after each fire. Allowing for the possibility of about ten successive fires in the same area in a 200-year period (a conservative estimate, Chapter 10), a continuing decline in nutrient content already amounting to perhaps 25% was envisaged. This suggestion together with an apparent decline in the productivity and 'carrying capacity' of the upland areas during this period, attracted a good deal of attention.

However, no complete assessment of the impact of management upon nutrient status is possible in the absence of basic facts of the nutrient budget in this type of ecosystem. Some of these facts have been supplied by recent research and a tentative approach can be made towards the construction of input and output balance sheets for some of the major nutrients.

The main pathways of input and loss to a heathland system which have to be quantified for each important nutrient are shown in Fig. 50. Brief discussion follows of the nutrient fund in the system, the main inputs and losses and the effects of management practices. Reference will then be made to attempts to draw up balance sheets, followed by tentative conclusions in respect of each of the main nutrients.

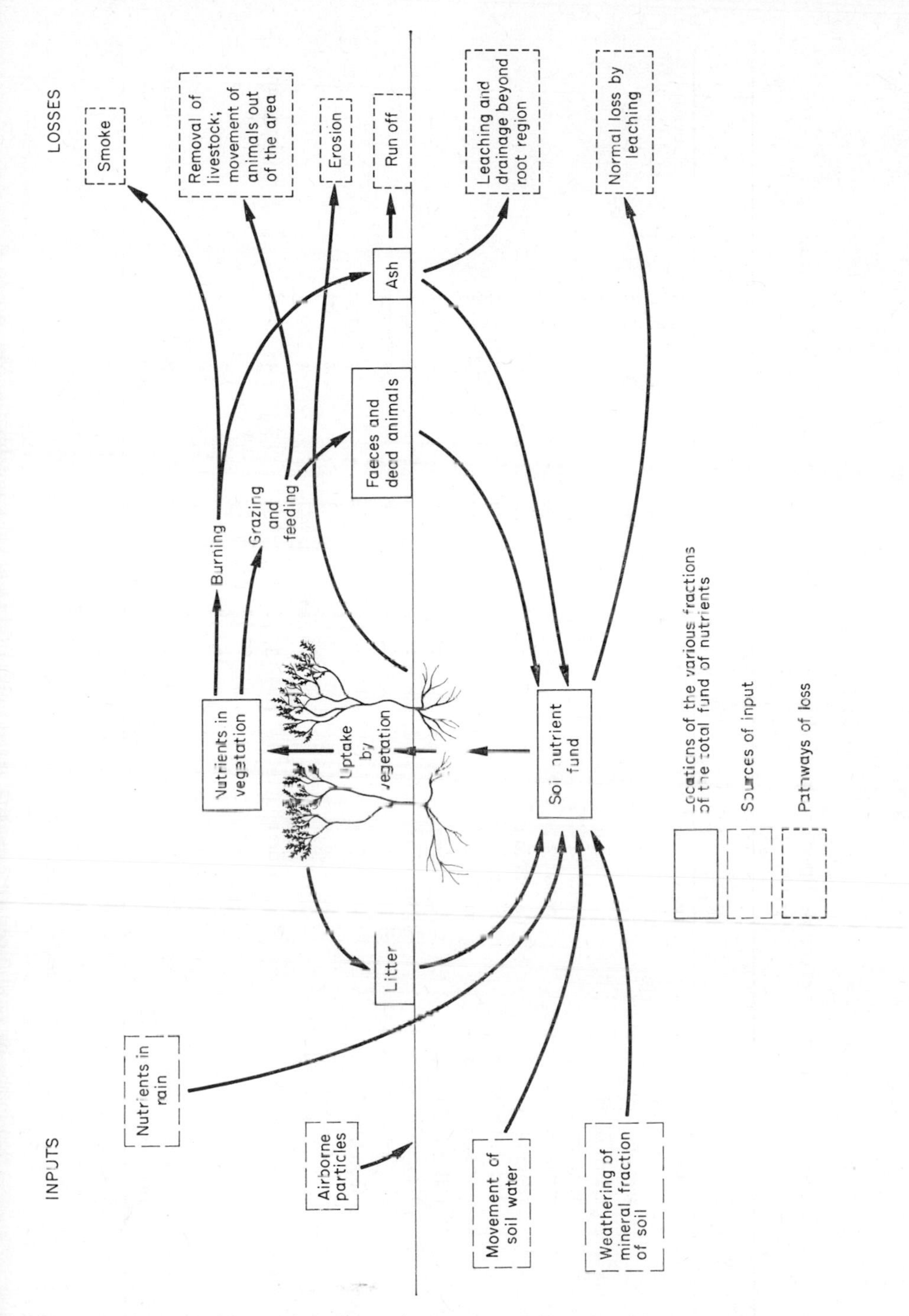

Fig. 50. Diagram illustrating the main sources of input and loss to the nutrient fund in heathland vegetation and soils.

TABLE 26

The nutrient fund in vegetation litter and the rooting region of the soil in heath stands of varying age
kg ha^{-1}

Age since burning	Location		Ca	Mg	K	Na	P	N	Source
2 years	Blanchland	Vegetation	18	7	25		4	55	Robertson
	Moor, Hexham,	Litter	Nil	Nil	Nil		Nil	Nil	and Davies
	northern England	Soil*	111	61	238		432	6396	(1965)
3 years	Cairn o' Mount,	Vegetation	14	6	22		4	59	,,
	Kincardineshire,	Litter	5	1	2		1	28	
	north-east Scotland	Soil†	605	376	189		207	5794	
5 years	Blanchland	Vegetation	32	12	45		6	89	,,
	Moor, Hexham	Litter	6	2	4		2	54	
	northern England	Soil*	108	62	209		362	5675	
8 years	Cairn o' Mount,	Vegetation	19	9	34		5	92	,,
	Kincardineshire,	Litter	11	3	3		3	61	
	north-east Scotland	Soil†	662	401	216		207	5951	
> 10 years	Blanchland	Vegetation	34	11	45		5	102	,,
	Moor, Hexham,	Litter	6	2	5		2	63	
	north-east England	Soil*	66	19	127		334	1991	
12 years	Poole Basin,	Vegetation	33	13	34	5	4	108	Chapman
	Dorset,	Litter	15	4	5	1	4	75	(1967)
	southern England	Soil‡	229	236	288	84	37	2210	
> 15 years	Cairn o' Mount,	Vegetation	39	17	56		8	192	Robertson
	Kincardineshire,	Litter	13	4	3		3	58	and Davies
	north-east Scotland	Soil†	591	321	178		166	5394	(1965)

* Depth Sampled – 15 cm. Exchangeable (in 1 N ammonium acetate) Ca, Mg, K. Total P and N.

† Depth sampled – 17.5 cm. Total Ca, Mg, K, P and N.

‡ Depth sampled – 20 cm. Total Ca, Mg, K, Na, P and N (perchloric acid digests).

Trends with age in the nutrient fund stored in the vegetation, and contributed to the litter, can be followed despite considerable variation in the content of certain elements in the soils of different sites, notably Ca, Mg, K and P.

The fund of nutrients in vegetation and soil

Measurement of the total fund of major nutrients in the vegetation, litter and rooting region of the soil have been made in different parts of Britain by Robertson and Davies (1965) and Chapman (1967). Extracts from their results are quoted in Table 26. Changes in the distribution of major nutrients between the above ground biomass, litter and soil as stands increase in age are illustrated by these figures. Owing to the fact that on managed heaths stands are seldom allowed to exceed 15 years of age, data for older examples are scarce, while comparative figures for unmanaged stands of uneven age-structure are lacking. All the published analyses have been made in connection with studies of the effects of management.

Immediately after burning, the total fund of nutrients in the system is located in or on the soil, but this is drawn upon by the regenerating vegetation. Table 27 shows changes with age of stand in the ratios of

TABLE 27

Approximate ratios of nutrients contained above ground (vegetation plus litter) to those in the rooting region of the soil in heath stands of varying age.

(Calculated from Table 26)

[Nutrient fund above ground taken as 1]

Age since burning	Location	Ca	Mg	K	Na	P	N
2 years	Blanchland Moor, Hexham, northern England	6·2	8·7	9·5		108·0	116·3
3 years	Cairn o' Mount, Kincardineshire, north-east Scotland	31·8	53·7	7·9		41·4	66·6
5 years	Blanchland Moor, Hexham, northern England	0·8	1·1	1·3		45·3	39·7
8 years	Cairn o' Mount Kincardineshire, north-east Scotland	22·1	33·4	5·8		25·9	38·9
> 10 years	Blanchland Moor, Hexham, northern England	1·7	1·5	2·5		47·7	12·1
12 years	Poole Basin, Dorset, southern England	4·8	13·9	7·4	14·0	4·6	12·1
> 15 years	Cairn o' Mount, Kincardineshire, north-east Scotland	11·4	15·3	3·0		15·1	21·6

Ratios are affected by differences in the content of certain elements, notably Ca, Mg and P, in the soils of different sites, but the influence of the developing stands of vegetation can also be detected.

quantities in vegetation and litter to quantities in the rooting region of the soil, for several nutrients. The actual values depend upon soil type, but the trends are repeated in each set of samples. At first no part of the total nutrient fund is contained in litter (except any which may remain from beforet he fire), but as shown by Chapman (1970) this becomes an increasingly important location of nutrient accumulation. The effects of fire, to be discussed later, will clearly vary greatly according to the age of the stand. If stands are burnt while still young, a high proportion of the nutrient content of the system remains unaffected in the soil, but with increasing age (at least up to 20 years) larger fractions will be affected because they have been taken up by the vegetation.

Inputs to the ecosystem

Theoretically, weathering of the soil parent material or breakdown of the mineral particles of the soil may yield nutrients to the system. This, however, may be discounted as a significant contribution to the nutrient fund. In many types of heathland ecosystem most of the plants root exclusively or largely in organic matter (peat, or the mor humus horizons of podsols). Even where organic layers are thin and there is appreciable rooting in mineral horizons, these are generally derived from slow-weathering nutrient-poor materials which have been subject to leaching for long periods.

Locally, lateral movement of soil water into a heathland area may provide a more important input of nutrients, especially if passage through richer strata or soils has occurred on the way. This is often the case where heaths occupy the lower slopes of hills, or valley floors. Normally, however, the flushed area is localized and recognizable on account of replacement of the heath species by grass, sedge or rush communities. Heaths subject to enrichment in these ways have not been considered in the studies under review.

It is likely that the deposition of air-borne particles of various origin (Tamm, 1958) may contribute a significant input of nutrients to an area of heath. Unfortunately little is known of the magnitude of this source, and until more estimates have been made no allowance can be made for it.

One further means by which substantial quantities of nutrients can be delivered to the ecosystem is in solution in rainfall. Throughout the heath region rainfall is high; moreover the location of heaths on the western margin of the European continent, subject to a prevailing westerly air-stream arriving from the Atlantic, implies that precipitation will be relatively highly charged with certain ions. Table 28 incorporates analyses of rainfall arriving at certain heathland sites and shows that there is an appreciable content of a number of elements, notably potassium, calcium and nitrogen, also sodium but only to a lesser extent phosphorus. (There is much variation in the reported concentrations of phosphorus in rainfall samples, but these are always small and errors are liable to be large in proportion.)

TABLE 28

Input of nutrients in rainfall at three localities in Britain
kg ha^{-1}year^{-1}

	Ca	Mg	K	Na	P	N	S	Source
Scotland: Means from 3 recording stations	7	4	3			6		Robertson and Davies (1965)
Kerloch, Kincardineshire	6·7	3·6	3·9	28	0·19			Allen Evans and Grimshaw (1969)
Means from Moor House, north Pennines and Grizedale, east Lancashire	9	5	3–6	25·5	0·4–0·7	8–9	12	Allen (1964) Crisp (1966)
Broxa Forest, north-east Yorkshire	9·8	4·5	4·8	36	0·77			Allen, Evans and Grimshaw (1969)
Merlewood, Lancashire	12	3·6	3·7	22	0·97			
Abbot's Moss, Cheshire	14	2·9	5·4	14	0·81			
Dorset, southern England	4·7	5·6	1·2	25·4	0·01	5·2		Chapman (1967)

The nutrients in rainfall are important to the ecosystem only if they are either taken up directly by the vegetation or retained in the soil of the rooting region. The proportion absorbed directly by the plants is likely to be small, except perhaps where lichens or mosses, especially *Sphagnum* (Allen, 1964, Clymo, 1963), are prominent, particularly in view of the fact that much of the rain falls in winter when the vegetation is inactive. Allen (1964) carried out some experiments to determine the quantities of nutrients that can be held in heathland soils and found that peat, and a clay soil from the edge of a heath, retained important quantities from solutions applied to them, particularly in the cases of potassium and calcium, whereas this did not apply to a sandstone soil. Small proportions only of magnesium and phosphorus were retained, and the soils were not tested for nitrogen retention. Other work reported by Allen indicated that peat may be capable of taking up in a year about 0·1 g m^{-2} potassium and 0·3 g m^{-2} nitrogen from rainfall in a year but only 0·005 g m^{-2} phosphorus. However, the actual proportion of the nutrient input in rainfall which is retained against leaching and becomes available to the ecosystem remains one of the less well documented of the processes concerned.

Losses to the ecosystem

The loss of nutrients by leaching in rainwater draining through the soil profile represents a normal feature of the ecosystem where podsols rather than peat are concerned, but also one which has not been effectively quantified. Some of the dissolved nutrients may disappear from the area altogether in drainage water, others will be depositied in the B horizons below the rooting region of most of the plants comprising the vegetation. It is generally assumed that the rate of loss by this means is slow and that it can therefore be ignored. While this may be acceptable in the case of peat soils, it may constitute an appreciable error elsewhere, particularly since the breakdown products of *Calluna* are undoubtedly important in mobilizing cations. The downward movement and subsequent deposition of iron and aluminium are well known, and the same process can be demonstrated for most nutrient ions. It is very probable that for this reason the continued dominance of *Calluna*, or its effective monoculture, may contribute towards a continuing process of nutrient loss by leaching. Further research on this hypothesis is urgently needed. Nutrients may also be lost in solution in water which runs off over the soil surface. The amount of run-off depends on the amount and intensity of rainfall, and on the angle of slope of the ground.

It is possible that losses through these pathways are increased after a fire, when nutrients may be taken into solution in rainwater from the ash deposited on the soil surface. This question will be discussed below when the effects of burning are further considered.

Losses due to erosion probably represent a very variable quantity depending on topography and management. Erosion is liable to be most serious on peat, but can be rapid whatever the nature of the soil if the slope is appreciable and vegetation cover is destroyed or disrupted. It follows that if regeneration of the vegetation is for any reason slow or delayed, significant losses due to erosion may occur (Chapter 10). Losses of this kind have been quantified only in one study, that by Crisp (1966) working on the peat ecosystem of the study area at Moor House (for details see p. 144). By filtering off the particles of peat being washed down the drainage streams, he estimated that a mean depth of 0·8 mm, or 1120 kg ha^{-1}, of peat was lost over the whole catchment per year, though most of this was probably derived from an actively eroding 10–20% of the area. Chemical analysis of the filtered particles indicated an appreciable loss of nutrients to the system, as shown in Table 29. In proportion to the total fund of nutrients, the loss of phosphorus was particularly important, while considerable quantities of nitrogen, calcium and potassium were also removed.

Herbivores increase the rate at which nutrients are cycled through the ecosystem, but cause loss to the system only when they move away from an area or are taken off as a crop. Apart from cropping, the movement of herbivores, whether vertebrate or invertebrate, out of any area is presumably more or less equated by inward movements. However, as shown by Crisp (1966) where stream flow is concerned there is a loss of animal

TABLE 29

Outline nutrient balance sheet for an 83 ha catchment at Moor House National Nature Reserve, north Pennines, England.
(From Crisp, 1966 – modified)
kg $ha^{-1}year^{-1}$

	Ca	K	Na	P	N
Outputs					
In stream water	53·7	9·0	45·2	0·4	2·9
By peat erosion	4·8	2·1	0·3	0·4	14·6
Drift of fauna in stream	Trace	Trace	Trace	Trace	0·001
Drift of fauna on surface of stream	Trace	0·005	0·001	0·005	0·055
Sale of sheep and wool	0·02	0·005	0·002	0·01	0·053
Total output	58·5	11·1	45·5	0·8	17·6
Input in precipitation	9·0	3·1	25·5	0·5 – 0·7	8·2
Difference = net loss	49·5	8·0	20·0	0·1 to 0·3	9·4

material in suspension in stream water and also on the water surface. Approximate estimates, based on the contents of filters and surface traps, indicated about 1·4 kg dry weight passing downstream in the water and about 49·8 kg on the surface, from a catchment of 83 ha. Chemical analyses of similar animal material indicates losses of nutrients to the ecosystem of the order shown in Table 29.

The removal of larger grazing animals such as sheep, deer, hares and grouse, may be of greater importance. Each of these may to a greater or lesser degree contribute to a 'crop' removed from heathland ecosystems (Chapter 8). For any particular locality, the dry weight of animal products removed per unit area can usually be calculated from estate records. Since the carcases and wool of sheep constitute by far the largest proportion, other grazing animals have generally been ignored in estimating the quantities of nutrients removed. Using analyses of the chemical composition of sheep carcases and wool available in Lawes and Gilbert (1859, 1883), Sears (1951), Esminger (1956), Spector (1956) and Millar, Turk and Foth (1958), approximate figures can be obtained for the annual loss of nutrients, as shown in Table 29. In view of the relatively low productivity of grazing animals in heathland ecosystems the quantities are not large. The output of nitrogen and phosphorus may, however, be of some significance. The latter, especially, is important in relation to the generally low levels of this element in the system and in the sources of input.

Such estimates refer to the general effects of removal of sheep from relatively large areas. The impact of the animal, whether sheep or grouse or any other grazer, upon a more restricted area varies according to the age of the stand. The shoots of young heather plants are both more palatable and richer in nutrient concentration than those of older plants. Kenworthy (1964) showed that the percentage of the standing crop of heath vegetation in Glen Clova, in the east of Scotland, removed by sheep declined in an approximately linear relationship with age of stand (Fig. 46). Since, however, production per unit area increases during the earlier part of the period concerned, the greatest total loss to the system due to grazing occurs between about seven to ten years after burning; 70% of this loss consists of young shoots of *Calluna*, the rest comprising grasses and other species. At this stage in stand development, young shoots still constitute a high proportion of the total biomass of *Calluna* (Chapter 8) and the total loss of the standing crop of *Calluna* is of the order of 50%. The proportion of nutrients removed is higher than this, because their concentration is greater in the young shoots than in the woody branches. At this stage the output of nutrients from the ecosystem, therefore, represents a higher-than-average loss.

The effects of heath fires on the redistribution of certain nutrient elements has been touched on in Chapter 10. It has been shown that losses in smoke are closely related to fire temperature, and therefore indirectly to the age of the stand at the time of burning. Smoke contains a relatively high proportion of the nitrogen, and lower proportions of the other nutrients which were located in the vegetation at the time of burning. These amounts are generally treated as losses to the system, though some return may be received by deposition when smoke drifts in from other areas. However, upland heaths are often located on the higher ground or on slopes from which smoke is more often lost than received from adjacent sites. Lowland heaths are much dissected and there is strong probability that any nutrients deposited from smoke will benefit land other than the heath. It therefore seems reasonable to regard the nutrients taken up in smoke as in the main lost to the ecosystem, though in so doing some over-estimation may result.

The remainder of the nutrients is deposited on the soil surface in ash (Plate 33), some such as potassium in a relatively high soluble form, others such as phosphorus apparently in a rather insoluble form (Chapter 10). Measurements of the total amounts of most nutrients in the upper horizons of the soil (the rooting region) show marked increases immediately after a fire. It is their subsequent fate which is of importance in relation to the effects of fire on the nutrient budget of the ecosystem. The problem can be approached experimentally using soil blocks, excavated to the depth of active rooting and removed to the laboratory. Controls, on which the vegetation remains intact, can be compared with blocks on which it has been burnt. Alternatively, the vegetation may be removed and replaced by the ash and charred debris produced by burning a known weight of vegetation equal to the average above-ground biomass. Known volumes of water are then sprayed on at intervals to simulate rainfall, and allowed to drain through to a collecting vessel below, ensuring that none can pass directly down the sides of the blocks. Finally, the leachates are analysed for nutrient content. In a preliminary experiment of this kind, Elliott (1953) found that the quantities of calcium and potassium leached were considerably increased as a result of burning. Measurements of total exchangeable bases in the leachate gave the results quoted on p. 206, which could, it appeared, amount to a serious depletion of nutrients. However, further experimentation along similar lines has modified this conclusion. Allen (1964), using synthetic rainwater (i.e. distilled water with ions added in concentrations equivalent to mean values for incoming rainfall) to leach soil blocks, found that, with the exception of calcium, the addition of ash did not increase the quantity of nutrients removed by leaching. Instead, the quantity recovered in subsequent analyses of the litter and upper peat layers indicated that most of the added nutrients were retained there. Herein lies a difference from the circumstances of Elliott's experiments, which were carried out on podsols derived from Millstone Grits, having only shallow litter and raw humus layers. It may be concluded that potential losses are minimized in the more organic soils characteristic of the majority of heaths, as they would be also on clay soils.

This, however, involves the assumption that all the water containing dissolved nutrients is draining through the soil. In fact, percolation is often retarded after burning because fine ash particles are washed into the spaces on the surface layers. This may impede drainage and increase run-off, with its attendant loss of nutrients. Taking all these points into consideration, Allen (1964) estimates the losses after a single fire on the peat ecosystem of Moor House as follows:

K	1 $kg\ ha^{-1}$	(leaching and smoke)	P	0·1 $kg\ ha^{-1}$	(leaching)
Ca	$<0·1$ $kg\ ha^{-1}$	(leaching)	N	45 $kg\ ha^{-1}$	(smoke)
Mg	$<0·2$ $kg\ ha^{-1}$	(leaching)	S	5 $kg\ ha^{-1}$	(smoke)

In a subsequent paper, Allen, Evans and Grimshaw (1969) report the results of similar tests on soils from seven heathland areas. In these experiments *Calluna* ash was spread on the soil, after cutting away the vegetation and briefly burning the surface with a flame gun. Instead of leaching with artificial rainwater the soil blocks or experimental plots were left exposed to natural rainfall for twelve months. At the end of this time on five of the soils the nutrients derived from the ash were almost wholly retained in the surface organic layer. The degree of retention appeared to relate to the thickness of the organic layer which in these soils was over 5 cm, but even in the remaining two a high proportion of the nutrients was retained in an organic layer of *c.* 1 cm overlying a sandy profile. Some movement down the profile was recorded in the latter, but income in precipitation greatly exceeded the amounts carried down. However, nutrients disappeared from the surface rather more rapidly in undisturbed soils than in experimental blocks, perhaps as a result of vertical channelling or surface run-off. Although uptake by regenerating vegetation could be detected, particularly on the soils with the lowest mineral status, it was not possible to account fully for the fate of the nutrients deposited in ash. (Nitrogen was not included in these experiments, which were concerned with sodium, potassium, calcium, magnesium, phosphorus, manganese, zinc and copper.)

Nutrient budgets

If nutrient losses following burning were to exceed compensation from input sources, a stepwise decline in the nutrient fund might be expected to follow from regular burning management. In practice, however, this proves difficult to test, firstly, because the differences at each step would be very small in relation to variation among samples and, secondly, because more than a lifetime would be required to follow trends throughout several burning cycles on one site. Elliott (1953) produced circumstantial evidence of cumulative losses by comparing the nutrient status of heaths of known history with that of relict woodland assumed to represent the ecosystem from which they had been derived. The differences were of an order compatible with his view that repeated burning would cause losses of about 2·5% of the total nutrient fund for each fire. Kenworthy (1964), on the other hand, found little evidence for such losses, except to some extent for phosphorus and perhaps also potassium. His approach was to compare the soil nutrient content in near-by heath stands constituting an age-series of up to 16 years from burning, on the grounds that differences associated with increasing length of the post-fire period should reflect trends in the system as a whole between one fire and the next.

However, reconstruction of a temporal sequence from simultaneous measurements on a number of stands, despite close proximity and apparent similarity of habitat, is always subject to unknown error arising from local environmental variation or unsuspected differences in past development. More informative are attempts to draw up a complete balance sheet for each nutrient concerned. Full or partial balance sheets have been worked out for several different types of heath ecosystems. Although inevitably there is a considerable element of approximation, these make a valuable contribution to an understanding of ecosystem dynamics as far as nutrients are concerned, and of the effects of management.

Table 29 is taken from Crisp (1966) and concerns a catchment of 83 ha on the Moor House National Nature Reserve. This area, already referred to in other connections, is occupied largely by wet hill peat. Much of the vegetation is dominated by *Calluna* and *Eriophorum vaginatum* and so may be regarded as illustrating conditions at one extreme of the range of heath communities. Management here does not involve burning, but the land is used for sheep grazing. The peat surface suffers from active erosion over a considerable proportion of the area (p. 212).

In the absence of burning, the main losses to the system are those resulting from leaching, as shown by the output of nutrients in stream water, and those consequent upon peat erosion. The removal of sheep and wool has a negligible effect on the nutrient balance, although as Crisp points out, sheep may affect the system in other ways. They may, for example, alter the distribution of nutrients and the composition of the vegetation, or contribute to the causes of erosion. The only source of input of nutrients to the ecosystem shown in the Table is rainfall. This, however, is sufficient to replace a very large proportion of the amounts lost. The residual deficits for sodium, potassium and calcium are probably made up by an unknown input from rocks and mineral soils in parts of the area, leaving net losses only in phosphorus and nitrogen. The output of these two nutrients appears to be substantial, but the greatest losses may be to that part of the fund which is of least significance to the vegetation. About 50% of the losses of phosphorus and 80% of those of nitrogen are due to erosion of peat, which has accumulated over many years but probably releases these nutrients in available form only very slowly. When viewed against the total fund, net losses are, in any case, small.

In the ecosystem described by Crisp, changes from year to year are slight and progressive, so that within a reasonable period mean annual figures may be used to build up the balance sheet. When, however, burning is introduced at intervals of several years this is no longer possible. Chapman (1967) tackles this problem for an area of lowland heath in Dorset, on soil derived from Bagshot sands, by stating the total nutrient fund in a stand aged 12 years when burnt, calculating the direct losses on burning and setting against these the sum of the inputs over the whole 12 years (Table 30). He draws attention to the lack of any estimate of losses by

TABLE 30

A nutrient balance sheet for an area of lowland heath, 12 years old when burnt. Dorset, southern England. (From Chapman, 1967)
kg ha^{-1}

	Na	K	Ca	Mg	P	N
Output						
Loss of nutrients on burning (in smoke)	1·5	8·3	12·5	4·0	2·2	173·1
Input						
12 years' input in rainfall	305	14	56	67	0·12	62
Difference						
Gain or loss over 12 year cycle	+303	+5·7	+43·5	+63	−2·08	−111

leaching and run-off during this period, but despite this it is clearly established that the quantities of calcium, potassium, sodium and magnesium lost on burning could be made good from precipitation in the course of a few years. On the other hand, the losses of phosphorus and nitrogen could not be fully compensated from this source. Although the actual amount of nitrogen lost over the whole period is considerably greater than that of phosphorus it represents about the same proportion of the total fund in the 12-year-old stand (4·6% and 4·8% respectively). While these elements are lost at much the same rate as in the Moor House catchment, they are derived entirely from that part of the fund which is being actively cycled through vegetation and soil, whereas losses caused by peat erosion include in large measure ions fixed in organic complexes unavailable to plants. Further data of a type suitable for drawing up balance sheets are provided by Robertson and Davies (1965) and Kenworthy (1964) working in north-east Scotland. Again, these comparisons lack any estimate of a regular output by solution in water leaching through the soil or running off the surface. However, despite differences in detail, the results reinforce conclusions that a large proportion of the possible nutrient loss due to burning is replenished by rainfall during the redevelopment of the stand. This is particularly evident in calcium and magnesium for which the input appears to be more than adequate both to make up for losses and to supply the quantities accumulated in a ten-year-old stand of vegetation. For potassium the levels may be more nearly balanced, but the figures in Robertson and Davies indicate that rainfall is almost adequate to compensate for 'maximum possible losses', while those of Kenworthy suggest some excess. Approximate estimates are given of the removal of nutrients contained in the carcases of animals sold for an area of heath, affecting nitrogen, calcium and phosphorus more than other nutrients. Of these, only for phosphates is the amount lost important in relation to other outputs, and to input levels. Although the requirements of the vegetation for this element are lower than those for any other major nutrient, rainfall is a very poor source of supply and there is evidence of a net loss over a ten-year period.

There is no doubt that losses of nitrogen in burning are substantial, although differences in methods of sampling and estimation have led to a wide range in the figures quoted. As a result it is not easy to determine the extent to which input in rainfall, also considerable, might be adequate to replenish the loss. A sizeable store of nitrogen is contained in the soil but, as Kenworthy points out, little of this is soluble and the vegetation may depend to a large extent upon inputs during the post-fire period. Further support for these conclusions is provided by Allen (1964) from the north of England. Since the effects of burning were the prime concern of this work and no reference is made to removal of animal products, the losses of phosphorus are less apparent. However, Allen's conclusion agrees with those of other authors that, although for most nutrients the amounts arriving in rain are well in excess of losses, the quantities of nitrogen and phosphorus lost as a result of burning can be replaced only if the vegetation and soil retain almost the entire income throughout the normal period between fires. Reference is made to determinations of the concentrations of potassium, phosphorus and nitrogen in stream water, representing the amounts removed in the drainage system. From this it emerges that, although peat is relatively highly retentive of nutrient ions, quite large proportions of the annual input disappear in the drainage water. Although the amounts retained may be about adequate to sustain growth, these findings imply a need for caution in assessing the input due to rainfall. In this instance, for example, a net loss of nitrogen may be occurring. The nutrients in rain represent an effective input only if they are retained in the rooting region or are taken up by the vegetation, and a large proportion of the annual rainfall takes place in winter when root uptake is inactive. Further research on the availability of nutrients contained in rainwater is required before the balance sheets can be interpreted with confidence.

Much still requires to be done before the data are fully adequate, but they demonstrate that it is impossible to generalize about the effects of grazing and burning management on the nutrient status of heathland ecosystems. While it seems that the concern felt at one time about the possibility of serious depletion of resources caused by regular burning cannot be sustained, its effects on the quantities and distribution of each element must be considered separately. Although the input and loss accounts for each nutrient are still far from complete, and the results of different studies are not always in agreement, it is possible to summarize the main findings as follows:

Calcium and magnesium

Although the concentration of magnesium in the vegetation is less than that of calcium these nutrients are alike as far as the nutrient budget is concerned. Most heath soils are not rich in calcium and magnesium, but the vegetation takes them up efficiently and contains relatively large amounts, at least of calcium. On burning, the greater part of this is deposited in ash in a moderately soluble form. Losses due to leaching are not excessive, but the removal of calcium in animal carcases may be significant. However, these elements (especially calcium) are supplied rather liberally in rainfall, and can be effectively retained by organic soils (or by *Sphagnum* when present). There is no sign that burning or any other form of management causes depletion of these nutrients.

Potassium

Potassium is also delivered in relatively large quantities in rainfall, but on the other hand losses due to burning are high. This nutrient is very freely dissolved from ash. While this does not mean that the ions taken up into solution are all lost to the system, especially where the more retentive organic substrata are concerned, the exchange capacity of the soils is low and undoubtedly an appreciable quantity disappears in run-off and drainage water. The demand of the developing stand of vegetation for potassium is high, depleting the amount contained in the soil. Because it is also readily leached out of litter there is little accumulation of insoluble potassium in the organic horizons. There may, indeed, be a slight decline in the total potassium content of the soil in the period following a burn, though at the same time the amount contained in the vegetation increases. Some studies show a small net loss but in general it seems unlikely that any significant decline is occurring as a result of management.

Sulphur

Such evidence as there is suggests a similar conclusion for sulphur. Losses of sulphur in smoke may exceed 50% of the amount contained in the vegetation, but on the other hand the concentration of sulphate in rain may be high, though variable.

Phosphorus

The findings for phosphorus agree in establishing a completely different pattern. Where animal produce is removed from a heathland grazing, a significant loss of phosphorus may be involved, but depletion due to burning is not particularly heavy. This is partly because there is little loss in smoke, while in ash the phosphorus seems to be fixed in an insoluble form. However, as no long-term accumulation of insoluble phosphorus has been found to result from burning, it seems that it must be released over the years following a fire. It is also released rather rapidly from the litter which is deposited in increasing quantities by the developing vegetation. Although some of this available phosphorus is recycled by uptake into the plants, doubtless some is lost by leaching. Wherever soil erosion or loss of peat occurs large proportions of the phosphorus fund in the soil may disappear rapidly.

The incoming amounts in rainfall, despite considerable variation between records from different parts of the country, are never large and could compensate for losses only if retained in the system in their entirety. A net loss over 10 or 12 year periods is in fact suggested by several authors: this in part may be inherent in a *Calluna*-ecosystem, but has probably been aggravated by management for herbivores. Burning itself might be regarded as of temporary benefit by causing immobilization of phosphorus in ash, but this seems to be offset by later losses from both soil and litter. Most heathland soils are notoriously deficient in phosphorus for agricultural or forestry purposes and although application of phosphatic fertilizers does not necessarily stimulate the growth of *Calluna* there is a marked response in increased phosphorus concentration in the foliage and young shoots. It seems probable that the amounts of available phosphorus in heath ecosystems may in many cases have fallen below limiting levels and, while they may never have been large, animal production with its attendant burning management has probably contributed to this situation, particularly if, as is often the case, this has exposed the soil surface to erosion.

Nitrogen

In contrast to the foregoing nutrients, no very clear conclusions can be drawn regarding nitrogen. Estimates of the total nitrogen contained in the vegetation and soil reveal a sizeable fund against which the losses due to burning, although large in comparison with other nutrients, seem to represent a small proportion. The greater part of the nitrogen lost goes in smoke, but estimates of the actual quantity by different workers vary considerably, according to the biomass of the vegetation before burning, the ratio of young shoots to woody stems, and the figure used for the percentage of the total nitrogen driven off. This varies from about 45% to 95%, according to whether determinations have been carried out in the laboratory or the field. There are also quite large variations in estimates of the nitrogen content of rainfall in different parts of the country, so it is hardly surprising that when losses are set against inputs there has been little consistency in the inferences drawn about the balance. Some authors have concluded that the losses could not be made good by rainfall alone over a period of 10–12 years, others that they might be replaced from this source if the entire input were retained by the soil and vegetation (which is probably not the case), and yet others that there is some, or even a considerable, excess of input over loss.

However, in the case of nitrogen the direct balancing of losses against input is perhaps less appropriate than for other nutrients, since most of the nitrogen contained in the vegetation would in the normal course of events be returned to the soil in an insoluble, unavailable form. *Calluna* litter releases nitrogen only very slowly, and most of the quite large fund of total nitrogen in the soil is locked up in an insoluble form in organic matter which has accumulated sometimes over many years. Furthermore, the rates of nitrification and nitrogen fixation are generally stated to be low, although this point has not received much investigation.* It follows that the demands of the vegetation, which are considerable, must be met from the limited amounts of soluble nitrogen in soil which are largely stocked from rainfall. It is the extent to which the rainfall input can meet the demands of a developing stand of vegetation which is important, and most authors have agreed that over the 10–12 year period the input is

* Account must also be taken of the possible influence of root-nodule-bearing species in heathland communities, particularly *Ulex* spp.

likely to be insufficient. It is therefore not surprising that the soils show every sign of nitrogen deficiency, as indicated by immediate responses in the growth of *Calluna* and many other plants to the addition of nitrogenous fertilizers, and the need to supply them to young trees on these soils. The available nitrogen added in this way invariably disappears rapidly, as demonstrated by the speedy exhaustion of the nitrogen fertilizers applied.

Clearly under these circumstances the heavy losses of nitrogen due to burning must be the cause of some concern. The real effects, however, lie not so much in the actual removal of given amounts of nitrogen, which in any case would probably in time have been deposited in an unavailable form on the soil, but in the frequent regeneration of the vegetation, producing new young stands of plants which create a demand for available nitrogen probably heavier than the rainfall input can supply.

The nitrogen balance of heathland ecosystems raises many interesting problems. As yet research has not proceeded much beyond the stage of identifying the important spheres for future inquiry. This is a field which will certainly repay more intensive investigation.

Although many aspects of nutrient cycling in heath ecosystems are not yet fully understood, recent investigations have made possible a more balanced judgement on the effects of burning. It seems unlikely that, under suitable conditions and with proper control, fire causes serious depletion of fertility. At the same time, there may be net losses in certain nutrient elements associated directly or indirectly with rotational burning. Particularly in the case of available phosphorus and, perhaps, of nitrogen, burning may intensify deficiencies already inherent in many heathland soils. However, fire cannot be considered in isolation, as the only factor affecting the productivity of heathland areas. The systems of land use which require prolonged monoculture of *Calluna* have numerous effects on the soil and habitat generally. In the final chapter consideration is given to these influences and to the outlook for improved forms of land use in heathland. This amounts to an appraisal of heathland conservation in its widest sense.

12 Land use and conservation

General

The history of west European heaths, apart from those of high altitudes, is bound up with use and management. The original forest clearances, to which in large measure they owe their origin, may have been undertaken for a variety of reasons, but heathland has been maintained since then only because it served a useful purpose. Generally this purpose has been to provide grazing for domestic animals, mainly cattle and sheep, on soils unsuitable for grass pasture. Productivity in such systems is low (Chapter 8), and for this reason the input of capital and labour has been minimal. Extensive, free range grazing has been the rule.

This pattern of land use still applies in the more northerly and upland parts of Britain, where hill sheep farming and, to a lesser extent, cattle grazing retain a prominent role. The valuable revenues from letting the shooting on grouse moors have contributed an additional strong incentive, unique to Britain, for retaining large tracts of heathland. Fire offers an inexpensive management tool which serves the needs both of sheep farming and grouse rearing. Elsewhere, however, generally intensified farm production has made obsolete the use of heaths as grazing land. In the absence of any valid reason for maintaining them, the areas occupied by heath have declined rapidly during the past 100 years in Sweden, Denmark, Germany, the Netherlands, Belgium, France and southern England. Sometimes they have been abandoned and left to revert to scrub or woodland, but wherever an injection of capital has been possible they have been reclaimed and put to new uses in agriculture or forestry.

A broad view of conservation embraces both the maintenance of representative examples of important types of community, and an approach to problems of land use and management which seeks to combine productivity with the ecological health of the system. In regions such as northern and upland Britain where heaths are still extensive it has been argued that traditional forms of land use, as currently practised, are not compatible with ecological health. Hill farming is insufficiently profitable to accumulate or attract capital and the land is subject to slow exploitation by continued production of herbivores on a minimum-input basis. A first essential therefore is an ecological appraisal of the nature and consequences of these systems of land management as at present practised, followed by an assessment of the possibilities of applying improved methods to the same objectives, so achieving a better use of resources. Only then is a position reached from which a comparison can be made, on an ecological basis, between existing objectives and alternatives which require replacement of heath by other types of vegetation. In the event, economic and sociological factors play a decisive part in determining whether land use patterns survive or change. None the less, account must be taken of the known ecological consequences of different systems of use and management, and of the extent to which any practice is compatible with the broad aims of conservation as here understood. Much of the scientific information necessary for sound judgements is already available, as shown in previous chapters of this book. Certain aspects of its application will be discussed below.

There may well be a continuing useful role for heathlands in Britain but it is probable that here, as elsewhere, improved management and changing demands may considerably reduce their area. In view of this, the complementary aspect of conservation must be invoked and action taken to maintain viable examples of this type of ecosystem, especially in regions where it is rapidly disappearing. The fact that heathland owes its origin to human activity in no way reduces its claim for such treatment; on the other hand this makes it the more vulnerable when human cultures and land use patterns change. Heaths are of value aesthetically and in connection with various forms of recreation; they are also of great importance for scientific research and as a reservoir of wild life. Heathland reserves have already been set up in several countries where without them this formation, which has been so characteristic a part of Western Europe, would have completely disappeared. Heaths survive only under regular management, which therefore plays a vital part in such reserves. The later part of this chapter is concerned with defining the aims of this more limited aspect of heathland conservation and outlining the means of achieving them.

Use for grazing

Hill sheep farming has constituted a major component of land use in upland Britain for well over 200 years. The fortunes of this industry have fluctuated greatly during its history. Highly profitable at first, the returns began to decline between 1850 and 1880. Between 1920 and 1930 there was a period of moderate prosperity, but this was temporary and during the past 30 years there has been heavy reliance on Government subsidy. This is not to say that hill sheep farming is unimportant in the general agricultural economy. Its contribution to the gross agricultural product of Britain may not be great (one thirtieth, McVean and Lockie, 1969), but it has additional value as a source of hardy breeding stock.

During the same period there has been a decline both in the area devoted to hill sheep farming and the stock carried. Estimates of the numbers of hill sheep vary considerably and it is not easy to obtain a clear picture of changes in productivity of the land. In addition to fluctuations in numbers, there have been changes in the proportions of different breeds and in the average live weight of animals. Despite these complications, there seems to be agreement that there has been a real decline in the carrying capacity of the upland grazings. In some areas this began shortly after the middle of the nineteenth century, in others not until after 1900, but it seems generally to have continued until about 1950. In Scotland at least, the numbers of hill sheep appeared to stabilize at this time, perhaps with some further decline in recent years (Cunningham, Smith and Doney, 1971).

The reasons for this fall in productivity have been the subject of much discussion. McVean and Lockie (1969) conclude that they are partly economic and partly ecological. Prior to the advent of the large sheep flocks the land had been only lightly grazed, mainly by cattle. Since then sheep have been in the ascendancy, and their influence upon the ecosystem, coupled with the effects of the necessary vegetational management by burning, can be judged on the basis of information contained in earlier chapters. The view has been expressed that this represents 200 years of misuse of the land (McVean and Lockie, 1969).

Undoubtedly the original clearance of forest and its replacement by heath was 'a management effect of profound ecological significance' (Nicholson, 1964), but as far as possible hill sheep farming as a system of land use must be assessed on the basis of its own impact on the ecosystem, not on the consequences of the preceding forest destruction. Grazing by sheep to the exclusion of other herbivores has undoubtedly caused unwanted changes in community composition (p. 184). To a limited extent the removal of animal produce causes depletion of certain mineral nutrients. Burning, although apparently causing little habitat deterioration when properly controlled, may aggravate nitrogen and phosphorus deficiencies, and always introduces a risk of erosion or spread of weed species if regeneration of the vegetation is delayed. The resulting monoculture of *Calluna* may promote podsolization and loss of fertility in the upper soil horizons.

However, although the evidence of these changes may amount to a condemnation of present practices this does not necessarily imply that a soundly based system of using heathland for herbivore production is impossible. Present practices, as Nicholson points out, are to a large extent a by-product of accepted demands of animal husbandry which have often been formulated for more hospitable environments and may be ill-adapted to upland grazings. The results of research should be able to suggest the necessary improvements in practice. If these can be contained within the limits set by economics the use of hill land for pastoral purposes can stand comparison with other systems of land management without negating the principles of conservation.

In an extensive grazing system such as that on heathlands, the only variable which can be closely controlled with minimum expense is the domestic animal, the numbers and to some extent the distribution of which can be varied at will. Vegetation management has to be largely confined to burning, the cheapest way of maintaining it in a productive condition. For this reason, there is a tendency to regard the animal as the resource and to pay little or no attention to long-term effects on the real resources, namely the soil and vegetation. It is this attitude which fundamentally conflicts with the aims of conservation, and attention should therefore be directed to the possibilities of improving the productive capacity of the habitat.

Burning is bound to remain an integral component of heathland management wherever a high proportion of *Calluna* in the vegetation is required. The conditions which must be observed to achieve good control of burning, and thereby minimize damage to the habitat, are now well understood (Chapter 10) and do not require reiteration here. The chief objectives of burning are to maintain the dominance of *Calluna*, and to ensure that the feeding value of the stand in terms of quantity and nutrient concentration is kept high. However, on soils capable of carrying a more productive community such as *Agrostis-Festuca* grassland, a course of more frequent burning associated with rather intensive grazing may be used to bring about a change from heath to pasture.

Attention to the soil may also improve the productive capacity of the habitat. The response of heath vegetation to fertilizers has been briefly considered in Chapter 9, and it is clearly possible to increase production by this means. A long series of investigations has been devoted to this subject (Greig, 1910; Fagan and Provan, 1930; Heddle and Ogg, 1933; Daves and Jones, 1932; Milton, 1934, 1938, 1940; Clouston, 1943; Milton and Davis, 1947; Watson and Gregor, 1956; Wannop, 1958, 1959 *a and b*; Fairbairn, 1959; Robertson and Nicholson, 1961; Nicholson, 1964; Miller, 1968). Application of fertilizer can be undertaken either with a view simply to improving the feeding value of existing heath vegetation, or in association with measures directed towards producing changes in the botanical composition of the community. For the former purpose, the quickest response in terms of increased growth is achieved by the addition of nitrogen, while the immediate effects of phosphatic fertilizer are seen in some increase of the concentration of phosphorus in the edible shoots and a marked stimulation of flowering. Lime, if applied at rather high rates, produces a substantial rise in the calcium content of the green shoots.

Little firm evidence has been produced to support the view that the addition of lime will kill out *Calluna,* but some decline may be shown after liming, perhaps as a result of changes in the competitive balance with other species and in differential grazing intensity.

Interest has been expressed in the possibility of increasing the grazing value of heath vegetation by the use of lime and fertilizer, in relation both to sheep farming and grouse shooting. Increases in the breeding stock of grouse on fertilized *Calluna* heath have been reported, and the importance to sheep of a good supply of nutritious *Calluna* needs no emphasis. However, the wisdom of pursuing this approach is questionable, mainly because of the expense of distributing large quantities of fertilizer in difficult country. This factor alone has precluded the use of aircraft for the purpose (Nicholson, 1960). Furthermore, there is an element of waste in supplying fertilizer to a perennial woody plant in which a proportion of any dry-weight increment is located in the non-edible, lignified branch system. None the less, Robertson and Nicholson (1961) conclude that there is sufficient advantage to be gained from the use of comparatively low rates of lime and phosphate application to warrant consideration of upgrading some hill grazings in this way, provided a satisfactory measure of stock control can be ensured.

Under certain conditions, particularly those associated with wet heath vegetation, direct improvement of the habitat by drainage may result in increased production. This is now widely used wherever surface water creates a problem, and can be effective in preventing water tables rising to the soil surface or above. It reduces the possibility of damage by excessive surface run-off, and thereby increases the capacity of the soil profile to retain precipitation water (Nicholson, 1964; Robertson, Nicholson and Hughes, 1963) and to benefit from the accompanying input of nutrients. The rooting region is maintained in an aerated, oxidising condition rather than being subject to repeated waterlogging. These changes contribute to improved performance of species such as *Calluna*, and may lead to gradual successional changes in the community as a whole, although in high rainfall areas and on peaty soils these may be rather slight unless accompanied by changes in grazing management. On wet ground drainage is, in any case, desirable from the point of view of the health of the animal stock.

Greater progress in improving the productivity of heathland can be made by turning attention to controlled change in community composition, at least over parts of the available area. Hunter (1954, 1962) has shown that where both mor and mull soils occur on the one hill grazing, as is usually the case, vegetation on the former (*Calluna, Molinia, Nardus,* etc.) receives only one-third to one-fifth the grazing intensity of that on the latter (*Agrostis, Festuca*). This suggests that the overall utilization of a grazing is affected by the proportions in which these two categories are represented, and that where mull swards are limited in extent there is much to be gained by upgrading the vegetation on selected areas of mor soil (Hunter, 1958*a*).

Heddle and Ogg (1933) and Gregor (1961) have shown that on a localized scale this can be achieved at relatively low cost by cutting channels from springs or streams to extend flushed areas. The resulting addition of nutrients, reduction in acidity, and improved water supply are all concerned in increasing the representation of *Agrostis* spp., *Festuca rubra, Poa trivialis, Trifolium repens,* etc., with an accompanying decrease in heath species. It is pointed out that no improvement in the quantity of available phosphorus is obtained, and that the potential for improvement might be increased if appropriate additives could be supplied to the water.

As already indicated, changes of this kind can be initiated over much wider areas by fertilizer treatment, commonly in the form of ground limestone and basic slag. Consideration has also been given to altering the botanical composition of the community by the use of herbicides, for example 2, 4, D or Paraquat, which have been found to kill *Calluna* provided it is in a condition of active growth, but this has never been attempted on a large scale. A more practical means of changing vegetational composition is by manipulating the type and intensity of grazing. However, before leaving the subject of improvement of the basic resources of vegetation and soil, reference must be made to the success with which patches of heath have been converted into grass-dominated communities by reseeding.

Suitable ground may first be ploughed, but a valuable low-cost improvement can be achieved by seeding with or even without surface cultivation (Clouston, 1943; Roberts, 1960). Existing vegetation is generally first cleared by burning and the area is limed and fertilized before the grass-clover mixture is sown. It is usual to select a number of small areas in the more freely-drained parts of a hill farm for improvement along these lines, and to leave these unfenced so that grazing animals have free access.

The success with which the introduced grasses and clovers are established depends to a considerable degree upon the amount of cultivation possible. This also determines the extent to which the native vegetation is destroyed and so prevented from regenerating and competing with the developing grass sward.

A common practice has been to sow the 'cleanings' from *Lolium perenne* and *Trifolium repens* seed crops, which contain in addition high percentages of *Holcus lanatus, Cynosurus cristatus* and other species. These are simply a by-product of seed crops grown on fertile lowland areas (Jones, 1967). More lasting improvements have been made with seed mixtures containing strains of *Lolium perenne* (e.g. Aberystwyth S23 type) and *Trifolium repens* selected for their performance in upland grazing areas, often in association with other species such as *Poa trivialis, Poa pratensis, Phleum pratense, Cynosurus cristatus,* etc. However, despite the progress made at the Welsh Plant Breeding Station, seed mixtures are still sometimes based on plants bred for lowland conditions. Relatively little attention has been paid to the breeding of hardy and persistent varieties which will yield well throughout the year in the less favourable environments, and grasses which contribute naturally to the heath communities in such areas have been little used. These, such as *Agrostis* spp., *Festuca rubra* and others, inferior on fertile lowland pastures and less palatable than the species commonly sown, might contribute effectively to a lasting conversion of heath to grassland.

Spectacular improvements have, however, been achieved along these lines, for example in Wales and parts of Scotland. The importance of *Trifolium repens* in maintaining an improved nitrogen status in the soil, and of the grasses in altering soil structure and contributing a mull type of humus cannot be overemphasized. These effects, together with those of fertilizer treatment, are not confined to the areas selected for improvement, because nutrients are transferred to surrounding land by the movements of the grazing animals. That there are problems in achieving lasting benefit, however, cannot be denied. The increases in herbage production are greatest in summer when the heath vegetation itself usually offers more than adequate forage. If the improved grassland is undergrazed it will rapidly revert towards the original vegetation. More commonly, animals concentrate upon the more palatable patches of improved ground, so accentuating the habitual summer under-grazing of the remainder. Furthermore, the increased stocking which becomes possible as a result of improvements aggravates the problem of providing sufficient winter-feed, although this question is beyond the scope of the present discussion. While changes in vegetation which represent great improvement from the viewpoint of livestock production can be accomplished in the ways outlined (Fig. 51), their full and lasting value can be realized only if they are accompanied by improved control of the grazing factor. This implies fencing which involves considerable expenditure, thus raising important economic considerations (Nicholson, Currie and Paterson, 1968).

Management of the grazing regime, even in the absence of reseeding, can cause far-reaching changes in botanical composition. These can be advantageous both in an agricultural and an ecological sense. Equally, they can be retrograde where, for example, over-grazing or too intensive grazing immediately after burning cause the spread of *Pteridium aquilinum*, *Nardus stricta* or *Molinia caerulea*. On heath vegetation, management designed to produce desirable plant succession can involve control both of the intensity and type of grazing. Considerations of economics and husbandry may dictate practice, but the principles upon which it should be based are well known. Where heath occurs on soils and in climates suitable for *Agrostis-Festuca* grassland, the change can be effected in the course of 10–12 years, or even less, by appropriate grazing management using

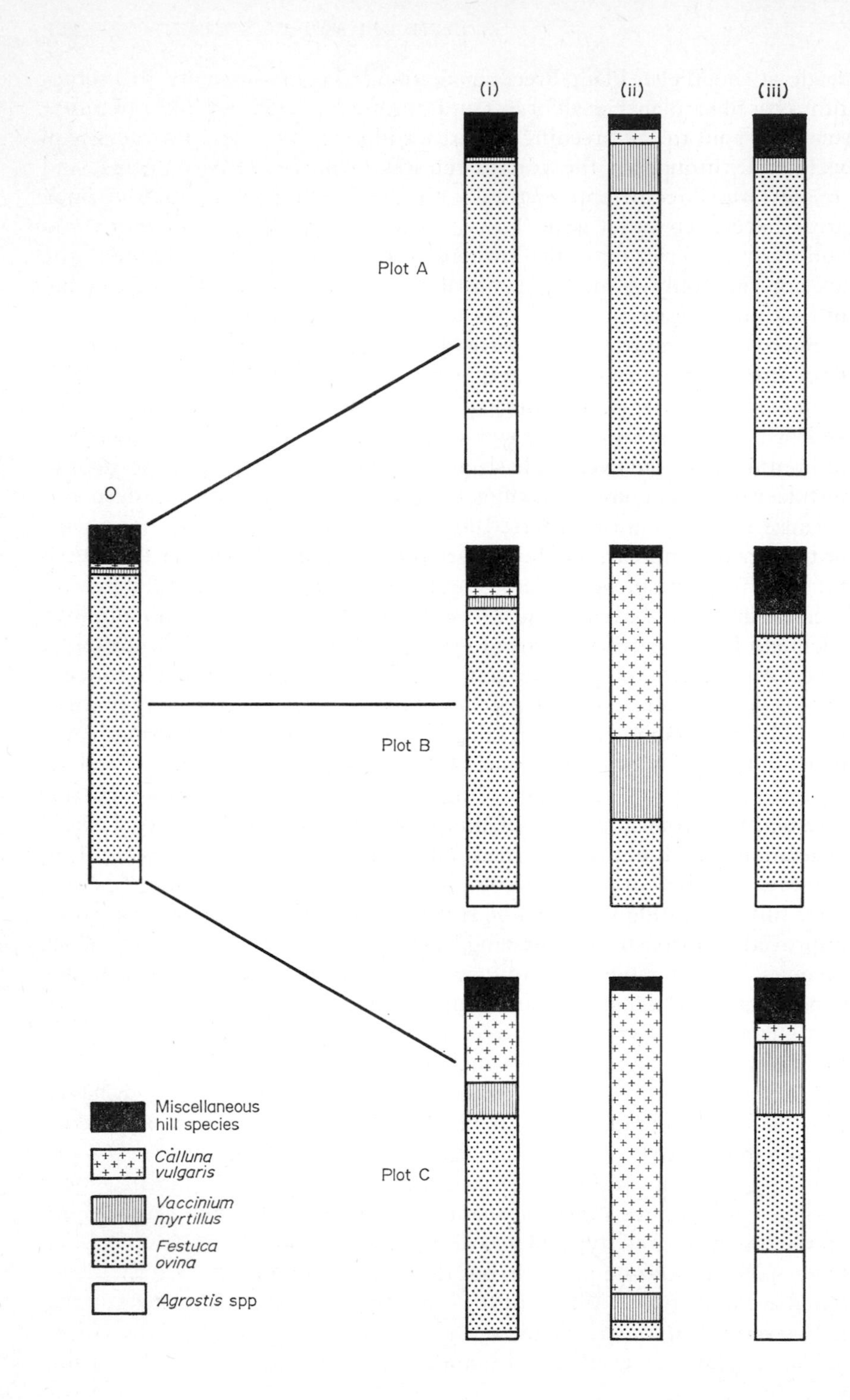
(i)
(ii)
(iii)
Plot A
O
Plot B
Plot C
Miscellaneous hill species
Calluna vulgaris
Vaccinium myrtillus
Festuca ovina
Agrostis spp

sheep alone (Fig. 51). The importance of introducing a degree of control into the utilization of upland grazings is shown by Hunter's (1954, 1960, 1962) work on the behaviour of hill sheep. Particularly with the Blackface breed, utilization of a hill grazing is determined not only by the preferences of individual sheep but also by their habitual formation of 'home range groups' of 20–50 sheep whose range is restricted to 40–50 ha (Hunter and Eadie, 1962). Under a free-range system of grazing the utilization of available herbage may therefore be very uneven, and this is accentuated by seasonal concentration on different plant communities. If control and rotation of grazing on different parts of an area could be accomplished inexpensively, beneficial effects on the vegetation would follow.

Even greater possibilities exist when the type of grazing is altered by returning to the practice of former times of combining cattle grazing with sheep. Hunter (1958*b*) gives a table of the respective grazing preferences of cattle and sheep, showing that where the grazing includes elements for which their preferences differ the two types of animal can be complementary. This makes for better utilization of available herbage, while the fact that cattle are less selective in grazing than sheep and more ready to take fibrous material results in a better control of botanical composition. The pressure of palatable species is reduced and unwanted plants are kept in check. The advantages of mixed grazing have been stressed by many writers (Clouston, 1943; Wannop, 1958, 1959 *a and b*; Taylor, 1961; Peart, 1963; Hunter, 1958 *b*; Nicholson, 1964 and others). Peart (1963) concludes: 'Higher levels of animal production from hill pastures would appear to be dependent on the evolution of grazing systems having a balance between cattle and sheep and possibly a form of grazing control.' Unfortunately, progress along these lines has, up to the present, been limited to rather few areas.

The use of fertilizer to improve the soil and vegetation, as described earlier, is effective only where accompanied by appropriate grazing management and this is even more true of improvement by reseeding. Although the benefits are appreciable where free access is allowed to patches

---

Fig. 51. Changes in the botanical composition of an area of upland grazing, produced by grazing management. (After Jones 1967.)

O – Composition at the start of the experiment

Plot A – (*i*) 1932. After frequent close grazing, 1930/31
(*ii*) 1944. After 13 years protection from grazing.
(*iii*) 1956. After 12 years open grazing.

Plot B – (*i*) 1932. After light grazing, 1930/31.
(*ii*) 1944. After 13 years protection from grazing.
(*iii*) 1956. After 12 years open grazing.

Plot C – (*i*) 1932. After protection from grazing.
(*ii*) 1944. After a further 13 years protection from grazing.
(*iii*) 1956. After 12 years open grazing.

of reseeded ground, in theory these could be much increased if their utilization could be planned and controlled according to the production of edible material as the season proceeds. Unfortunately, owing to expense, this is seldom possible except where improvements are adjacent to 'in-bye' land on low ground or close to the farm.

Use for sport

A large proportion of the heathland in Britain is used for sport, either exclusively or in conjunction with sheep farming. Over much of the area this takes the form of grouse shooting, and the relationship between the populations of grouse and burning management have been fully discussed in Chapter 10. As mentioned there, it has been estimated that grouse use only about 5% of the annual production of edible material by *Calluna,* so it is unlikely that there is significant competition with sheep as far as food supply is concerned. In large measure, management by burning is appropriate for both purposes. Because the food of grouse consists so largely of *Calluna* the possibilities for improvement are limited to improved burning management and improvement in the nutrient content of the *Calluna* shoots. The former, as shown by Piccozzi (1968) and Miller, Watson and Jenkins (1970) can be achieved by greater attention to the burning programme which should be planned to ensure, first, that very little of the total *Calluna*-dominated vegetation is allowed to pass beyond the building phase and, second, that this is achieved as far as possible by means of numerous small fires (Chapter 10, Plate 34). Each grouse territory must contain both young *Calluna* for food and older stands for cover. More territories can therefore be set up where burning of small areas has produced a patchwork of stands of various age than where even-aged

TABLE 31

Number of grouse per km² in spring and number reared per old bird in August, on burnt and unburnt areas.
(From Miller, Watson and Jenkins, 1970)

Year	Burnt area		Unburnt area	
	Grouse/km² in spring	Young: 1 old in August	Grouse/Km² in spring	Young: 1 old in August
1962	49	0·5	52	0·1
1963	49	1·1	44	0·8
1964	51	1·0	54	1·1
1965	78	0·3	57	1·1
1966	82	0·2	51	0·2
1967	73	0·1	53	0·4
1968	51	0·5	53	0·4

Burning was carried out in the following seasons: 1961–2; 1962–3; 1963–4 and 1964–5.

stands are extensive. Experimental evidence that it is possible to increase the numbers of breeding grouse in an area by appropriate burning management has been given by Miller, Watson and Jenkins (1970). At Kerloch, Kincardineshire, grouse populations had shown a long-term decline prior to 1961. During the four years following this date numerous small fires were burnt in an area of 49 ha, covering a total of about 30% of the *Calluna* in patches of 0·4 ha or less. From 1965 the density of grouse almost doubled for about three years, in contrast to an adjacent unburnt area of similar size. (Breeding success was unaffected and was generally low in the neighbourhood.) Subsequently the grouse density again declined (Table 31). On a wider scale, a survey by Piccozzi (1968) of a large number of grouse moors revealed that, with few exceptions, the areas with high densities of grouse were those most systematically managed by burning.

The possiblity of improving breeding density and breeding success in grouse by using fertilizer to increase the concentration of nutrients in *Calluna* shoots has also been tested (Chapter 9). There is no doubt that this is feasible, though it is perhaps of limited significance owing to the difficulty of applying fertilizer in rough and often remote country, as discussed above in connection with sheep grazing. The greatest success has been achieved with nitrogenous fertilizer, but its effects are short-lived, and only short-term improvements can be expected from its use.

'Reclamation' of heathland: farmland or forest

The wisdom of harvesting a protein crop produced by herbivores from vegetation on acid, nutrient-poor soils has been called in question by several writers including Albrecht (1957) and McVean and Lockie (1969). Attention is drawn to the quantities of nitrogen and phosphorus removed in protein, elements which are notably deficient in these ecosystems. While this is doubtless important, it should be assessed in the context of the generally low rates of production and extraction of animal carcases. Calculations summarized in Chapter 11 indicate that the loss of nutrients resulting from the removal of animal produce is a relatively minor component of the input-output balance sheet. None the less it may be argued that, as at present practised, herbivore production on heathland is ecologically an inappropriate form of land use, although capable of improvement If this is accepted, two further possibilities remain to be considered: first, the wholesale replacement of heath vegetation by different forms of productive vegetation, and second the preservation of heath for purposes other than production.

The first of these presents the alternatives of substituting cultivated farmland for heath, or afforestation. The former has been effective chiefly on the lowland heaths in milder climates, notably in Denmark and the south of England. The process of reduction and fragmentation of heathland in Dorset, southern England, resulting from reclamation for agricultural purposes has been documented by Moore (1962). Today the total area of heathland in east Dorset and Hampshire west of the River Avon is only one-third of its extent in 1811.

Elsewhere in the south of England similar conversions have taken place, especially in Hampshire, Sussex, Surrey, Kent and East Anglia (the Breckland). Heath survives today only in a fraction of the area it occupied about 100 years ago. The most spectacular transformation of heath into cultivated land took place in Denmark, following the formation in 1866 of the Danish Heath Society as mentioned in Chapter 2. With the aid of State subsidies, finance was organized for bringing large areas under cultivation, at first manually and later by machinery. Between 1866 and 1955 about 700 000 ha of heath and bog were cultivated, drained where necessary, limed and fertilized, some for forest and shelter belt plantation, and some for agricultural purposes. Virtually the only heathland surviving in Jutland is that now preserved in reserves. The original soils were acid and nutrient-poor, being derived in large part from fluvio-glacial outwash, but for this reason the terrain is generally level, free from boulders, and amenable to ploughing by specially designed giant ploughs capable of breaking up hardpan at depths of about 90 cm. The success of the scheme, however, depended upon the existence of an organization able to attract and administer strong financial aid from the State, and to undertake the technical oversight of the enterprise. Smallholders alone could not have attempted anything on this scale.

In most other countries, especially where access is more difficult and population densities low, forestry has been the chief rival of the grazing animal on heathland. In the past, there were considerable technical problems in the way of afforestation on heath soils, especially where a hardpan was present in the profile, or a very wet peaty layer at the surface. These have been solved by the development of appropriate machinery for deep ploughing and draining, and by choosing tree species appropriate for the soils concerned. Chief among these are *Pinus contorta* for soils with hardpan or other indurated horizons and *Picea sitchensis* for peaty soils. There is a tendency, as discussed in Chapter 5, for the latter to enter a condition of check when planted among *Calluna,* but this can generally be alleviated by the addition of fertilizer. Some additional expense may be incurred as a result, but otherwise there are now few obstacles in the path of establishing successful plantations on heathland.

In southern Sweden almost all available heath has been converted into forest and some fine, productive stands have been grown. On the dry heaths of Denmark, north Germany and parts of the Netherlands, where the soils often lack indurated horizons and deep ploughing is easy, much planting has been accomplished. There is also an extensive forestry programme on British heathlands, now concentrating particularly on Scotland. Until recently several species were used including *Larix leptolepis, Larix decidua, Pinus sylvestris, Picea sitchensis* and *Pinus contorta*, but at present only the last two are widely planted. To a considerable extent in the past planting has been undertaken with little reference to the purposes for which the adjacent unplanted heathland is used. Large blocks of forest trees have reduced the areas available for sheep grazing, and in some cases have occupied ground regarded as of good quality for grazing. Access for animals from the glens to summer grazing grounds at higher levels has on occasion been restricted by continuous belts of forest occupying intermediate altitudes. However, much more attention is now paid to the integration of forestry with other uses, and under some conditions the sheep farmer may reap positive benefit from the shelter provided by the forest.

Successful establishment of forest trees depends, firstly, on the techniques of ploughing and planting. On the drier soils furrow-planting is common, while in wetter habitats the young trees are usually planted in a notch on the side of the plough-ridge. Secondly, the appropriate provision of fertilizer is essential. This normally includes a supply of phosphate at the time of planting. Decay of the organic matter exposed by the plough may increase the amount of available nitrogen, but additional supplies may be needed and are then incorporated in the fertilizer. Later, if the trees show signs of going into check a further dose of fertilizer is given. Nutrient deficiencies may develop even at a later stage, when it is sometimes necessary to supply additional potassium, and on occasion aircraft may be required to spread the fertilizer.

These treatments undoubtedly contribute to a general improvement of the soil, and as the plantation becomes older the tree roots penetrate rather more deeply than those of the heath community. Nutrients otherwise unavailable to the vegetation are brought back into the cycle, and eventually returned via the shed needles. However, it is unwise to assume too readily that all the effects of afforestation are beneficial and that because woodland constituted the original vegetation cover, return to forest in any form must necessarily upgrade the habitat. Trees of commercial value are conifers, and in the main exotic species. They do not normally root deeply, as indicated by damage due to wind-throw (Edwards, Atterson and Howell, 1963), and therefore draw nutrients from the lower soil horizons only to a limited degree. Their leaf litter is acid, adding to the layer of mor humus (Rennie, 1961). This is slow to break down, and does little to halt the processes of podsolization.

However, many of these undesirable effects are the legacy of the widespread planting of *Pinus sylvestris* with little cultivation up to 80 years ago. At the present time, foresters are fully aware of potential dangers to the habitat. The more extensive use of spruce has required progressively deeper cultivation, up to 1 m in some instances. As a result the trees root more deeply and nutrients are more freely circulated. At the same time, species producing contrasting types of litter are more often planted in mixtures, so avoiding the accumulation of a uniform raw humus. These are steps which may contribute to improvement of the habitat, or at least may avoid deterioration.

Some nutrients are eventually removed from the system in the timber crop, but since this consists largely of lignin the drain on the nutrient fund is not so great as that caused by removal of whole animals.

The conversion of heathland to forest is clearly a practicable undertaking, again provided the necessary capital is available. There is a continuing demand for the produce, though since there is a long time-interval between the investment of capital and the final return it is difficult to predict the economic success of the venture. This may be much affected by changes in interest rates during the period of tree growth, and in the prices which can be realized on harvesting. In the short term certain ecological improvements are made to the habitat, but much less is known about the long term effects. Indeed, reafforestation of heaths is still too recent an event for the full effects upon the soil yet to have emerged. Local changes in micro-climate are also to be expected, most of which will probably be beneficial.

Preservation of heathland for purposes of recreation and wild-life conservation

It has been shown that over much of the heathland region forces favouring conversion of heathland to other forms of vegetation have been predominant, except where reserves have been established. In Britain both sheep farming and grouse shooting have acted powerfully for the retention of heath, at least in the upland areas. The present indications are that even if hill sheep farming is not on a very sound economic foundation, the return which can be had from grouse moors will ensure that large areas will be retained for this purpose, at least for the forseeable future. For example, on the Airlie Estates (east central Scotland) the income from the game account exceeds that of the total agricultural rental, while on one farm of 1396 ha the ratio of the contribution of sport to that of agriculture in calculating the capital value of the land is about 13 : 3 (Earl of Airlie, 1971).

However, even in northern Britain where there seems to be a continuing role for heathland, its area will doubtless be diminished in future years as a result of the forestry programme and other demands for land. The need to preserve examples of heath vegetation is therefore just as real in this country as in others where reclamation has already proceeded almost to the limit. Enough is known of heathland ecology, as outlined in this book, to serve as a basis to assess the management requirements for conservation, but it is essential to recognize that these will vary according to the exact purposes of any conservation programme. Apart from grazing, retention of heathland may be desired for two main reasons: firstly for amenity, recreation and sport; and secondly for scientific or educational purposes and wild-life protection. Unfortunately the management appropriate for the former is in the main incompatible with the latter.

The management appropriate for grouse moors has already been outlined and serves, with minor modifications, for sheep or cattle grazings depending on *Calluna*. This, at the same time, has the effect of maintaining quite large tracts of country as open, virtually tree-less heath. This landscape, to be found for example in Dartmoor, Exmoor, parts of Dorset, Hampshire, Sussex, Surrey, East Anglia, the Pennines, Wales and Scotland, also has great aesthetic appeal. It is, in fact, an asset of considerable value as a tourist attraction for purposes of general recreation as well as sport. The essential aim of management, if all these demands are to be satisfied, is to keep as much of the area as possible covered with dense stands of building *Calluna,* which combine good feeding for herbivores with free flowering in late summer. The means to achieve this is a well planned burning programme.

Earlier chapters have detailed the ecological consequences of this system of management. While habitat deterioration can probably, with careful control, be reduced to a minimum, the outcome is a monoculture of *Calluna* in its most vigorous and competitive phase, producing uniformity of the micro-environment and severe restriction of associated species. This biological monotony is not a necessary feature of heath ecosystems, many of which can be quite rich in animal and plant life: it is a consequence of management. Conservation of the scientific and wild-life interest of heath ecosystems therefore requires a different approach to management, and it may be emphasized that even where extensive areas of heath remain devoted to sport and amenity there is still urgent need for smaller reserves to safeguard at least representative examples of the main types. These must be selected on the basis of their scientific and historical value as well as their flora and fauna, taking their place among the main categories of habitat and community requiring representation in an international conservation programme.

Because the majority of heaths owe their origin and survival to human activity, it follows that active management is required in most heath reserves. Without this, plant succession will proceed and the ecosystem which is the object of preservation will be lost. The formulation of a management policy which will prevent tree and shrub colonization while retaining maximum variety in specific composition presents a considerable problem. Its solution can be found only by discovering the fundamental characteristics of the richer heath communities, and determining management aims in the light of knowledge which has been acquired about the dynamic processes involved.

The most striking characteristic of heaths in which the plant and animal communities have not been impoverished is an intense vegetational pattern on a rather small scale. Investigations on pattern in heath communities

(Chapters 6 and 7), and particularly on the cycle of growth phases in *Calluna* as one of the main causes of pattern, therefore acquire a special significance in this context. In contrast to the uniformity of large areas of even-aged building *Calluna,* relieved only by recently-burnt patches passing quickly through the pioneer phase, some heaths show an uneven-aged structure with a patchwork of all phases of *Calluna* and gaps occupied by other plants. The canopy is uneven and interrupted, resulting in heterogeneity of micro-environmental factors and providing a variety of habitats suitable for a considerable diversity of plants and animals. The conditions associated with individuals of *Calluna* in each of the main growth phases and the cyclical progression of change from one to another have been described in Chapter 7, together with the accompanying fluctuations in associated species. The greatest variety in composition occurs where a considerable proportion of the area is occupied by plants either in the degenerate or pioneer phases, in which there are gaps in the canopy and the competitive vigour of the *Calluna* is least.

In reserves, management should therefore aim at encouraging pattern by means of maintaining, as far as possible, an uneven-aged community structure (Gimingham, 1971 *b*). This may require freedom from burning for periods of considerable length, although in most areas recourse to occasional burning will always be essential to prevent succession to scrub and woodland (unless this can be achieved by hand labour). It may be difficult to strike the necessary balance between the management needed to prevent the establishment of unwanted plants and the freedom from disturbance required for the development of an uneven-aged structure. After burning, from 20 to more than 30 years may have to pass before a reasonable proportion of individuals has reached the degenerate phase, while even longer periods are necessary before all stages of cyclical change are represented. The frequency of burning will vary according to local conditions, but is bound to be less than that required for sheep farming or grouse. During recovery, the stand is even-aged for a number of years and it is therefore all the more important to burn in very small patches rather than in large blocks. Once irregular die-back of the central branches of *Calluna* or even of whole bushes begins, gaps are opened up to colonization by other species and eventually to the cyclic return of *Calluna* itself. Under these conditions the stand rather rapidly becomes uneven-aged, and the characteristic small-scale pattern develops, encouraging the entry of a variety of plants and animals. This must be the aim of heathland reserve management.

In the special case of rare species belonging to heath communities, their survival may depend upon the representation of a particular phase of *Calluna* or on an abundance of some other plant. Management must then aim at a balance which secures these conditions. For example, Moore (1962) demonstrated in Dorset heaths the dependence of the Dartford warbler

(*Sylvia undata*) on adequate supplies of old *Calluna* and *Ulex europaeus*, as produced by a mosaic of not too frequent fires.

A further important question arising in connection with reserves concerns their size. Very little attention has been paid to the criteria upon which to estimate the minimum area required for the maintenance of viable examples of heathland ecosystem. No hard and fast rules can be laid down, because populations of different species require very different areas for survival. It is necessary first to recognize the key species of the ecosystem (Moore, 1962). These are species essential to the relative stability of the ecosystem which, if they are removed, undergoes radical change. Some plants or animals which do not satisfy this definition may also, as a group, constitute a 'key force' – for example, species of the bryophyte or lichen strata, or a group of herbivores or predators. The size of habitat for a viable ecosystem, and therefore of a reserve, is equivalent to the smallest area which will support a viable population of the weakest key species. For many purposes, however, this minimum size may be inadequate because the area, although large enough to maintain a functioning heath ecosystem, may be insufficient for the survival of species which characterize a particular community-type or are of special interest in other ways. An additional list of species of conservation interest is required, as indicated by Moore. The relationships between occurrence and the size of available areas may then be investigated for each of these, as well as for the key species. Progress along these lines has, however, been very limited although Moore obtained valuable data on the requirements of a number of rarities and indicator species for Dorset heaths. He also found evidence that, in addition to size of area, distance from other suitable habitats is an important factor. The more isolated fragments of heath showed, for example, some impoverishment of the fauna, particularly in respect of four indicator species which are wholly confined to heath: *Ceriagrion tenellum* (Odonata), *Plebeius argus* (Lepidoptera), *Lacerta agilis* (sand lizard) and *Sylvia undata* (Dartford warbler). It is pointed out that this effect is bound to be emphasized where local communities are repeatedly exterminated by fire, and depend for re-establishment on neighbouring populations.

The foregoing discussion does not exhaust the possible uses to which heathland has been put, or which may be devised for it in the future. *Calluna* has been harvested in the past for thatch, bedding or dye-extraction and is still used in bales as a foundation for road building. The surface humus has been removed from the soil for fuel or to act as a supply of organic material for sandy fields, while the excellence of *Calluna* honey still rewards bee-keepers for locating their hives on the heaths. At the other extreme, lowland heaths have suffered extensively from the spread of urban development, roads, golf courses and military training grounds.

With so many competing claims on heathland, at least some of which can be shown to be reasonably compatible with a broad view of conservation, it remains to consider whether any general principles can be enumerated as a guide in assigning priorities. High priority should clearly be given to the establishment of reserves for scientific research, education and wild-life protection, since their value is out of all proportion to their size, and the areas required for this purpose will deny little to other, productive forms of land use. Beyond this, it is much more difficult to lay down clear guide lines as so much depends upon economic and social change. However, it is possible to apply some form of land capability classification to heathland areas, provided that vegetation and its past management are taken into account as well as soil. A scheme of this kind prepared by McVean and Lockie (1969) for the Scottish Highlands may be taken as an example which might with little modification be widely applicable to heathlands. A valuable feature is the specification of possible secondary uses which can appropriately be carried on in addition to or alongside the primary use recommended for each category (Table 32).

There is probably less scope now than formerly for complete conversion of heaths to cultivated crop-land since this has already been accomplished on the more fertile ground, and general agricultural over-production in Europe has removed the incentive for breaking-in additional marginal land. Forestry undoubtedly has a good case for reclaiming further areas of heathland, particularly in Britain. In addition to providing a valuable product for which there will be continuing demand, probably on balance most heathland habitats benefit ecologically from afforestation. However, strong cases can be made out for retaining a good proportion of our remaining heathland on account of its value for grazing, sport and recreation. Much of it, indeed, might already have disappeared if it had not been for the farmers and landowners. At present hill farming plays a significant part in the pattern of agriculture in this country even if, on economic grounds alone, its future may seem unsure. In this field, improvements in practice are essential if further habitat deterioration is to be avoided, but there are few technical difficulties in the way of this. On the other hand, the importance of sporting revenues to the finances of estates which incorporate heathland cannot be denied, and this discourages major change in land use while stimulating efforts towards the best possible management of *Calluna,* within the limits of available resources and methods.

A major drawback of the present arrangement is that it encourages the commitment of very large tracts to the one form of management, itself geared to the creation of biological uniformity. Such diversity as exists is provided by forestry plantations which at least until recently have not added conspicuously to variety. Certain types of land use cannot be combined on one and the same area of land: commercial forestry in particular does not mix readily with other uses, while the interests of grouse shooting

TABLE 32

Land capability classification for the Scottish Highlands
(From McVean and Lockie, 1969)

Land capability and precautions in use	Primary uses	Secondary uses
I. Suitable for retention or reclamation as crop land. This would not necessarily imply that reclamation was economically feasible but only that it would be ecologically valid.	Agriculture	Recreation Wildlife Grazing
II. Suitable for improvement as grazing by cultivation and reseeding. Sub-class (*a*) mineral soils (*b*) peats.	Grazing	Recreation Wildlife
III. Suitable for improvement as grazing by methods other than cultivation. Sub-class (*a*) mineral soils (*b*) peats.	Grazing	Recreation Wildlife
IV. Suitable for retention as unimproved rough grazing in association with classes I–III. Position with respect to surrounding areas of the other classes would have to be taken into account. Careful moor burning permissible.	Grazing	Recreation Wildlife Watershed management
V. More suitable for commercial afforestation than grazing.	Commercial forestry	Recreation Wildlife Grazing Watershed management
VI. Suitable mainly for protection afforestation and wildlife. Moor burning not permissible.	Protection afforestation Wildlife	Recreation Grazing Watershed management
VII. Suitable only for wildlife. Moor burning not permissible.	Wildlife	Recreation Watershed management
VIII. Any areas requiring urgent counter-erosion works, including areas of severe peat hagging, badly gullied or sheet eroded slopes, landslips and river bank failures.	Erosion control Watershed management	

can be reconciled with little except sheep grazing. McVean and Lockie make a convincing plea for greater ecological diversity in upland land use generally, and this may be supported in the special context of heathlands. However, if too many different types of land use were to be imposed on an area in the name of diversity this might be self-destructive, for each category of use requires a certain minimum of land to support it. Nevertheless, diversity can effectively be encouraged by introducing, alongside land retained for traditional purposes, more intensive improvement of selected areas for livestock including cattle as well as sheep, and additional, more varied afforestation along lines currently being developed by the Forestry Commission to combine recreational use and wild-life interest with timber production. The intensification of production in parts of the heathland would provide the justification for setting aside others as reserves, perhaps also allowing a further fraction to revert naturally to *Betula*, *Betula-Pinus* or other mixed woodland communities including *Alnus glutinosa* in the wetter sites. This could lead to improved nutrient cycling and the counteraction of podsolization processes, as well as enrichment of wild-life. Experiments to investigate this kind of effect have been started by the Nature Conservancy on the Island of Rhum, where small areas have been planted with tree and shrub species such as *Quercus petraea*, *Pinus sylvestris*, *Corylus avellana*, *Fagus sylvatica*, *Betula pubescens*, *Alnus glutinosa*, *Sorbus aucuparia*, *Crataegus monogyna*, *Populus tremula*, *Prunus padus*, *Salix atrocinerea*, *Sarothamnus scoparius* and *Ulex europaeus*, in an attempt to reproduce examples of woodland and scrub of types formerly native to the area (Wormell, 1968).

The introduction of experiments of this kind, together with the widely recognized need for habitat improvement in heathland areas and the ambitious programmes of afforestation, point inescapably to one conclusion. This is that destruction of native forest on so wide-spread a scale in an oceanic climate and on acid, generally infertile soil was a most unfortunate exploitation of a natural resource of great economic and ecological value. It set in train a series of consequences, both as regards land use and management and as regards developments in vegetation and soils, which have been inclined towards deterioration of the habitat. But at the same time, the spread of heathlands has provided an object lesson of the utmost value and a landscape of much delight. Much of both scientific and practical importance has been revealed by research converging from numerous avenues of approach upon the problems presented by this unique, largely man-made type of ecosystem. Possibilities now exist for the redemption of what has been damaged or even destroyed. Current changes in patterns of land use and the variety of claims on heathland make it a test case for acceptance or rejection of the standpoint of conservation, understood in its broadest sense and based upon thorough ecological analysis. Acceptance will secure a good prospect of accommodating increased production with facilities for recreation and adequate provision of reserves. In this way a satisfactory future can be envisaged for a type of vegetation which has provided fascinating insights into the repercussions of man's impact upon natural systems.

References

AHLGREN, F. and AHLGREN, C. E. (1960). 'Ecological effects of forest fires, including temperature, fertility and chemical composition'. *Bot. Rev.*, **26**, 483-533.

AIRLIE, EARL OF (1971). 'Making full use of an upland estate'. *Scottish Landowner*, No. 142, 3–6.

ALBRECHT, W. A. (1957). 'Soil fertility and biotic geography'. *Geogrl. Rev.*, **47**, 86–105.

ALLEN, S. E. (1964). 'Chemical aspects of heather burning.' *J. appl. Ecol.*, **1**, 347–67.

ALLEN, S. E., EVANS, C. C. and GRIMSHAW, H. M. (1969). 'The distribution of mineral nutrients in soil after heather burning.' *Oikos*, **20**, 16–25.

ANDERSON, D. J. (1961). 'The structure of some upland plant communities in Caernarvonshire. II The pattern shown by *Vaccinium myrtillus* and *Calluna vulgaris*,' *J. Ecol.*, **49**, 731–8.

ANDERSON, D. J. (1963). 'The structure of some upland plant communities in Caernarvonshire. III The continuum analysis', *J. Ecol.*, **51**, 403–14.

ARMSTRONG, J. I., CALVERT, J. and INGOLD, C. T. (1930). 'The ecology of the Mountains of Mourne with special reference to Slieve Donard', *Proc. R. Ir. Acad.* (*B*)., **39**, 440–52

ATLESTAM, P. O. (1942). 'Bohusläns Ljunghedar.' *Medd. Göteborgs Högsk. Geogr. Instn.*, **30**, 1–132.

BANNISTER, P. (1964*a*). 'Stomatal responses of heath plants to water deficits'. *J. Ecol.*, **52**, 151–8.

BANNISTER, P. (1964*b*). 'The water relations of certain heath plants with reference to their ecological amptitude. I Introduction: germination and establishment'. *J. Ecol.*, **52**, 423–32.

BANNISTER, P. (1964*c*). 'The water relations of certain heath plants with reference to their ecological amplitude. II Field studies'. *J. Ecol.*, **52**, 481–97.

BANNISTER, P. (1964*d*). 'The water relations of certain heath plants with reference to their ecological amplitude. III Experimental studies: general conclusions'. *J. Ecol.*, **52**, 499–509.

BANNISTER, P. (1965). 'Biological Flora of the British Isles. *Erica cinerea L*'. *J. Ecol.*, **53**, 527–42

BANNISTER, P. (1966). 'Biological Flora of the British Isles. *Erica tetralix L*'. *J. Ecol.*, **54**, 795–813.

BARCLAY-ESTRUP, P. (1970). 'The description and interpretation of cyclical processes in a heath community. II Changes in biomass and shoot production during the *Calluna* cycle'. *J. Ecol.*, **58**, 243–9.

BARCLAY-ESTRUP, P. (1971). 'The description and interpretation of cyclical processes in a heath community. III Microclimate in relation to the *Calluna* cycle'. *J. Ecol.*, **59**, 143–66.

BARCLAY-ESTRUP, P. and GIMINGHAM, C. H. (1969). 'The description and interpretation of cyclical processes in a heath community. I Vegetational change in relation to the *Calluna* cycle.' *J. Ecol.*, **57**, 737–58.

BEIJERINCK, W. (1940). '*Calluna:* a monograph on the Scotch heather'. *Verh. Akad. Wet. Amst.* (3rd Sect), **38**, 1–180.

BELLAMY, D. J. and HOLLAND, P. J. (1966). 'Determination of the net annual aerial production of *Calluna vulgaris* (*L*) Hull, in northern England'. *Oikos,* **17,** 272–5.

BELLAMY, D. J., BRIDGEWATER, P., MARSHALL, C., and TICKLE, W. M. (1969). 'Status of the Teesdale rarities'. *Nature, Lond.,* **222,** 238–43.

BERGLUND, B. E. (1962). 'Vegetation på ön Senoren. I Vegetations historia.' *Bot. Notiser,* **115,** 387–420.

BIRSE, E. L. (1968). *Hill-land vegetation in Scotland.* Proc. Symp. Hill-land Productivity – European Grassland Federation 4 pp. July 1968.

BJÖRKMAN, E. (1945). 'Studier över ljusets betydelse för föryngringens höjdtillväxt på Norrländska tallhedar.' *Meddn. St. SkogsförsAnst.,* **34,** 497–542.

BLISS, L. C. (1956). 'A comparison of plant development in the microenvironments of arctic and alpine tundras.' *Ecol. Monogr.,* **26,** 303–37.

BØCHER, T. W. (1937). 'Om Udbredelsen af Ericaceae, Vacciniaceae og Empetraceae i Danmark.' *Bot. Tidsskr.,* **44,** 6–40.

BØCHER, T. W. (1943). 'Studies on the plant geography of the North-Atlantic heath formation. II Danish dwarf shrub communities in relation to those of Northern Europe.' *K. Danske vidensk. Selsk., Biol., Skr.,* **2,** 1–129.

BOGGIE, R., HUNTER, R. F., and KNIGHT, A. H. (1958). 'Studies of the root development of plants in the field using radioactive tracers.' *J. Ecol.,* **46,** 621–39.

BOND, G. and SCOTT, G. D. (1955). 'An examination of some symbiotic systems for fixation of nitrogen.' *Ann. Bot., Lond. N.S.,* **19,** 67–77.

BRAID, K. W. and TERVET, I. W. (1937). 'Certain botantical aspects of the dying-out of heather.' *Scott. J. Agric.,* **20,** 365–72.

BRAUN-BLANQUET, J. (1964). *Pflanzensociologie.* Wien & Berlin.

BRIDGEWATER, P. B. (in preparation). Phytosociological studies in the British heath formation I.

CAMERON, A. E., MCHARDY, J. W. and BENNETT, A. H. (1944). *The heather beetle* (Lochmaea suturalis)*: its biology and control.* British Field Sports Society. Petworth.

CHAPMAN, S. B. (1967). 'Nutrient budgets for a dry heath ecosystem in the south of England.' *J. Ecol.,* **55,** 677–89.

CHAPMAN, S. B. (1970). 'The nutrient content of the soil and root system of a dry heath ecosystem.' *J. Ecol.,* **58,** 445–52.

CHRISTOPH, H. (1921). 'Untersuchungen uber die mykotrophen Verhaltnisse der "Ericales" und die Keimung van Pirolaceen.' *Beih. bot. Zbl.,* **38,** 115–57.

CLAPHAM, A. R., TUTIN, T. G. and WARBURG, E. F. (1962). *Flora of the British Isles.* 2nd Ed. Cambridge.

CLOUSTON, D. (1943) 'The improvement of heath and moorland grazings'. *Scott. J. Agric.,* **24,** 104–12.

CLYMO, R. S. (1963). 'Ion exchange in *Sphagnum* and its relation to bog ecology.' *Ann. Bot., Lond. N.S.,* **27,** 309–24.

CONWAY, V. M. (1947). 'Ringinglow Bog near Sheffield.' *J. Ecol.* **34,** 149–81

COOMBE, D. E. and FROST, L. C. (1956). 'The nature and origin of the soils over the Cornish Serpentine.' *J. Ecol.* **44,** 605–15.

COPPINS, B. J. and SHIMWELL, D. W. (1971). 'Cryptogam complement and biomass in dry *Calluna* heath of different ages'. *Oikos,* **22,** 204–9.

CORMACK, E. and GIMINGHAM, C. H. (1964). 'Litter production by *Calluna vulgaris* (L.) Hull.' *J. Ecol.,* **52,** 285–97.

COSTIN, A. B. (1959). Vegetation of high mountains in Australia in relation to land use. In: *Biogeography and Ecology in Australia* (Ed: Keast, A., Crocker, R. L. and Christian, C. S.). The Hague.

COSTIN, A. B. (1967). Alpine ecosystems of the Australasian region. In: *Arctic and Alpine Environments* (Ed: Wright, H. E. and Osburn, W. H.). Indiana.

CRAMPTON, C. B. (1911). *The vegetation of Caithness considered in relation to the geology*. Committee for the survey and study of British vegetation. Cambridge.

CRAMPTON, C. B. and MACGREGOR, M. (1913). 'The plant ecology of Ben Armine (Sutherlandshire).' *Scott. Geogr. Mag.*, **29**, 256–66.

CRISP, D. T. (1966). 'Input and output of minerals for an area of Pennine moorland: The importance of precipitation, drainage, peat erosion and animals'. *J. appl. Ecol.*, **3**, 327–48.

CUNNINGHAM, J. M. M., SMITH, A. D. M. and DONEY, J. M. (1971). *Trends in livestock populations in hill areas in Scotland.* Hill Farming Research Organization, Edinburgh. Fifth Report, 1967–70; 88–95.

DAHL, E. (1956). *Rondane: mountain vegetation in south Norway and its relation to the environment.* Oslo.

DAMMAN, A. W. H. (1957). 'The south-Swedish *Calluna* heath and its relation to the Calluneto-Genistetum, *Bot. Notiser.* **110,** 363–98.

DARLING, F. F. (1947). *Natural history in the Highlands and Islands.* London.

DAUBENMIRE, R. (1968). Ecology of fire in grasslands, In: *Advances in Ecological Research* (Ed: Cragg, J. B.). **5**, 209–66.

DAVIES, W. and JONES, T. E. (1932). 'The yield and response to manures of contrasting pasture types'. *Welsh J. Agric.* **8**, 170–92.

DELANY, M. J. (1953). 'Studies on the microclimate of *Calluna* heathland'. *J. Anim. Ecol.*, **22**, 227–39.

DIMBLEBY, G. W. (1952). 'Soil regeneration on the north-east Yorkshire moors'. *J. Ecol.*, **40**, 331–41

DIMBLEBY, G. W. (1962). 'The development of British heathlands and their soils'. *Oxf. For. Mem.*, **23**, 1–121.

DIMBLEBY, G. W. (1965). 'Post-glacial changes in soil profiles'. *Proc. R. Soc. B.* **161**, 355–62.

DUCHAUFOUR, P. (1948). 'Récherches écologiques sur la chenaie atlantique française'. *Ann. Ec. natn. Eaux Forêts, Nancy*, **11**, 1–332.

DUCHAUFOUR, P. (1950). 'Observations sur "la faim d'azote" de l'opicea'. *Rev. for Franc.*, **2**, 1–4.

DUCHAUFOUR, P. (1956). *Note sur les phases de la podzolisation sur grès vosgien* Trans. 6th Int. Cong. Soil Sci. E., pp. 367–70.

DU RIETZ, G. E. (1931). 'Life forms of terrestrial flowering plants. *Acta phytogeogr, suec.*, **3**, 1–95.

DURNO, S. E. (1957). 'Certain aspects of vegetational history in North-East Scotland'. *Scott. Geogr., Mag.*, **73**, 176–84.

DURNO, S. E. (1958). 'Pollen analysis of peat deposits in Eastern Sutherland and Caithness'. *Scott. geogr., Mag.*, **74**, 127–35.

DURNO, S. E. (1965). 'Pollen analytical evidence of 'landnam' from two Scottish sites'. *Trans. Bot. Soc. Edinb.*, **40**, 13–19.

EADIE, J. (1967). *The nutrition of grazing hill sheep; utilization of hill pastures.* Hill Farming Research Organization, Fourth Report 1964–7; pp. 38–45.

EDWARDS, M. V., ATTERSON, J., and HOWELL, R. S. (1963). 'Wind-loosening of young trees on upland heaths'. *Forest Rec. Lond.*, **50**, 1–16.

ELGEE, F. (1914). 'The vegetation of the eastern moorlands of Yorkshire'. *J. Ecol.*, **2**, 1–18.

ELLENBERG, H. (1954). 'Steppenheide und Waldweide'. *Erdkunde*, **8**, 1 88–94.

ELLENBERG, H. and MUELLER-DOMBOIS, D. (1966 *a*). 'Tentative physiognomic-ecological classification of plant formations of the earth'. *Ber. geobot. Forsch Inst. Rübel* **37**, 21–55.

ELLENBERG, H. and MUELLER-DOMBOIS, D. (1966 *b*). 'A Key to Raunkiaer plant life forms with revised subdivisions'. *Ber. geobot. Forsch Inst., Rübel,* **37**, 56–73.

ELLIOTT, R. J. (1953). *Heather burning.* Ph.D. Thesis, University of Sheffield.

ESMINGER, M. E. (1956). *Sheep husbandry.* Danville, U.S.A.

EVANS, C. C. and ALLEN, S. E. (1971). 'Nutrient losses in smoke produced during heather burning'. *Oikos,* **22**, 149–54.

FAEGRI, K. (1940). 'Quartärgeologische Untersuchungen im westlichen Norwegen. II Zur spätquartären Geschichte Jaerens'. *Bergens Mus. Årb. naturvid. rekke,* **7**, 1–201.

FAGAN, T. W., and PROVAN, A. L. (1930). 'The effect of manures on the nitrogen and mineral content of upland and lowland pastures'. *Bull. Welsh Pl. Breed. Stn. Series H,* **11**, 27–37.

FAIRBAIRN, C. B. (1959). *The composition and probable nutritive value of some plants indigenous to Northern England.* Ph.D. Thesis, Univ. of Durham.

FARROW, E. P. (1917). 'On the ecology of the vegetation of Breckland. III General effects of rabbits on the vegetation'. *J. Ecol.,* **5**, 1–18.

FARROW, E. P. (1925). 'On the ecology of the vegetation of Breckland. VIII Views relating to the probable former distribution of *Calluna* heath in England'. *J. Ecol.,* **13**, 121–37.

FENTON, E. W. (1936). 'The problem of moor mat grass'. *Scott. J. Agric.,* **19**, 143–8.

FENTON, E. W. (1937 *a*). 'Some aspects of man's influence on the vegetation of Scotland'. *Scott. geogr. Mag.,* **53**, 16–24.

FENTON, E. W. (1937 *b*). 'The influence of sheep on the vegetation of hill grazings in Scotland'. *J. Ecol.,* **25**, 424–30.

FENTON, E. W. (1940). 'The influence of rabbits on the vegetation of certain hill-grazing districts of Scotland'. *J. Ecol.* **28**, 438–49.

FITZPATRICK, E. A. (1964). The Soils of Scotland. In: *The Vegetation of Scotland* (Ed: Burnett, J. H.). Edinburgh and London.

FLOWER-ELLIS, J. G. K. (1971). 'Age structure and dynamics in stands of bilberry (*Vaccinium myrtillus* L.)'. *Avdelningen för Skogsekologi, Stockholm,* **9**, 1–108.

FORREST, G. I. (1971). 'Structure and production of North Pennine blanket bog vegetation'. *J. Ecol.,* **59** 453–79.

FOSBERG, F. R. (1967). Classification of Vegetation. In: *Guide to the check sheet for I.B.P. areas.* I.B.P. Handbook. No. 4 (Ed. Peterken, G.F.). Oxford and Edinburgh.

FREISLEBEN, R. (1935). 'Weitere Untersuchungen über die Mykotrophie der Ericaceen'. *Jb. wiss. Bot.,* **82**, 413–59.

FRITSCH, F. E. (1927). 'The heath association on Hindhead Common, 1910–1926'. *J. Ecol.,* **15**, 344–72.

FRITSCH, F. E. and SALISBURY, E. J. (1915). 'Further observations on the heath association of Hindhead Common', *New Phytol.,* **14**, 116–38.

GIMINGHAM, C. H. (1949). 'The effects of grazing on the balance between *Erica cinerea* L. and *Calluna vulgaris* (L.) Hull in upland heath, and their morphological responses'. *J. Ecol.,* **37**, 100–19.

GIMINGHAM, C. H. (1960). 'Biological flora of the British Isles. *Calluna vulgaris* (L.) Hull'. *J. Ecol.,* **48**, 455–83.

GIMINGHAM, C. H. (1961). 'North European heath communities a "network of variation"'. *J. Ecol.,* **49**, 655–94.

GIMINGHAM, C. H. (1964 *a*). Dwarf-shrub Heaths. In: *The Vegetation of Scotland* (Ed. Burnett, J. H.). Edinburgh and London.

GIMINGHAM, C. H. (1964 *b*). The composition of the vegetation and its balance with environment. In: *Symposium on Land Use in the Scottish Highlands. Advmt. Sci., Lond.*, **21**, 148–52

GIMINGHAM, C. H. (1969). 'The interpretation of variation in north-European dwarf-shrub heath communities'. *Vegetatio*, **17**, 89–108.

GIMINGHAM, C. H. (1971 *a*). 'British heathland ecosystems: the outcome of many years of management by fire'. *Proc. 10th Annual Tall Timbers Fire Ecology Conference*, 293–321.

GIMINGHAM, C. H. (1971 *b*). *Calluna* heathlands: use and conservation in the light of some ecological effects of management. In: *Scientific management of plant and animal communities for conservation*. Brit. Ecol. Soc. Symposium No. 11 (Ed. Duffey, E. A. G. and Walt, A. S.). Oxford and Edinburgh.

GIMINGHAM, C. H. and MILLER, G. R. (1968). Measurement of the primary production of dwarf shrub heaths. In: *Methods for the measurement of primary production of grassland*. I.B.P. Handbook No. 6 (Ed. Milner, C. and Hughes, R. E.), 43–51. Oxford and Edinburgh.

GODWIN, H. (1944 *a*). 'Neolithic forest clearance'. *Nature, Lond.*, **153**, 511–12.

GODWIN, H. (1944 *b*). 'Age and origin of the Breckland heaths of East Anglia'. *Nature, Lond.*, **154**, 6.

GODWIN, H. (1948). 'Studies of the post glacial history of British vegetation. X. Correlation between climate, forest composition, prehistoric agriculture and peat stratigraphy in Sub-boreal and Sub-atlantic peats of the Somerset levels.' *Phil. Trans. R. Soc. B.*, **233**, 275–86.

GODWIN, H. (1956). *History of the British flora*. Cambridge.

GRACE, J. and WOOLHOUSE, H. W. (1970). 'A physiological and mathematical study of the growth and productivity of a *Calluna-Sphagnum* community. I Net. photosynthesis of *Calluna vulgaris* (L.) Hull'. *J. appl. Ecol.*, **7**, 363–81

GRAEBNER, P. (1901, 1st Edition; 1925 2nd Edition). *Die Heide Norddeutschlands* (Die Vegetation der Erde, V. Ed. Engler, A. and Drude, O.) Leipzig.

GRANT, SHEILA A. and HUNTER, R. F. (1962). 'Ecotypic differentiation of *Calluna vulgaris* (L.) in relation to altitude'. *New Phytol.*, **61**, 44–55.

GRANT, SHEILA A. and HUNTER, R. F. (1966). 'The effects of frequency and season of clipping on the morphology, productivity and chemical composition of *Calluna vulgaris* (L.) Hull'. *New Phytol.*, **65**, 125–33.

GRANT, SHEILA A., and HUNTER, R. F. (1968). 'Interactions of grazing and burning on heather moors and their implications in heather management'. *J. Brit. Grassl. Soc.*, **23**, 285–93.

GREEN, F. H. W. (1964). 'A map of annual average potential water deficit in the British Isles'. *J. appl. Ecol.*, **1**, 151–8.

GREGOR, J. W. (1961). 'Hill grazings: irrigation as a possible improvement technique'. *Rep. Scott. Pl. Breed. Stn. for* 1961, 31–8.

GREIG, R. B. (1910). 'Report on the improvement of hill pasture as determined by effect on stock'. *Bull. Aberd. N. Scotl. Coll. Agric.*, **16**, 1–24.

GREIG-SMITH, P. (1952). 'The use of random and contiguous quadrats in the study of the structure of plant communities'. *Ann. Bot., Lond., N.S.*, **16**, 293–316.

GREVILLIUS, A. Y. and KIRCHNER, O. VON (1923). *Ericaceae*. In: *Lebensgeschichte der Blutenpflanzen Mitteleuropas*. 4 (Ed: Kirchner, O. von; Loew, E. and Schröter, C.). Stuttgart.

GREVILLIUS, A. Y. and KIRCHNER, O. VON (1925). *Empetraceae*. In: *Lebensgeschichte der Blutenpflanzen Mitteleuropas*. 4 (Ed: Kirchner, O. von; Loew, E. and Schröter, C.). Stuttgart.

GRUBB, P. J.; GREEN, H. E., and MERRIFIELD, R. C. J. (1969). 'The ecology of chalk heath: its relevence to the calcicole-calcifuge and soil acidification problems'. *J. Ecol.*, **57**, 175–212.

GUILLOTEAU, J. (1956). 'The problem of bush fires and burns in land development and soil conservation in Africa south of the Sahara'. *Sols. afr.*, **4**, 64–102.

HAGERUP, O. (1946). 'Studies on the Empetraceae'. *Dansk. vidensk. Selsk., Biol. Medd.*, **20**, 5–9.

HANDLEY, W. R. C. (1963). 'Mycorrhizal associations and *Calluna* heathland afforestation'. *Bull. For. Commn., Lond.*, **36**, 1–70.

HARRISON, C. M. (1970). 'The phytosociology of certain English heathland communities'. *J. Ecol.*, **58**, 573–89.

HEATH, G. H. and LUCKWILL, L. C. (1938). 'The rooting systems of heath plants'. *J. Ecol.*, **26**, 331–52.

HEDBERG, O. (1951). 'Vegetation belts of the East African mountains'. *Svensk. bot. Tidskr.*, **45**, 140–202.

HEDDLE, R. G. and OGG, W. G. (1933). 'Experiments in the improvement of hill pasture'. *Scott. J. Agric.*, **16**, 431–46.

HOPE-SIMPSON, J. F. (1941). 'Studies of the vegetation of the English chalk. VIII A second survey of the chalk grasslands of the South Downs'. *J. Ecol.*, **29**, 217–67.

HOPKINS, B. (1965). 'Observations on savanna burning in the Olokemeji Forest Reserve, Nigeria'. *J. appl. Ecol.*, **2**, 367–81.

HUNTER, R. F. (1954). 'The grazing of hill pasture sward types'. *J. Br. Grassld. Soc.*, **9**, 195–208.

HUNTER, R. F. (1958 *a*). 'The direction and problems of research in hill pasture improvement'. *J. Br. Grassld. Soc.*, **13**, 121–5.

HUNTER, R. F. (1958 *b*). 'Hill land improvement'. *Advmt. Sci., Lond.*, **15**, 194–6.

HUNTER, R. F. (1960). 'Aims and methods in grazing-behaviour studies on hill pastures'. *Proc. VIIIth Int. Grassld. Congress*, pp. 454–7.

HUNTER, R. F. (1962). 'Hill sheep and their pasture: a study of sheep-grazing in south-east Scotland'. *J. Ecol.*, **50**, 651–80.

HUNTER, R. F. and EADIE, J. (1962). Botanical Research at the Hill Farming Research Organization. *Proc. Europ. Conf. for Forage Production on Natural Grassland in Mountain Regions*, 1962. 7 pp.

IVERSEN, J. (1941). 'Landnam i Danmarks Stenalder'. *Danm. Geol. Unders IIR.*, **66**, 1–67.

IVERSEN, J. (1949). 'The influence of prehistoric man on soil fertility'. *Danm. geol. Unders. IV R.*, **6**, 1–25.

IVERSEN, J. (1964). 'Retrogressive vegetational succession in the post-glacial'. *J. Ecol.*, **52**, (Suppl.), 59–70.

IVERSEN, J. (1969). 'Retrogressive development of a forest ecosystem demonstrated by pollen diagrams from fossil mor'. *Oikos Suppl.*, **12**, 35–49.

IWANAMI, Y. (1969). 'Temperatures during *Miscanthus* type grassland fires and their effect on the regeneration of *Miscanthus sinensis*'. *Sci. Rep. Res. Insts. Tôhoku Univ., Series D.*, **20**, 47–88.

IWANAMI, Y. and IIZUMI, S. (1969). 'Report on the burning temperatures of Japanese lawn grass (*Zoisia japonica* Steud.)'. *Jap. J. Ecol.*, **16**, 40–41.

JENKINS, D., WATSON, A., and MILLER, G. R. (1963). 'Population studies on red grouse, *Lagopus lagopus scoticus* (Lath.) in north-east Scotland'. *J. Anim. Ecol.*, **32**, 317–76.

JENKINS, D., WATSON, A., and MILLER, G. R. (1967). 'Population fluctuations in the red grouse (*Lagopus lagopus scoticus*)'. *J. Anim. Ecol.*, **36**, 97–122.

JONASSEN, H. (1950). 'Recent pollen sedimentation and Jutland heath diagrams'. *Dansk. bot. Ark.*, **13**, 1–168.

JONES, H. E. (1971 *a*). 'Comparative studies of plant growth and distribution in relation to waterlogging. II. An experimental study of the relationship between transpiration and the uptake of iron in *Erica cinerea* L. and *E. tetralix* L.' *J. Ecol.*, **59**, 167–78.

JONES, H. E. (1971 *b*). 'Comparative studies of plant growth and distribution in relation to waterlogging. III. The response of *Erica cinerea* L. to waterlogging in peat soils of differing iron content'. *J. Ecol.*, **59**, 583–91.

JONES, H. E. and ETHERINGTON, J. R. (1970). 'Comparative studies of plant growth and distribution in relation to waterlogging. I The survival of *Erica cinerea* L. and *E. tetralix* L. and its apparent relationship to iron and manganese uptake in waterlogged soil'. *J. Ecol.*, **58**, 487–96.

JONES, H. E., FORREST, G. I. and GORE, A. J. P. (1971). First stage of a model for the growth and decay of *Calluna vulgaris* at Moor House. In: *Tundra Biome: working meeting on analysis of ecosystems, Kevo, Finland, September, 1970* (Ed: Heal, O. W.). I.B.P. London.

JONES, LD.I. (1967). *Studies on hill land in Wales*. Tech. Bull., Welsh Pl. Breed. Stn., 2, Aberystwyth.

JONES, W. N. and SMITH, M. L. (1928). 'On the fixation of atmospheric nitrogen by *Phoma radicis Callunae*, including a new method for investigating N fixation in micro-organisms'. *Brit. J. exp. Biol.*, **6**, 167–89.

KAYLL, A. J. (1966). 'Some characteristics of heath fires in north-east Scotland'. *J. appl. Ecol.*, **3**, 29–40.

KAYLL, A. J. and GIMINGHAM, C. H. (1965). 'Vegetative regeneration of *Calluna vulgaris* after fire'. *J. Ecol.*, **53**, 729–34.

KEEF, P. A. M., WYMER, J. J. and DIMBLEBY, G. W. (1965). 'A mesolithic site on Iping Common, Sussex'. *Proc. prehist. Soc.*, **31**, 85–92.

KENWORTHY, J. B. (1963). 'Temperatures in heather burning'. *Nature, Lond.*, **200**, 1226.

KENWORTHY, J. B. (1964). *A study of the changes in plant and soil nutrients associated with moorburning and grazing*. Ph.D. Thesis, Univ. of St. Andrews.

KERSHAW, K. A. (1957). 'The use of cover and frequency in the detection of pattern in plant communities'. *Ecology*, **38**, 291–9.

KERSHAW, K. A. (1967). 'Ecological methods and computers'. *Sci. Prog., Lond.*, **55**, 437–51.

KERSHAW, K. A. and TALLIS, J. H. (1958). 'Pattern in a high-level *Juncus squarrosus* community'. *J. Ecol.*, **46**, 739–48.

KING, J. (1960). Observations on the seeding establishment and growth of *Nardus stricta* in burned *Callunetum*'. *J. Ecol.*, **48**, 667–77.

KNAPP, R. (1942). *Zur Systematik der Walder, Zwergstrauch-heider und Trockenrasen des eurosibirischen Vegetationskreis*. Inaug. Diss. Freiburg. 12 Rd.Z. Hannover.

KNUDSON, L. (1928). 'Is the fungus necessary for development of seedlings of *Calluna vulgaris*?' *Am. J. Bot.*, **15**, 624–5.

KNUDSON, L. (1929). 'Seed germination and growth of *Calluna vulgaris*'. *New Phytol.*, **28**, 369–76.

KNUDSON, L. (1933). 'Non-symbiotic development of seedlings of *Calluna vulgaris*', *New Phytol.*, **32**, 115–27.

KÖPPEN, W. (1918). 'Klassification der Klimate nach Temperatur, Niederschlag und Jahresverlauf', *Petermann's Mitt.* **64**, 193–203 and 243–8.

KUBIËNA, W. L. (1953). *The Soils of Europe*. London.

LAUDER, A. and COMRIE, A. (1936). 'The composition of heather (*Calluna vulgaris*)'. *Scott. J. Agric.*, **19**, 148–52.

LAWES, J. B. and GILBERT, J. H. (1859). 'Experimental inquiry into the composition of some animals fed and slaughtered for human food'. *Proc. R. Soc.*, **149**, 496–600.

LAWES, J. B., and GILBERT, J. H. (1883). An additional note on 'Experimental inquiry into the composition of some animals fed and slaughtered for human food'. Composition of entire animals and of certain separated parts'. *Proc. R. Soc.*, **174**, 865–90.

LEBRUN, J., NOIRFALISE, A., HEINEMANN, P. and VANDEN BERGHEN, C. (1949). 'Les Associations Vegetales de Belgique'. *Bull. Soc. r. Bot. Belg.*, **82**, 105–207.

LEMÉE, G. (1938). 'Récherches écologiques sur la végétation du Perche'. *Revue gén. Bot.*, **50**, 547–63 and 671–90.

LEYTON, L. (1954). 'The growth and mineral nutrition of Spruce and Pine in heathland plantations'. *Inst. Pap. Commonw. For. Inst.*, **31**, 1–109.

LEYTON, L. (1955). 'The influence of artificial shading of the ground vegetation on the nutrition and growth of Sitka spruce (*Picea sitchensis* Carr.) in a heathland plantation'. *Forestry*, **28**, 1–6.

LINTON, A. (1918). *The grazing of hill pastures*. Selkirk.

LOACH, K. (1966). 'Relations between soil nutrients and vegetation in wet-heaths. I Soil nutrient content and moisture conditions'. *J. Ecol.*, **54**, 597–608.

LOVAT, LORD (1911). Heather Burning. In: *The grouse in health and disease* (Ed: Leslie, A.S.). 392–412. London.

MACDONALD, J. A. (1949). 'Heather rhizomorph fungus in Scotland'. *Proc. R. Soc. Edinb. B.*, **63**, 230–41.

MCLEAN, R. C. (1935). 'An ungrazed grassland on limestone in Wales, with a note on plant "dominions"'. *J. Ecol.*, **23**, 436–42

MCINTOSH, R. (1967). 'The continuum concept of vegetation'. *Bot. Rev.*, **33**, 130–87.

MACLEOD, A. C. (1955). 'Heather in the seasonal dietary of sheep'. *Proc. Br. Soc. Anim. Prod.*, 1955, 13–17.

MCVEAN, D. N. (1964). *Grass Heaths and Moss Heaths*. In: *The Vegetation of Scotland* (Ed: Burnett, J. H.). Edinburgh and London.

MCVEAN, D. N. (1969). 'Alpine vegetation of the central Snowy Mountains of New South Wales'. *J. Ecol.*, **57**, 67–81.

MCVEAN, D. N. and LOCKIE, J. (1969). *Ecology and land use in upland Scotland.* Edinburgh.

MCVEAN, D. N. and RATCLIFFE, D. A. (1962) *Plant communities of the Scottish Highlands*. London.

MALLOCH, A. J. C. (1971). 'Vegetation of the maritime cliff-tops of the Lizard and Land's End peninsulas, west Cornwall'. *New Phytol. 70*, 1155–97.

MALMER, N. (1965). 'The South-western dwarf-shrub heaths'. In: The plant cover of Sweden. *Acta Phytogeogr. Suec.*, **50**, 123–30.

MALMSTRÖM, C. (1937). 'Tönnersjöhedens försökspark i Halland'. *Meddn. St. Skogsförs Anst.*, **30**, 323–528.

MALMSTRÖM, C. (1939). 'Hallands skogar under de senaste 300 aren'. *Meddn. St. Skogsförs Anst.*, **31**, 171–300.

MANLEY, G. (1942). 'Meteorological observation on Dun Fell'. *Q. Jl. R. met. Soc.*, **68**, 251–65.

MATTHEWS, J. R. (1937). 'Geographical relationships of the British flora'. *J. Ecol.*, **25**, 1–90.

METCALFE, G. (1950). 'The ecology of the Cairngorms. Part II. The mountain Callunetum'. *J. Ecol.*, **38**, 46–74.

METEOROLOGICAL OFFICE (Air Ministry) (1958). *Tables of Temperature, Relative Humidity and Precipitation for the World.* H.M.S.O., London.

MILLAR, C. E., TURK, L. M., and FOTH, H. D. (1958). *Fundamentals of soil science.* New York.

MILLER, G. R. (1964). 'The management of heather moors'. In: Symposium on Land use in the Scottish Highlands. *Advmt. Sci*, **21**, 163–9.

MILLER, G. R. (1968). 'Evidence for selective feeding on fertilized plots by red grouse, hares and rabbits'. *J. Wildl. Mgmt.*, **32**, 849–53.

MILLER, G. R. (1971). Grazing and the regeneration of shrubs and trees. In: *Range. Ecology Research. 1st Progress Report*, pp. 27–40. The Nature Conservancy, Edinburgh.

MILLER, G. R. and MILES, A. M. (1969). Productivity and management of heather. In: *Grouse research in Scotland. 13th Progress Report*, pp. 31–45. The Nature Conservancy, Edinburgh.

MILLER, G. R. and MILES, J. (1970). 'Regeneration of heather (*Calluna vulgaris* (L.) Hull) at different ages and seasons in north-east Scotland'. *J. appl. Ecol.*, **7**, 51–60.

MILLER, G. R., JENKINS, D. and WATSON, A. (1966). 'Heather performance and red grouse populations. I Visual estimates of heather performance'. *J. appl. Ecol.*, **3**, 313–26.

MILLER, G. R., WATSON, A. and JENKINS, D. (1970). Responses of red grouse populations to experimental improvement of their food. In: *Animal populations in relation to their food resources.* Brit. Ecol. Soc. Symposium No. 10 (Ed: Watson, A.). Oxford and Edinburgh.

MILTON, W. E. J. (1934). 'The effect of controlled grazing and manuring on natural hill pastures'. *Welsh J. Agric.*, **10**, 196–211.

MILTON, W. E. J. (1938). 'The composition of natural hill pastures under controlled and free grazing, cutting and manuring'. *Welsh J. Agric.*, **14**, 182–95.

MILTON, W. E. J. (1940). 'The effect of manuring, grazing and cutting on the yield, botanical and chemical composition of natural hill pastures. I Yield and Botanical Section'. *J. Ecol.*, **28**, 326–56.

MILTON, W. E. J. and DAVIS, R. O. (1947). 'The yield, botanical and chemical composition of natural hill herbage under manuring, controlled grazing and hay conditions'. *J. Ecol.*, **35**, 65–95.

MITCHELL, G. F. (1951). 'Studies in Irish Quaternary deposits, No. 7'. *Proc. R. Ir. Acad.*, **53**, B, 111–206.

MITCHELL, G. F. (1956). 'Post Boreal pollen diagrams from Irish raised bogs'. *Proc. R. Ir. Acad.*, **57**, B., 185–251.

MOHAMED, B. F. and GIMINGHAM, C. H. (1970). 'The morphology of vegetative regeneration in *Calluna vulgaris*'. *New Phytol.*, **69**, 743–50.

MOORE, J. J. (1968). A classification of the bogs and wet heaths of northern Europe (Oxycocco-Sphagnetea Br. – Bl. et Tx. 1943). In: *Pflanzensoziologische Systematik* (Ed: Tüxen, R.). pp. 306–20. Den Haag.

MOORE, N. W. (1962). 'The heaths of Dorset and their conservation'. *J. Ecol.*, **50**, 369–91.

MORAN, T. and PACE, J. (1962). 'A note on the amino acid composition of the protein in heather shoots'. *J. agric. Sci., Camb.*, **59**, 93–4.

MORISON, G. (1963). *The Heather beetle* (Lochmaea suturalis, *Thomson*). North of Scotland College of Agriculture, Aberdeen.

MORK, E. (1946). 'Om Skogsbunnens lyngvegetasjon'. *Meddn. Norsk. Skogsforsøksv.*, **9**, 269–356.

MORRISON, M. E. S. (1959). 'Evidence and interpretation of "landnam" in the North-East of Ireland'. *Bot. Notiser.*, **112**, 185–204.

MOSS, C. E. 1913. *Vegetation of the Peak District*, Cambridge.

MOSS, R. (1967). Probable limiting nutrients in the main food of red grouse (Lagopus lagopus scoticus). In: *Secondary Productivity of Terrestrial Ecosystems* (Ed. Petrusewicz, K.) pp. 369–79. Warsaw.

MOSS, R. (1968). Food selection and nutrition in ptarmigan (*Lagopus mutus*). In: *Symp. Zool. Soc. Lond. No.*, **21**, (Ed: Crawford, M. A.). pp. 207–16. London.

MOSS, R. (1969 *a*). 'A comparison of red grouse (*Lagopus l. scoticus*) stocks with the production and nutritive value of heather (*Calluna vulgaris*)'. *J. Anim. Ecol.*, **38**, 103–22.

MOSS, R. (1969 *b*). Nutrition in red grouse and ptarmigan. In: *Grouse Research in Scotland. 13th Progress Report*, pp 18–24. The Nature Conservancy, Edinburgh.

MUIR, A. and FRASER, G. K. (1940). 'The soils and vegetation of the Bin and Clashindarroch forests'. *Trans. R. Soc. Edinb.*, **60**, 233–341.

MÜLLER, P. E. (1884). 'Studier over skovjord, som bidrag til skovdyrkningens theori. II Om muld og mor i egeskov og paa heder'. *Tidsskr. f. Skovbrug.*, **7**, 1–232.

MÜLLER, P. E. (1924). 'Bidrag til de jydske hedesletters naturhistorie'. *Kgl. Danske Vidensk. Selsk. Biol. Medd.*, **4**.

NICHOLSON, I. A. (1960). 'Aircraft in agriculture'. *Scott. Agric.*, **39**, 127–32.

NICHOLSON, I. A. (1964). The influence of management practices on the present day vegetational pattern, and developmental trends. In: Symposium on Land use in the Scottish Highlands. *Advmt. Sci., Lond.*, **21**, 158–63.

NICHOLSON, I. A., CURRIE, D. C. and PATERSON, I. S. (1968). 'Hill grazing management and increased production'. *Scott. Agric.*, **47**, 123–31.

NICHOLSON, I. A., PATERSON, I. S., and CURRIE, A. (1970). A study of vegetational selection by sheep and cattle in *Nardus* pasture. In: *Animal populations in relation to their food resources*, Brit. Ecol. Soc. Symposium No. 10 (Ed: Watson, A.). Oxford and Edinburgh.

NICHOLSON, I. A. and ROBERTSON, R. A. (1958). 'Some observations on the ecology of an upland grazing in north-east Scotland with special reference to Callunetum'. *J. Ecol.*, **46**, 239–70.

NORDHAGEN, R. (1920). 'Vegetationsstudien auf der Insel Utsire im westlichen Norwegen'. *Bergens Mus. Årb. 1920 naturvid. rekke*, **1**, 1–149.

NORDHAGEN, R. (1928). *Die Vegetation und Flora des Sylenegebeites*. Oslo.

NORDHAGEN, R. (1936). 'Versuch einer neuen Einteilung der subalpinen – alpinen Vegetation Norwegens'. *Bergens Mus. Årb. naturvid. rekke*, **7**, Bergen.

NORDHAGEN, R. (1937). 'Studien uber die monotypische Gattung *Calluna* Salisb'. I. *Bergens Mus. Årb.*, 1937, *naturvid. rekke*, **4**, 1–55.

OVERBECK, F. and SCHMITZ, H. (1931). 'Zur Geschichte der Moore, Marschen und Walder Nordwestdeutschlands'. I. *Mitt. d. Provinzialstelle f. Naturdenkmalpflege Hannover*, **3**, 1–179.

PEARSALL, W. H. (1934). 'Woodland destruction in Northern Britain'. *The Naturalist for 1934*, 25–8.

PEARSALL, W. H. (1938). 'The soil complex in relation to plant communities. III Moorlands and bogs'. *J. Ecol.*, **26**, 298–314.

PEARSALL, W. H. (1950). *Mountains and Moorlands*. London.

PEART, J. N. (1963). 'Increased production from hill pastures: Sourhope trials with cattle and sheep'. *Scott. Agric.*, **42**, 147–51.

PICCOZZI, N. (1968). 'Grouse bags in relation to the management and geology of heather moors'. *J. appl. Ecol.*, **5**, 483–8.

PITOT, A. and MASSON, H. (1951). 'Quelques données sur latemperature au cours des feux de brosse aux environs de Dakar'. *Bull. Inst. fr. Afr. noire,* **13,** 711–32.

POEL, L. W. (1948). 'Effects of aeration on bracken and heather grown in nutrient solution'. *Nature, Lond.,* **162,** 115.

POEL, L. W. (1949). 'Germination and development of heather and the Hydrogen ion concentration of the medium'. *Nature, Lond.,* **163,** 647.

POORE, M. E. D. (1955). 'The use of phytosociological methods in ecological investigations. III Practical applications'. *J. Ecol.,* **43,** 606–51.

POORE, M. E. D. and MCVEAN, D. N. (1957). 'A new approach to Scottish mountain vegetation.' *J. Ecol.,* **45,** 401–39.

PRAEGER, R. L. (1911). 'A biological survey of Clare Island. 10. Phanerogamia and Pteriodphyta'. *Proc. R. Ir. Acad.,* **31,** 1–112.

PREISING, E. (1949). 'Nardo-Callunetea. Zur Systematik der Zwergstrauch-Heiden und Magertriften Europas mit Ausnahme des Mediterrangebeites, der Arktis und der Hochgebirge'. *Mit. Flor. – Soz. Arbeitsgem., N.F.* **1,** 12–25.

RAINS, A. B. (1963). 'Grassland research in Northern Nigeria 1952–62'. *Misc. Pap. Inst. agric. Res. Ahmadu Bello Univ.* No. 1.

RANKINE, W. F., RANKINE, W. M., and DIMBLEBY, G. F. (1960). 'Further excavations at a Mesolithic site at Oakhangar, Selborne, Hants'. *Proc. Prehist. Soc.* **26,** 246–62.

RAUNKIAER, C. (1907). The life-forms of plants and their bearing on geography. In: *The life forms of plants and statistical plant geography* (English Trans. 1934). Oxford.

RAUNKIAER, S. (1910). Investigation and statistics of plant formations. In: *The life forms of plants and statistical plant geography* (English trans., 1934). Oxford.

RAUNKIAER, C. (1913). Statistical investigation of the plant formations of Skagens Odde (The Skaw). In: *The life forms of plants and statistical plant geography* (English trans. 1934). Oxford.

RAWES, M. and WELCH, D. (1969) 'Upland productivity of vegetation and sheep at Moor House National Nature Reserve, Westmorland, England'. *Oikos, Suppl.,* **11,** 7–72.

RAYNER, M. C. (1913). 'The ecology of *Calluna vulgaris*'. *New Phytol.,* **12,** 59–77.

RAYNER, M. C. (1915). 'Obligate symbiosis in *Calluna vulgaris* I. *New Phytol.,* **12,** 59–77.

RAYNER, M. C. (1921). 'The ecology of *Calluna vulgaris.* II The calcifuge habit'. *J. Ecol.,* **9,** 60–74.

RAYNER, M. C. (1922 *a*). 'Mycorrhiza in the Ericaceae'. *Trans. Br. mycol. Soc.,* **8** 61–6.

RAYNER, M. C. (1922 *b*). 'Nitrogen fixation in Ericaceae'. *Bot. Gaz.,* **73,** 226–35.

RAYNER, M. C. (1923). 'Contributions to the biology of mycorrhiza in the Ericaceae'. *Rep. Br. Ass. Advmt. Sci., Liverpool.* 1923.

RAYNER, M. C. and SMITH, M. L. (1929). 'Phoma radicis Callunae. A physiological study'. *New Phytol.,* **28,** 261–90.

RENNIE, P. J. (1957). 'Effects of the afforestation of catchment areas upon water yield'. *Nature, Lond.,* **180,** 663–4.

RENNIE, P. J. (1961). 'Some long-term effects of tree growth on soil productivity'. In: *Recent Advances in Botany,* (pp. 1636–40). Toronto.

RITCHIE, J. C. (1955). 'Biological Flora of the British Isles. *Vaccinium vitis-idaea* L'. *J. Ecol.,* **43,** 701–8.

RITCHIE, J. C. (1956). 'Biological flora of the British Isles. *Vaccinium myrtillus* L'. *J. Ecol.,* **44,** 291–9.

ROBERTS, H. W. (1960). 'Chemical renovation of hill land in the north of Scotland'. *Trans. R. Highld. agric. Soc. Scotl.,* **5,** 29–37.

ROBERTSON, R. A. and DAVIES, G. E. (1965). 'Quantities of plant nutrients in heather ecosystems'. *J. appl. Ecol.*, **2**, 211–19.

ROBERTSON, R. A. and NICHOLSON, I. A. (1961). 'The response of some hill pasture types to lime and phosphate'. *J. Br. Grassld. Soc.*, **16**, 117–25.

ROBERTSON, R. A., NICHOLSON, I. A. and HUGHES, R. (1963). *Runoff studies on a peat catchment,* Internat. Peat Congress, Leningrad, 1963, pp 1–8.

ROBINSON, R. K. (1971). Importance of soil toxicity in relation to the stability of plant communities. In: *Scientific management of plant and animal communities for conservation.* pp. 105–13. Brit. Ecol. Soc. Symposium No. 11 (Ed: Duffey, E. A. G. and Watt, A. S.). Oxford and Edinburgh.

ROFF, W. J. (1964). *An analysis of competition between* Calluna vulgaris *and* Festuca ovina. Dissertation for the Degree of Ph.D., University of Cambridge.

ROMELL, L. G. (1951). Liens landskap och mulens. In: *Sver. Nat.* 42. Stockholm.

ROMELL, L. G. (1952). Heden. In: *Natur i Halland.* Stockholm.

RUTTER, A. J. (1955). 'The composition of wet-heath vegetation in relation to the water-table'. *J. Ecol.*, **43**, 507–43.

SALISBURY, E. J. (1921). 'Stratification and hydrogen – ion concentration of the soil in relation to leaching and plant succession, with special reference to woodlands'. *J. Ecol.*, **9**, 220–40.

SALISBURY, E. J. (1942). *The Reproductive Capacity of Plants.* London.

SEARS, P. D. (1951). 'Plant and animal nutrition in relation to soil and climatic factors'. *Proc. Spec. Conf. Agric. Br. Commonw. Sci. Off. Aust. 1949.*, **409**.

SHEIKH, K. H. (1969). 'The effects of competition and nutrition on the inter-relations of some wet-heath plants'. *J. Ecol.*, **57**, 87–99.

SHEIKH, K. H. (1970). 'The responses of *Molinia caerulea* and *Erica tetralix* to soil aeration and related factors. III Effects of different gas concentrations on growth in solution culture; and general conclusions'. *J. Ecol.*, **58**, 141–54.

SHEIKH, K. H. and RUTTER, A. J. (1969). 'The responses of *Molinia caerulea* and *Erica tetralix* to soil aeration and related factors. I Root distribution in relation to soil porosity'. *J. Ecol.*, **57**, 713–26.

SJÖRS, H. (1950). 'Regional studies in north Swedish mire vegetation'. *Bot. Notiser* **2**, 173–222.

DE SMIDT, J. T. (1967). 'Phytogeographical relations in the North West European heath.' *Acta bot. Neerl.*, **15**, 630–47.

SMITH, A. G. and WILLIS, E. H. (1962). 'Radiocarbon dating of the Fallahogy landnam phase'. *Ulster J. Archaeol.*, **24–5**, 16–24.

SMITH, R. (1900). 'Botanical survey of Scotland. I. Edinburgh district'. *Scott. geogr. Mag.*, **16**, 385–416. II North Perthshire district. *Scott. geogr. Mag.*, **16**, 441–67.

SMITH, W. G. (1902). 'The origin and development of heather moorland'. *Scott. geogr. Mag.*, **18**, 587–97.

SMITH, W. G. (1905). 'Botanical survey of Scotland. III and IV Forfar and Fife'. *Scott. geogr. Mag.*, **21**, 1–20 and 57–83.

SMITH, W. G. (1911). Scottish Heaths. In: *Types of British Vegetation* (Ed. Tansley, A. G.). Cambridge.

SPECTOR, W. S. (1956). *Handbook of biological data.* Philadelphia.

SPENCE, D. H. N. (1960). 'Studies on the vegetation of Shetland. III Scrub in Shetland and in South Uist, Outer Hebrides'. *J. Ecol.*, **48**, 73–95.

STOCKER, O. (1923). 'Die Transpiration und Wasserökologie nordwestdeutscher Heide-und Moorpflanzen am Standort'. *Z. Bot.*, **15**, 1–41.

STOCKER, O. (1924). 'Beitrage zum Halophytenproblem. I Oekologische Untersuchungen an Strand-und Dunenpflanzen des Darss (Vorpommern)'. *Z. Bot.*, **16**, 289–330.

STOUTJESDIJK, P. (1959). 'Heaths and inland dunes of the Veluwe'. *Wentia,* **2,** 1–96.

TAMM, O. (1948). *Influence exercée par la végétation forestiere et les bruyères sur les sols de la partie méridionale de la Suede.* C. R. Conf. Pedol. medit. 1947, pp. 206–9.

TAMM, C. O. (1958). Die Mineralstoffquellen der Pflanze. c. The atmosphere. In: *Encyclopedia of plant physiology.* (Ed: Ruhland, W.) IV. pp. 233–42. Berlin-Gottingen-Heidelberg.

TANSLEY, A. G. (1911). (Editor). *Types of British Vegetation.* Cambridge.

TANSLEY, A. G. (1939). *The British Islands and their Vegetation.* Cambridge.

TANSLEY, A. G. and ADAMSON, R. S. (1926). 'Studies of the vegetation of the English chalk. IV A preliminary survey of the chalk grasslands of the Sussex downs'. *J. Ecol.,* **14,** 1–32.

TAYLOR, E. L. (1961). 'Control of worms in ruminants by pasture management'. *Outl. Agric.,* **3,** 139–44.

THOMAS, A. S. (1957). 'Nature of chalk heath soils'. *Nature, Lond.,* **179,** 545–6.

THOMAS, B. (1934). The composition of common heather. *J. agric. Sci., Camb.,* **24,** 151–5

THOMAS, B. (1937). 'The composition and feeding value of heather at different periods of the year'. *J. Minist. Agric. Fish.,* **43,** 1050–55.

THOMAS, B. (1956). 'Heather (*Calluna vulgaris*) as a food for livestock'. *Herb. Abstr.,* **26,** 1–7.

THOMAS, B. and DOUGALL, H. W. (1947). 'Yield of edible material from common heather'. *Scott. Agric.,* **27,** 35–8

THOMAS, B., ESKRITT, J. R. and TRINDER, N. (1945). 'Minor elements of common heather (*Calluna vulgaris*)', *Emp. J. Exp. Agric.,* **13,** 93–9.

THOMAS, B. and TRINDER, N. (1947). 'The ash components of some moorland plants', *Emp. J. Exp. Agric.,* **15,** 237–48.

THREN, R. (1934). 'Jahreszeitliche Schwankungen des osmotischen Wertes verschiedener ökologischer Typen in der Umgebung von Heidelberg'. *Z. Bot.,* **26,** 449–526.

TRACZYK, T. (1967). 'Studies on herb layer production estimate and the size of plant fall'. *Ekologia Polska. Se. A.,* **15,** 837–67

TÜXEN, R. (1937). 'Die Pflanzengesellschaften Norddeutschlands'. *Mitt. Flor.-Soz. Arbeitsgem,* **3,** 1–170.

TÜXEN, R. (1938). 'Von der nordwest-deutschen Heide'. *Natur und Volk,* **68,** 253–63.

TÜXEN, R. (1955). 'Das System der nordwest-deutschen Pflazengesellschaften'. *Mitt. Flor.-Soz. Arbeitsgem. N.F.,* **5,** 155–74.

VAARTAJA, O. (1949). 'High surface soil temperatures: on methods of investigation, and thermocouple observations on a wooded heath in the south of Finland'. *Oikos,* **1,** 6–8.

VANDEN BERGHEN, C. (1959). 'Etude sur la végétation des dunes et des landes de la Bretagne'. *Vegetatio,* **8,** 193–208.

WALLACE, R. (1917). *Heather and Moor Burning for Grouse and Sheep.* Edinburgh.

WANNOP, A. R. (1958). A review of hill farm problems. *Advmt. Sci., Lond.,* **59,** 189–94.

WANNOP, A. R. (1959). 'Progress in hill farm research'. *Scott. Agric.,* **38,** 111–14.

WARD, S. D. (1968). *A study of the distribution and vegetational composition of* Calluna-Arctostaphylos *heaths in north-east Scotland and of related Scandinavian communities.* Ph.D. Thesis, Univ. of Aberdeen.

WARD, S. D. (1970). 'The phytosociology of *Calluna*-Arctostaphylos heaths in Scotland and Scandinavia. I Dinnet Moor, Aberdeenshire'. *J. Ecol.,* **58,** 847–63.

WARMING, E. (1909). *Oecology of Plants* (English trans.). Oxford.

WATERBOLK, H. T. (1954). *De praehistorische mens en zijn milieu.* Groningen.

WATERBOLK, H. T. (1957). 'Pollenanalytisch onderzoek van twee noordbrabantse tumuli.' In *Twee grafheuvels in Noord-Brabant*, Beex, G. Bijdr. Studie. Brabantse Heem. **9**, 34–9.

WATSON, A. (1964). 'The food of Ptarmigan (*Lagopus mutus*) in Scotland'. *Scott. Nat.,* **71**, 60–6.

WATSON, A., MILLER, G. R. and GREEN, F. H. W. (1966). 'Winter browning of Heather (*Calluna vulgaris*) and other moorland plants'. *Trans. Bot. Soc. Edinb.,* **40**, 195–203.

WATSON, A. and MILLER, G. R. (1970). *Grouse Management.* The Game Conservancy, Booklet 12. Fordingbridge.

WATSON, P. J. and GREGOR, J. W. (1956). 'Reflections on hill-land improvement'. *Herb. Abstr.* **26**, 137–45

WATT, A. S. (1936). 'Studies in the ecology of Breckland. I Climate, soil and vegetation'. *J. Ecol.,* **24**, 117–38.

WATT, A. S. (1940). 'Studies in the ecology of Breckland. IV. The grass-heath'. *J. Ecol.,* **28**, 42–70.

WATT, A. S. (1945). 'Conributions to the ecology of bracken. III Frond types and the make up of the population'. *New Phytol.,* **44**, 156–78.

WATT, A. S. (1947 *b*). 'Pattern and process in the plant community'. *J. Ecol.,* **35**, 1–22.

WATT, A. S. (1955). 'Bracken versus heather, a study in plant sociology'. *J. Ecol.,* **43**, 490–506.

WATT, A. S. (1961). Ecology. In: *Contemporary Botanical Thought* (Ed: MacLeod, A. M. and Cobley, L. S.). Edinburgh and London.

WEATHERELL, J. (1953). 'The checking of forest trees by heather'. *Forestry,* **26**, 37–41.

WEBLEY, D. M., EASTWOOD, D. J. and GIMINGHAM, C. H. (1952). 'Development of a soil microflora in relation to plant succession on sand-dunes, including the 'rhizosphere' flora associated with colonizing species'. *J. Ecol.,* **40**, 168–78.

WHITTAKER, E. (1960). *Ecological effects of moor burning.* Ph.D. Thesis, Univ. of Aberdeen.

WHITTAKER, E. (1961). 'Temperatures in heath fires'. *J. Ecol.,* **49**, 709–15.

WHITTAKER, E. and GIMINGHAM, C. H. (1962). 'The effects of fire on regeneration of *Calluna vulgaris* (L.) Hull from seed'. *J. Ecol.,* **50**, 815–22.

WILLIAMS, W. T. and LAMBERT, J. M. (1959). 'Multivariate methods in plant ecology. I Association analysis in plant communities'. *J. Ecol.,* **47**, 83–101.

WILLIAMS, W. T. and LAMBERT, J. M. (1960). 'Multivariate methods in plant ecology. II The use of an electronic digital computer for association analysis'. *J. Ecol.,* **48**, 689–710.

WILSON, E. (1911). Observations on the food of grouse, based on examination of crop contents. In: *The grouse in health and disease.* (Ed: Leslie, A. S.) pp. 67–87. London.

WILSON, K. (1960). 'The time factor in the development of dune soils at South Haven Peninsula, Dorset'. *J. Ecol.,* **48**, 341–59.

WORMELL, P. (1968). 'Establishing woodland on the Ise of Rhum'. *Scott. For. J.,* **22**, 207–20.

WATT, A. S. (1947 *a*). 'Contributions to the ecology of bracken. IV The structure of the community'. *New Phytol.,* **46**, 97–121.

ZEHETMAYR, J. W. L. (1960). *Afforestation of upland heaths.* Bull. For. Commn. Lond., **32**, H.M.S.O., London.

Index

(Where reference is to an illustration the page number is shown in bold type)